Lecture Notes on Coastal and Estuarine Studies

Managing Editors:
Malcolm J. Bowman   Richard T. Barber
Christopher N.K. Mooers   John A. Raven

22

Bengt-Owe Jansson (Ed.)

Coastal-Offshore Ecosystem Interactions

Proceedings of a Symposium sponsored by SCOR, UNESCO,
San Francisco Society, California Sea Grant Program,
and U.S. Dept. of Interior, Mineral Management Service
held at San Francisco State University, Tiburon,
California, April 7–22, 1986

Springer-Verlag
Berlin Heidelberg GmbH

Managing Editors

Malcolm J. Bowman
Marine Sciences Research Center, State University of New York
Stony Brook, N.Y. 11794, USA

Richard T. Barber
Duke University, Marine Laboratory
Beaufort, N.C. 28516, USA

Christopher N.K. Mooers
Institute for Naval Oceanography
National Space Technology Laboratories
MS 39529, USA

John A. Raven
Dept. of Biological Sciences, Dundee University
Dundee, DD1 4HN, Scotland

Contributing Editors

Ain Aitsam (Tallinn, USSR) · Larry Atkinson (Savannah, USA)
Robert C. Beardsley (Woods Hole, USA) · Tseng Cheng-Ken (Qingdao, PRC)
Keith R. Dyer (Merseyside, UK) · Jon B. Hinwood (Melbourne, AUS)
Jorg Imberger (Western Australia, AUS) · Hideo Kawai (Kyoto, Japan)
Paul H. Le Blond (Vancouver, Canada) · L. Mysak (Montreal, Canada)
Akira Okuboi (Stony Brook, USA) · William S. Reebourgh (Fairbanks, USA)
David A. Ross (Woods Hole, USA) · John H. Simpson (Gwynedd, UK)
Absornsuda Siripong (Bangkok, Thailand) · Robert L. Smith (Covallis, USA)
Mathias Tomczak (Sydney, AUS) · Paul Tyler (Swansea, UK)

Editor

Bengt-Owe Jansson
University of Stockholm, Askö Laboratory
10691 Stockholm, Sweden

ISBN 978-3-540-19051-6   ISBN 978-3-642-52452-3 (eBook)
DOI 10.1007/978-3-642-52452-3

1-Jenne Zijlstra, 2-Bengt-Owe Jansson, 3-Mario Pamatmat, 4-Juanita Gearing, 5-Proserpina Gomez, 6-Eileen Hofmann, 7-John Field, 8-Job Dronkers, 9-Henk Postma, 10-Stephen Smith, 11-Charles Hopkinson, 12-Robert Twilley, 13-Bernt Zeitzschel, 14-Charles Epifanio, 15-Tom Kessler, 16-Alasdair McIntyre, 17-Reginald Uncles, 18-Ulrich Horstmann, 19-Thomas Pearson, 20-Scott Nixon, 21-Tom McClimans.
Not pictured: David Peterson, Peter Rothlisberg, Gary Sharp.

SCOR WORKING GROUP 65 CONFERENCE 1986

CONTENTS

ABSTRACT

Jansson, B.-O. (ed.). Coastal-offshore ecosystem interaction. Lecture
Notes on Coastal and Estuarine Studies. Springer Verlag 22:

Interactions between coastal and offshore ecosystems are considered,
focusing on four aspects. 1. Water exchange, crucial for most couplings
is classified for two types of system: shallow coastal areas and'
narrow, deep shelf areas. 2. Mass balance studies of tidal flats, salt
marshes, mangroves, fjord systems and coral reefs give a strong
indication of recirculation of nutrients and suggest that imported
organic material mostly remains in the nearshore areas. 3. Active
transport is demonstrated for fish and crustacean species occupying
coastal nurseries. Both crab and shrimp larvae are vertical migrators
which by reacting to fine-tuned hydrodynamics are retained in
favourable adult habitats. 4. Numerical modelling as a means of
synthesizing relevant physical and biological processes is analyzed for
several existing ecosystem models and recommendations for suitable
techniques are made.

Evaluation of present evidence shows that:
a) on a global scale and of the scale of years to decades, outwelling
is quantitatively insignificant in the biogeochemistry or productivity
of the sea b) productivity of many coastal systems are determined in
the short term more by recycling than by inputs, though the
relationship between the two remains to be determined c) "information
flows" in the form of oceanic populations using the coastal areas as
nursery grounds are important.

BACKGROUND AND ACKNOWLEDGEMENTS

In 1980 the Scientific Committee on Oceanic Reseach, in close collaboration with UNESCO and IABO, initiated the formation of SCOR Working Group 65 (Coastal-offshore ecosystems relationships) with the following terms of reference:

(i)   to review and compare the energetics of coastal (littoral and estuarine) and offshore pelagic and benthic populations.

(ii)   to suggest methods for improving knowledge of energy conversion between coastal and offshore pelagic migratory and benthic populations and to determine what further research is needed.

In consultation with IABO the Working Group decided to concern itself primarily with differences in the energetics of coastal and offshore ecosystems and with significant energy and material fluxes between such systems. These fluxes could include the exchange of organic material and plant-nutrients between the two systems. In addition it was recognized that fluxes might exist which are probably insignificant in terms of energy exchange, but are important in terms of quality and should therefore be considered. Such fluxes could include e.g. migrations of (juvenile) crustacea and fish from the coastal zone to offshore populations as well as fluxes of pollutants.

The Working Group, established in the course of 1980, had two meetings, the first in Bordeaux (France) from 5-7 September 1981 in conjunction with the International Symposium on Coastal Lagoons, 8-13 September 1981), the second at Texel (The Netherlands) from 12-15 September 1983. The membership of the W.G. and participants in the meetings were as follows:

	September 1981	September 1983
B.O. Jansson (Sweden)	x	x
B. Kjerfve (USA)		x
P. Lasserre (France)	x	
A.D. McIntyre (UK) secretary	x	x
R.C. Newell (UK)	x	
S.W. Nixon (USA)	x	x
M.M. Pamatmat (USA)	x	x
B. Zeitzschel (FRG)	x	x
J.J. Zijlstra (The Netherlands) chairman	x	x

B. Kjerfve was coopted by the W.G. after its first meeting to provide expertise on physical processes involved in the coastal-offshore relationships.

The group recognized the great diversity of the coastal zone, which might lead to local differences in the relationship between coastal and offshore ecosystems. It was therefore decided to exchange documented accounts of the situation with which each of the members was most familiar. These accounts covered areas as different as San Francisco Bay, North Inlet (South Carolina, USA), the Bermuda platform, a southern Benguela kelp community (South-Africa), sandy beaches in western Scotland, a Baltic coastal-offshore system and the Wadden Sea (The Netherlands).

These reports together with exchange of views during meetings and by correspondence, assisted in focussing attention on six aspects, which appear to be of general interest for all coastal-offshore situations (with the possible exception of tropical areas, for which no information was presented) and provide a background for the relationship between the two ecosystems.

1) coastal/offshore boundaries and water exchange.
2) nutrient exchange between coastal and offshore systems.
3) transport of matter across the coastal/offshore boundary.
4) coastal-offshore relations in terms of animal populations.
5) relative biological productivity in coastal and offshore systems.
6) effects of man-made disturbances.

The discussions within the working-group of these six aspects can be summarized as follows:

1) The boundary between coastal and offshore systems is highly dynamic and varies with river discharge, wave climate, wind stress, and other physical forces or events generated outside the immediate system. As used here, the term coastal includes estuaries and nearshore waters. Little is known about material exchange between nearshore and offshore areas. Studies have been published dealing with exchange between estuaries and adjoining nearshore waters but their results are not conclusive with regard to the exchange between nearshore and offshore waters.

2) It seems doubtful that coastal areas, in particular the estuaries, contribute significantly to the nutrient budget of offshore areas.

3) "Outwelling" of organic matter from coastal estuarine areas to offshore regions is probably much smaller than formerly postulated. In fact, there are indications of net organic matter import by some coastal waters from offshore areas.

4) migrations of nekton, especially large crustaceans and fish, across the coastal-offshore boundary is qualitatively established for commercially important species. Except perhaps for some migratory species like the salmon in some river systems, there are no reliable measurements or estimates of population movements in both directions.

5) Notwithstanding higher nutrient levels, higher rates of nutrient recycling, a higher potential energy and higher habitat diversity, coastal ecosystems may not always be as productive on all trophic levels as is generally postulated in comparison to offshore systems.

6) Most man-made disturbances are from point sources and will have localized effects. Impacts on coastal/offshore interactions can be expected especially at the mouth of large river systems or in areas bordering highly urbanized regions.

The group concluded that it did not seem possible from the kind of existing information to quantitatively evaluate the importance of estuaries, lagoons, mangrove swamps, or coastal waters in general, in the ecological energetics and productivity of offshore waters

To provide a more detailed basis for this possibly controversial conclusion the W.G. 65 proposed to SCOR that a workshop meeting be held to bring scientists together in order to develop an international consensus on gaps in our knowledge, the necessary approach and methodology.

After discussions between SCOR-representatives and members of the W.G.65 (Lasserre, Nixon) the program for the meeting was outlined. SCOR, the United Nations Educational and Scientific Council and San Francisco Bay Foundation were most helpful as co-sponsors of the workshop meeting and in the provision of financial support.

The meeting took place at Paul F. Romberg Tiburon Center for Environmental Studies, San Francisco State University, from April 7-12, 1986. The local arrangements were excellently directed by Prof. Mario Pamatmat who was also of great assistance during the first important editing phase of the manuscripts. Prof. M. Josselyn and his staff through their hospitality and efforts provided an inspiring background to the meeting.

Mrs Elizabeth Tidmarsh, Executive Secretary of SCOR shares greatly in the realization of this workshop through her constructive handling of the administration. I am also most grateful to Mrs Antoniella Cerri, Springer-Verlag, for her patience and support when editing processes were difficult. Maureen Moir undertook the Herculean task of typing the whole book.

GUIDE TO THE CONTENTS

The intention of this work is to summarize some of the present evidence for the interactions between coastal and offshore ecosystems, and at the same time to reveal gaps in knowledge and to make recommendations for future work. Not every author has chosen the state-of-the-art approach. Some have preferred to concentrate on, from their point of view, crucial problems which need further elucidation. A few give a detailed analysis and synthesis of the coastal/offshore exchange. These differences are probably significant for our knowledge today - it is patchy in both space and depth.

By examining different types of ecosystem it was our hope to scrutinize the generality of the six previously stated aspects of the coastal/offshore relationships. It was not possible, however, to obtain studies of all major systems. In particular, we did not include the tropical systems due mainly to the difficulty of assembling the necessary data.

Water exchange provides the most obvious and direct connection between coastal and offshore areas. Two types of systems and one extra methodological paper describe this type of coupling. DRONKERS discusses water exchange in shallow coastal systems, stressing tide, wind and buoyancy as principal agents. A classification system relating major mixing zones to mixing agents and geomorphological characteristics is presented. MCCLIMANS describes the coastal/offshore water exchange in narrow, deep shelf areas. He concludes that the common density front between coastal and offshore areas is a sufficient dividing line. Thanks to satellite images, the multiple fronts in this border areas can be assessed. Here filaments of highly productive regions are deformed in spirals and mixed through wind action. HORSTMAN advocates satellite remote sensing data from two or more consecutive days for estimating coastal-offshore fluxes with examples from the Baltic Sea.

Mass balance study is a classical tool for quantifying imports and exports of matter. GEARING presents a review of the promising technique of using stable isotope ratios for tracing transport of organic matter. Being more of an independent method, this technique might be used as a rough check on mass balance calculations. POSTMA summarizes the present knowledge for tidal flats stating that local production of organic matter is mostly insufficient, the system running on imports from outside. An unusually high percentage of this organic matter is metabolized by anaerobic bacteria which are fed through bioturbation. HOPKINSON starts with a critical evaluation of the direct flux and mass balance approaches for estimating the transfer of matter. Through a detailed analysis of five marsh/estuarine systems he then arrives at the overall conclusion that a substantial transfer of estuarine carbon to the nearshore region exists but that the source of "new" allochthonous nutrients cannot be defined on the basis of present information. TWILLEY concludes from studies of mangrove forests, that there is a more conclusive flux from forested wetlands than from salt marshes, partly due to the continuous litterfall in the former. Although present data on nutrients are scarce there are indications that nutrient recycling may vary along a hydrologic continuum. GOMEZ reports from her studies of Philippean mangrove areas that there is a net export of particulate material from the estuary to the open sea coinciding with the peak of litterfall of major mangrove species and the wet period. PEARSON, in his summary of boreal and polar fjord systems, states that boreal well-mixed fjords tend to export considerable amounts of nutrients to adjacent coastal waters whereas stagnant boreal fjords appear to be nutrient sinks. Stratified fjords are probably sinks for both nutrients and carbon throughout the year. SMITH interprets available metabolic data on coral reefs and stresses that reefs are not metabolically different from other shoal-water systems, but have very limited metabolic interaction with the surrounding ocean. Produced new carbon is only slightly higher than the new production of the surrounding plankton communities. PETERSON, HAGER, SCHEMEL and CAYAN take a global view of the riverine C, N, Si and P transport to the coastal ocean. They find that after aphotic and benthic mineralization the "leftovers" for eventual exchange are difficult to quantify by empirical methods. Large-scale onshore fluxes dominate the coastal nutrient budgets and the riverine/estuarine nutrient sources are secondary to the ocean, except locally.

Active transport between coast and ocean is studied through reviews of fish and crustacean case studies. ZIJLSTRA reviews the evidence of migrating fish as a coastal/offshore transport agent. He concludes that the diadromous fish play a minor role, partly due to the deterioration of rivers and estuaries. Fish using the coastal areas as nursery grounds are important, however, utilizing the favourable conditions in these areas such as: high production of food, high temperature in late spring and summer and scarceness of large predators including adults of the same species. Although there is a netflow of biomass from the coastal to the offshore areas this is important in terms of its quality rather than its quantity. ROTHLISBERG describes the flow of living matter between coast and offshore through the migration patterns of three crustaceans, all commercially important, from different oceanographic regimes and with different larval form and life span. All are active vertical migrators, showing three different larval transport trajectories due to the different hydrodynamic pattern. EPIFANIO shows how three species of crabs have evolved behavioural traits which allow control of horizontal advection with consequent larval retention in areas near favourable adult habitats. These traits can be summarized as: 1) constant maintenance of position deep in the water column; 2) downstream advection of surface-dwelling immature larvae; 3) tidally rhythmic vertical migration of larvae.

Numerical modelling as a tool for synthesizing physical and biological processes was discussed in the last session. UNCLES reviews the currently used techniques for coupling hydrodynamical and ecological models of large tidal estuarine ecosystems. He arrives at the conclusion that the fixed element, tidally-averaged model is the most suitable for ecosystem simulations.

The evidence of the separate reviews and the discussions during the meeting are summarized by the Working Group members. Several recommendations for future research and for suitable methods are made.

BENGT-OWE JANSSON
Stockholm, Sweden.
(Chairman of the workshop)

CONTRIBUTORS

Dr. D.R. CAYAN, Scripps Institution of La Jolla,
CA 92093, USA.

Dr. JOB DRONKERS, Ministerie van verkeer en waterstaat
rijkswaterstaat, Postbus 20907, v. Alkemadelaan 400,
2597 AT's-Gravenhage, The Netherlands.

Dr. CHARLES E. EPIFANIO, College of Marine Studies,
University of Delaware, Lewes, DW 19958, USA.

Dr. JUANITA GEARING, Department of Fisheries and Oceans,
Maurice Lamontagne Institute, Mont-Joli, Quebec G5H 3Z4,
Canada.

Dr. PROSERPINA GOMEZ, Mindanao State University,
Marawe City, Lanao del Sur, Philippines.

Dr. STEPHEN W. HAGER, U.S. Geological Survey,
345 Middlefield Road-MS 496, Menlo Park,
CA 94025, USA.

Dr. CHARLES HOPKINSON, Marine Institute, University
of Georgia, Sapelo Island, GA 31327, USA.

Dr. ULRICH HORSTMANN, Institut für Meereskunde, an der
Universität Kiel, Dusternbrooker Weg 20, D 2300 Kiel 1,
FRG.

Dr. BENGT-OWE JANSSON, Institute of Marine Ecology,
University of Stockholm, 106 91 Stockholm, Sweden.

Dr. TOM McCLIMANS, Institutet for Marin Byggeteknikk,
Alfred Getz Vei 3, N-7034 Trondheim-NTH, Norway.

Dr. ALASDAIR D. McINTYRE, Department of Agriculture and
Fisheries for Scotland, Marine Laboratory, Victoria Road,
Torry, Aberdeen, AB9 8DB, UK.

Dr. SCOTT NIXON, Graduate School of Oceanography,
University of Rhode Island, Kingston, RI 02881, USA.

Dr. MARIO M. PAMATMAT, Tiburon Center for Environmental
Studies, San Francisco State University, P.O. Box 855,
Tiburon, CA 94920, USA.

Dr. THOMAS H. PEARSON, Dunstaffnage Marine Research
Laboratory, P.O. Box 3, Oban, Argyll PA34 4AD, U.K.

Dr. DAVID PETERSON, U.S. Geological Survey,
345 Middlefield Road-MS 496, Menlo Park,
CA 94025, USA.

Dr. HENK POSTMA, Netherlands Institute for Sea Research,
Postbox 59, 1790 Ab Den Burg, The Netherlands.

Dr. PETER C. ROTHLISBERG, CSIRO Marine Laboratories,
233 Middle Street, Cleveland, QLD, Australia.

Dr. L.E. SCHEMEL, U.S. Geological Survey,
345 Middlefield Road-MS 496, Menlo Park,
CA 94025, USA.

Dr. STEPHEN V. SMITH, Department of Oceanography,
University of Hawaii at Manoa, 1000 Pope Road,
Honolulu, Hawaii 96822, USA

Dr. ROBERT TWILLEY, Department of Biology, University
of SW Louisiana, P.O. Box 42251, Lafayette, LA 70504, USA.

Dr. REGINALD J. UNCLES, Institute for Marine Environmental
Research, Prospect Place, The Hoe, Plymouth PL1 3DH, UK.

Dr. BERNT ZEITZSCHEL, Institut fur Meereskunde an der
Universitat Kiel, Dusternbrooker Weg 20, 2300 Kiel, FDR.

Dr. JENNE J. ZIJLSTRA, Netherlands Institute for Sea Research,
Postbox 59, 1790 Ab Den Burg, The Netherlands.

I. WATER EXCHANGE

INSHORE/OFFSHORE WATER EXCHANGE IN SHALLOW COASTAL SYSTEMS

J. Dronkers
Tidal waters Division, Rijkswaterstaat
The Hague, The Netherlands

Tide, wind and buoyancy are the principal agents for the exchange of
water and constituents (dissolved and particulate) between shallow
coastal areas and the offshore shelf zone. The mixing processes are
conditioned by topography and earth rotation. Mixing of inland and
ocean waters is accomplished partly in inshore basins (lagoons or
estuaries), partly around inlets and partly offshore on the coastal
shelf. Depending on the type of coastal system considered the
principal mixing zone is situated in one of these compartments. A
classification of shallow coastal systems is discussed which relates
major mixing zones to mixing agents and to geomorphological
characteristics. Along coasts with strong tides important mixing takes
place in inshore basins and inlet regions. Topographical structures
form an essential element in tidally induced mixing. For large rivers
mixing is accomplished essentially on the coastal shelf, often under
the influence of wind and buoyancy induced processes. Following the
lines indicated above a qualitative discussion of the most important
mixing processes is presented. Conclusions are put forward concerning
needs for future research.

1. INTRODUCTION

Shallow coastal shelf systems are widespread along the continental
boundaries. In these areas mixing of inland and offshore waters takes
place. Mixing zones are characterized by the presence of substantial
salinity gradients. They are located either in inshore basins or partly
offshore, but generally extend over both inshore and offshore regions.
Different types of coastal systems can be distinguished and several
classifications have been proposed, for instance by Hansen and Rattray
(1966), Schubel (1971) and Wanless (1976). These classifications are
based either on hydrodynamic properties, on geomorphological
properties, or on both. A common feature of all shallow coastal systems
is the existence of a definite relationship between hydrodynamic and
geomorphological properties, based on the dynamics of sediment
transport.

Lecture Notes on Coastal and Estuarine Studies, Vol. 22
B.-O. Jansson (Ed.), Coastal-Offshore Ecosystem. Interactions.
© Springer-Verlag Berlin Heidelberg 1988

Inshore/offshore water (and constituent) exchange is caused by a large number of processes, a few of which dominate in each particular type of coastal system. Processes which are very important in one type of system may have hardly any influence in another type. Therefore a general discussion of exchange processes requires a classification of coastal systems as a reference framework. The classification which is adopted, follows closely the one given by Wanless (1976). It is based on the parameters "river runoff", "tidal range" and "wave energy", see Table 1. The main arguments for the selection of these parameters are:

(a) River inflow is strongly related to landward input of fine cohesive sediment (clay and silt). This sediment is deposited in regions where the current slows down. If the landward sediment input is large, down stream storage basins will not exist; the river water is discharged directly on the coastal shelf.

(b) Oceanic tides amplified on the shelf cause a periodic inflow and outflow of large water masses through coastal inlets, which are consequently subject to scour. If tides are strong the channel cross sections increase downstream to the entrance and exhibit a funnel shape. Upstream an inshore system of braiding ebb and flood channels traversing vast tidal flat areas is generally present.

(c) Short wind-generated waves approaching the coast are distorted in such a way that in the near-shore zone a landward flux of bottom sediment is established. In this way coast-parallel barriers can be formed (Swift, 1976; Postma, 1980) protecting the inshore basin from direct oceanic influences. These areas are generally designated coastal lagoons.

For the inshore/offshore water exchange three compartments can be distinguished: the inshore basin, the inlet region and the coastal shelf area. In river dominated coastal systems (type I-III of Table 1) hardly any inshore seawater intrusion occurs, at least at high discharge. Mixing of inland and marine waters essentially takes place on the coastal shelf. For tide dominated coastal systems (type IV-VII) the mixing of inland and offshore waters is accomplished to a large degree inshore and around the inlet. When entering the coastal shelf inland waters and constituents are strongly diluted.

Mixing and flushing processes induced by tides will be dealt with in section 2, for each compartment. In section 3 the non-tidal exchange

type	river inflow	tide	waves	description	examples
I	+	–	–	river delta	Mississippi, Volga, Danube.
II	+	–	+	river delta with coastal barriers	Nile, Rhône, Pô
III	+	+	–	tidal river delta (funnel shaped inlets)	Ganga, Mekong, Amazon, Irrawaddy, Yangtze.
IV	o	+	–	coastal plain estuary (funnel shaped channels)	Thames, Western Scheldt, Ems, Weser, Elbe, Loire, Gironde, Potomac, Delaware, St. Lawrence.
V	–	+	+	tidal lagoon	Wadden Sea, Baie d'Arcachon.
VI	–	+	–	bay	Bay of Fundy, Wash.
VII	–	–	+	coastal lagoon	East Coast USA (Pamlico Sound, Indian River, Biscayne Bay), Texas Coast (Corpus Christi Bay), Baltic Sea (Haffens), Fleet.

Table 1. Classification of coastal systems, related to the relative importance of river inflow, tide and waves. For example, – + means: river discharge very small with respect to tidal discharge; sediment transport by waves significant compared to sediment transport by tidal currents.

type	description	Principal mixing zones and agents		
		inshore	inlet	offshore
I	river delta			wind, buoyancy
II	barrier river delta		buoyancy	wind, buoyancy
III	tidal river delta		tide, buoyancy	wind, buoyancy
IV	coastal plain estuary	tide, buoyancy	tide	wind, tide
V	tidal lagoón	tide, wind	tide	wind, tide
VI	bay	tide		
VII	coastal lagoon	wind, tide	wind, tide, buoyancy	

Table 2. Principal mixing zones and agents for water exchange

processes due to buoyancy and wind will be discussed. In section 4, attention will be paid briefly to the transport of particulate material, which presents some fundamental differences as compared to the transport of dissolved substances.

2.   TIDAL MIXING AND FLUSHING

2.1. The inshore coastal basin (coastal systems IV-VI)

The coastal basins considered here are characterized by a complex geometry, consisting of meandering and interlacing channels bounded by tidal flats. Ebb flow and flood flow are often concentrated in different channels or in different parts of the channel cross-section. The current velocity presents a strong cross-sectional variability; the tidal reversing of currents does not occur simultaneously in the cross- section. Due to the irregular current pattern, volumes of fluid which are initially far apart can be brought by tidal motion close to each other (Fig. 1). These volumes may exchange water and constituents due to turbulent water motions. This mechanism for large scale exchanges, known as "shear diffusion", has been described in detail for idealized tidal and geometrical conditions (Bowden, 1965; Okubo, 1967; Taylor III, 1974; Fischer et al., 1976).

Shoreline irregularities may also occasion large scale mixing of water masses (Okubo, 1973). The water retained in pockets or "dead zones" adjacent to the tidal channels can be exchanged with water masses moving along with the tide (Fig. 2). If the exchanges are caused by turbulent water motions or by geometrically induced eddies, one is dealing with a particular form of shear diffusion. However, exchanges between "dead zones" and the tidal channel can also be driven by other forces, for instance, density differences or tidal in- and outflow. The first process is important for the exchange between tidal rivers and harbour basins (Fig. 3; Abraham et al., 1986). The second process occurs for tidal flat areas and contributes to large-scale mixing if the longitudinal water motion in the channels is not in phase with the inflow and outflow of tidal flat areas (Fig. 4; Dronkers, 1978). This latter process can be viewed as belonging to a type of large-scale exchange mechanism related to the occurrence of spatial gradients in tidal phase lag. This mechanism operates especially in tidal basins with a branched channel system and an irregular geometry.

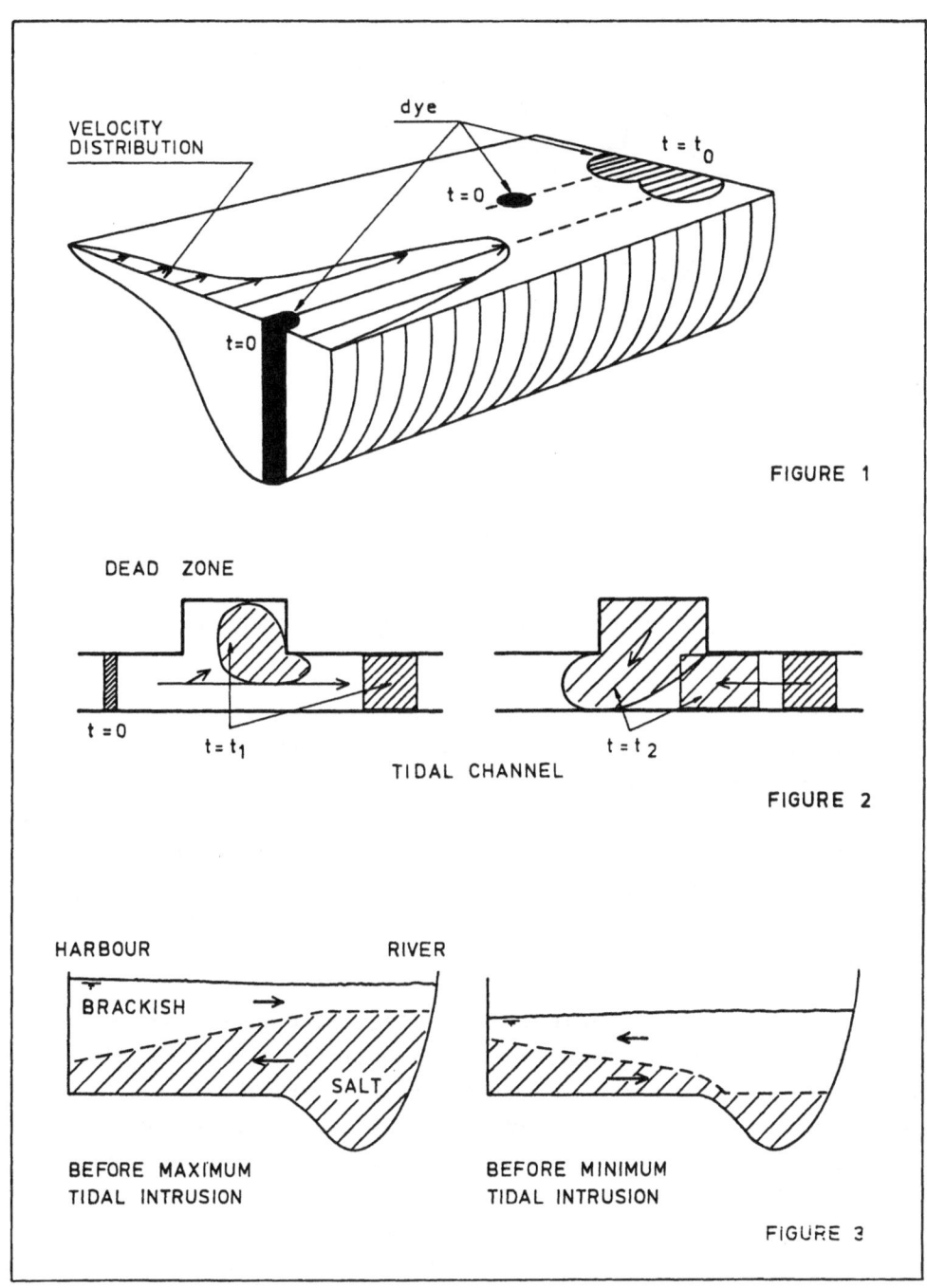

Fig. 1. Shear diffusion in a tidal channel
Fig. 2. Diffusion by dead zones
Fig. 3. Density driven exchanges between tidal river and harbour basins

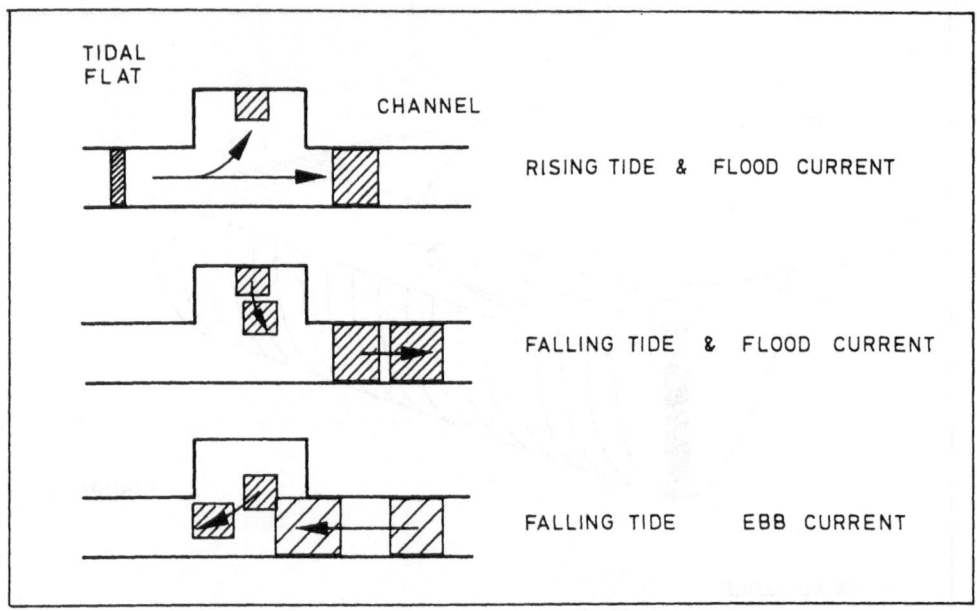

Fig. 4. Diffusion by tidal phase lag between channel flow and water storage on tidal flats

Lateral gradients in tidal phase lag arise from differences in bottom friction and differences in propagation and reflection of the tidal wave along opposite channel boundaries. As a consequence of these phase lag differences water parcels may experience a net Lagrangian displacement during a tidal period, even in the absence of a residual Eulerian velocity component (Fig. 5). The large dispersion coefficients measured at the junction of different channel systems can be explained in this way (Fig. 6).

In meandering channel systems the tidal flow generates an Eulerian residual flow pattern. This residual flow field has a three dimensional structure. In the plane perpendicular to the main tidal flow a rotating motion is superimposed, directed to the outer channel bend near the surface and to the inner bend near the bottom. In the horizontal plane a residual current is directed away from the outer bend and towards the inner bend (Fig. 7). This secondary flow field is thought to maintain or enhance the channel meander structure and to contribute to the building up of tidal flats (Pingree and Maddock, 1979; Heathershaw and Hammond, 1980). The origin of the residual eddy field can be described in terms of vorticity conservation (Zimmerman, 1981), but also in terms of centrifugal flow acceleration. Centrifugal

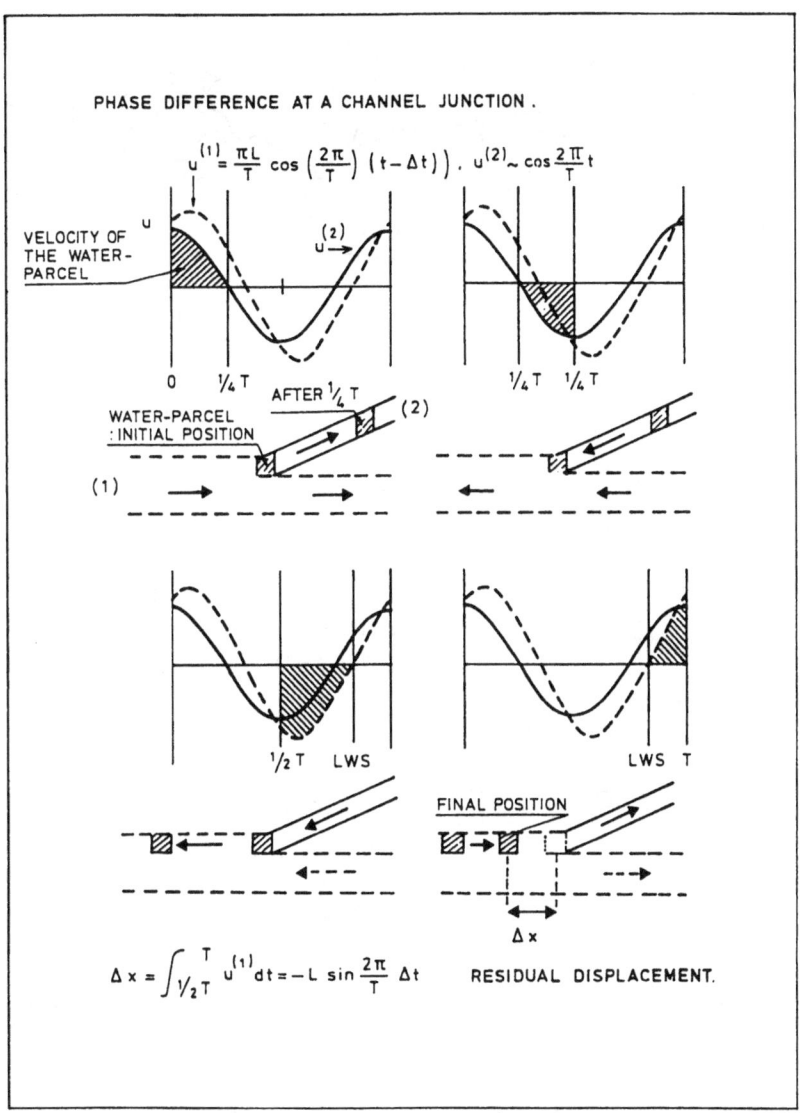

Fig. 5. Residual displacement of a water parcel in a branched channel system, caused by tidal phase lag between currents at a channel junction.

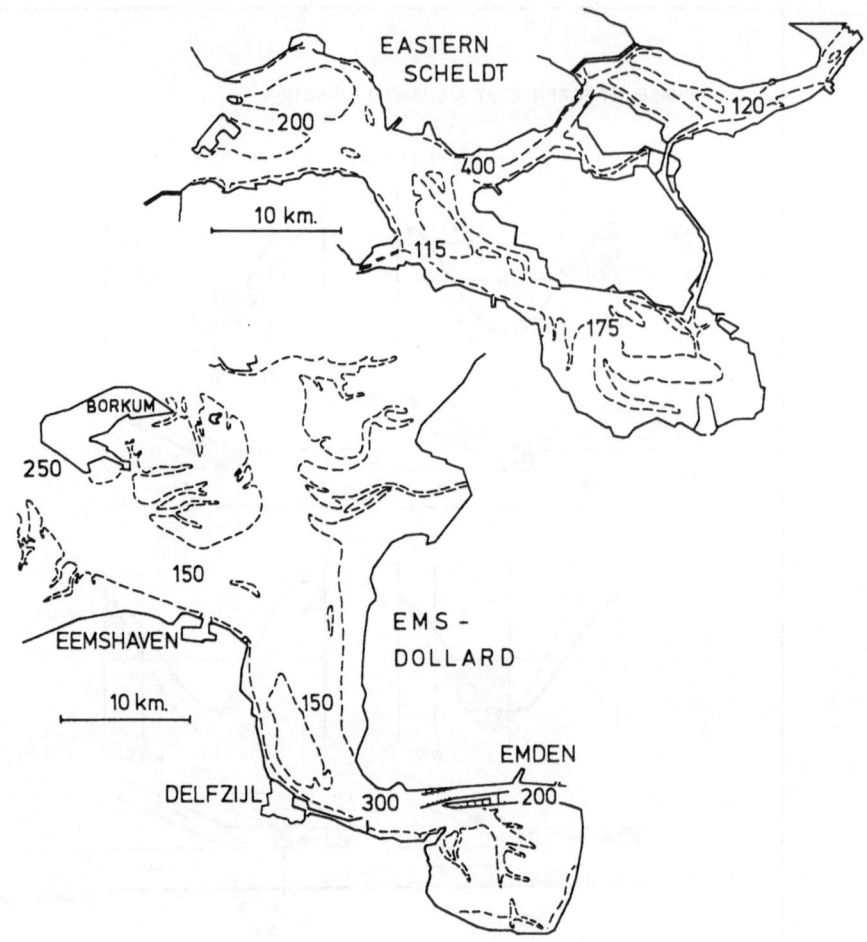

Fig. 6. Dispersion coefficients ($m^2 \cdot s^{-1}$) in branched tidal systems (Dronkers & Zimmerman, 1982; Helder & Ruarddij, 1983). Note the high values at channel junctions.

acceleration occurs if the flow curves when following the channel bend: the fluid velocity in the deep outer bend exceeds the velocity in the shallow inner bend. The radius of flow curvature is given by R = $u_1/(\delta u_1/\delta r)$, where u_1 is the amplitude of the tidal velocity, r the cross-channel coordinate. The transverse residual flow results from an imbalance in the vertical between centrifugal force u^2/R and lateral surface inclination δ /δr. This imbalance has a similar direction for ebb and flood. The horizontal residual flow results from the longitudinal surface inclination: the water level increases from

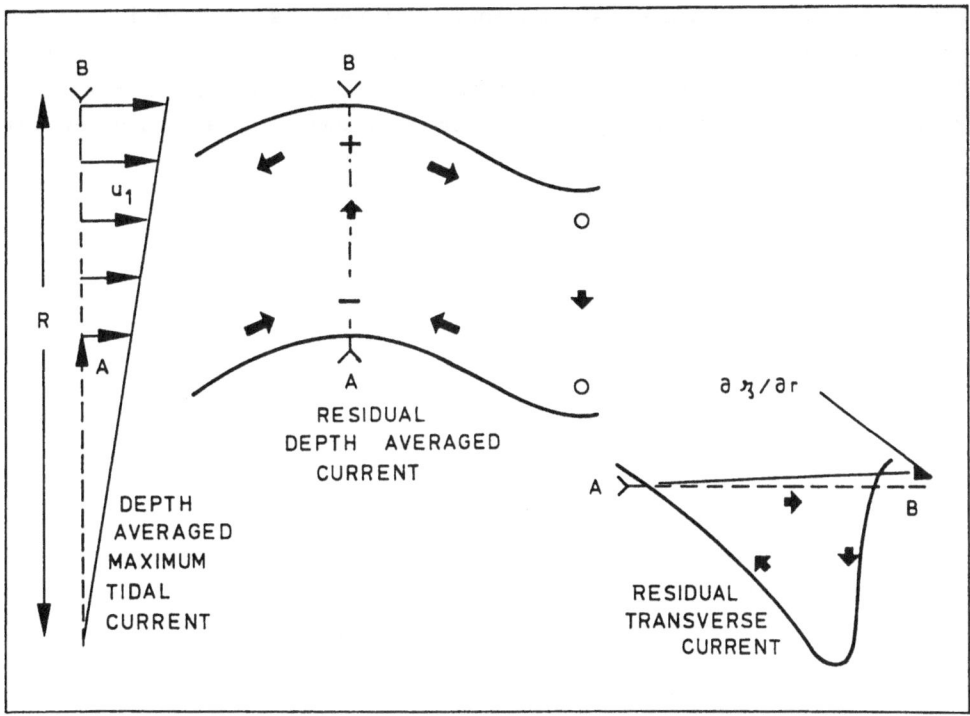

Fig. 7. Residual flow field in a channel bend.

either direction towards the maximum at the outer bend and decreases
from either direction towards the minimum at the inner bend.

An ebb-flood asymmetry in the transverse secondary flow will appear if
the Rossby number u_1/fR is of the order one or less. The two
circulation cells in Figure 7 become asymmetric: the flow will tend to
follow the boundaries at the right hand more closely. For an exact
computation of the residual flow structure the interaction between
transverse and horizontal eddies needs to be taken into account
(Zimmerman, 1986a).

The residual flow cells described above are generally smaller than the
mean tidal excursion. In a meandering estuary, water parcels are
transported during the tidal cycle through a field of such residual
flow cells. In this way a quasi-random motion is generated causing
mixing of sea and inland waters in the basin (Zimmerman, 1976). The
earlier described shear diffusion and phase induced Lagrangian drift

also contribute to the quasi-randomness of the residual motion of water parcels. Turbulent mixing, but especially geometrical irregularity (including channel meandering, braiding and tidal flats) are responsible for large quasi-random residual displacements of water parcels after a tidal cycle.

If the flow field is sufficiently irregular, neighbouring water parcels may experience already after a short excursion slightly different accelerations in magnitude and direction; from that moment they can rapidly diverge. In this way particle motions become uncorrelated and "chaotic", even if turbulent mixing is absent (Zimmerman, 1986b). This chaotic behaviour is enhanced by the aforementioned "deterministic" dispersion processes related to geometrical irregularity of the tidal basin. The time scale at which successive displacements of water parcels become uncorrelated is in the order of the time scale for transverse mixing $\tau_\perp$. If this time scale is much smaller than the flushing time of the basin τ_f , then the flushing can be described by a random walk with time steps $\delta\tau \approx \tau_\perp$. In that case the dispersive transport of any dissolved substance follows the same gradient-type law (Dronkers, 1982)

$$\dot{M} = QC - DA\frac{dc}{dx}$$

Here M, Q, A and C are the tidally averaged values of the transport, the discharge, the cross-section area and the cross-sectionally averaged concentration. The location dependent dispersion coefficient D(x) is a measure for the mixing efficiency. For stationary conditions the same value applies to all dissolved substances. Values typically range between 100 and 1,000 m^2 s^{-1} (Dronkers and Zimmerman, 1982).

2.2. The inlet region for tidal basins IV-VI

Certain coastal inlet shelf systems present a wide entrance (type IV and VI), others are more constricted (type V). In both cases an important exchange of inland and coastal waters through the inlet may occur. The underlying physical processes are similar in both situations, as they relate to the spatial variability of the tidal current field in both magnitude and direction.

The case of a constricted inlet has been studied numerically by Awaji, Imasato and Kunishi (1980). They conclude that the net water exchange is hardly influenced by turbulent diffusion but originates from a Lagrangian drift of water parcels induced directly by the tidal motion. Two mechanisms are identified, which contribute in approximately the same degree:

(a) Spatially rapid changes of amplitude, direction and phase lag of the demidiurnal tidal current. In such a flow-field, non-linear effects are important and particle motion becomes rapidly chaotic. Besides, flow separation causes a displacement of water parcels from high speed regions during flood to low speed regions during ebb and inversely;

(b) residual Eulerian eddies caused by centrifugal forces and flow separation, in a way similar as discussed in the previous section (Fig. 7). The effects of (a) and (b) are combined in Figure 8. The result is very similar to the classic picture of water exchange through inlets proposed by Stommel and Farmer (1952). The separation of the inflowing and outflowing water volumes illustrates the strong exchange through the inlet, which is estimated at 80-90% by Awaji et al. (1980).

Strong tidal currents parallel to the coast can occasion important exchanges between inland and sea water not only in narrow but also in wide tidal inlets. This is a situation which occurs in particular in partially enclosed seas, like the North Sea or the Yellow Sea, where the tide travels along the coast as a Kelvin wave. The tidal wave in the inlet generally has a standing wave character, due to inshore reflection. The tidal wave in the coastal zone, on the contrary, has a more progressive wave character. Therefore longshore coastal currents are out of phase with the cross-shore tidal flow through the inlet, as schematically indicated in Figure 9.As shown in this figure, water parcels flowing seaward through the inlet experience a net displacement after a tidal cycle as a consequence of the phase lag between the cross-shore and long-shore tidal flows. The direction is opposite to the propagation direction of the tidal wave along the coast. This is illustrated in Figure 10 for the southern Rhine-Meuse-Scheldt outflow region in the southern North Sea. The particle tracks have been computed from velocity data for average tidal conditions in a large number of locations (Terwindt, 1965). These data were obtained from moored current meters deployed during several months.

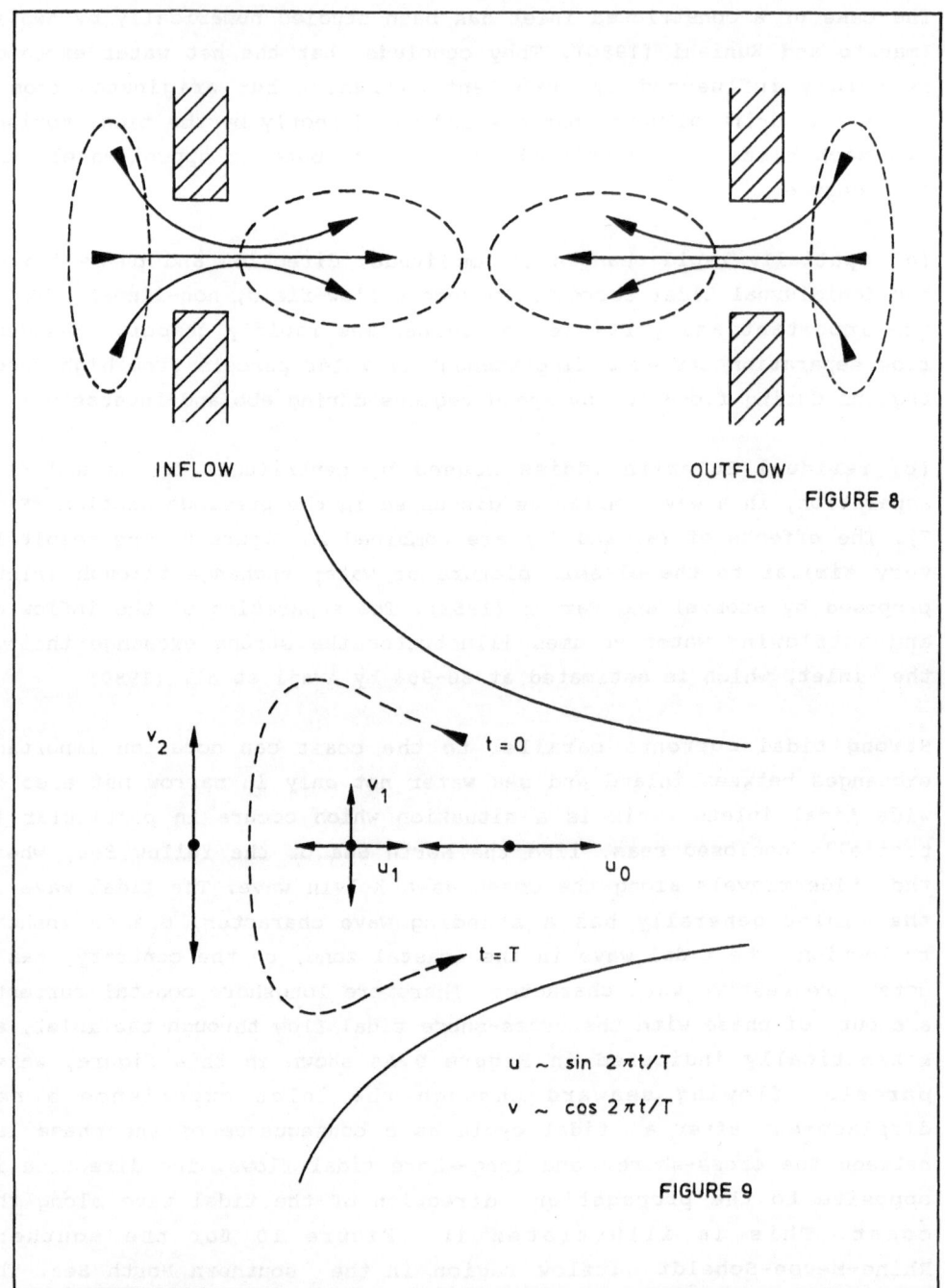

Fig. 8. Water exchange through a constricted tidal inlet.

Fig. 9. Schematic representation of the Lagrangian drift of a water parcel at the transition region of estuarine and coastal tidal flow.

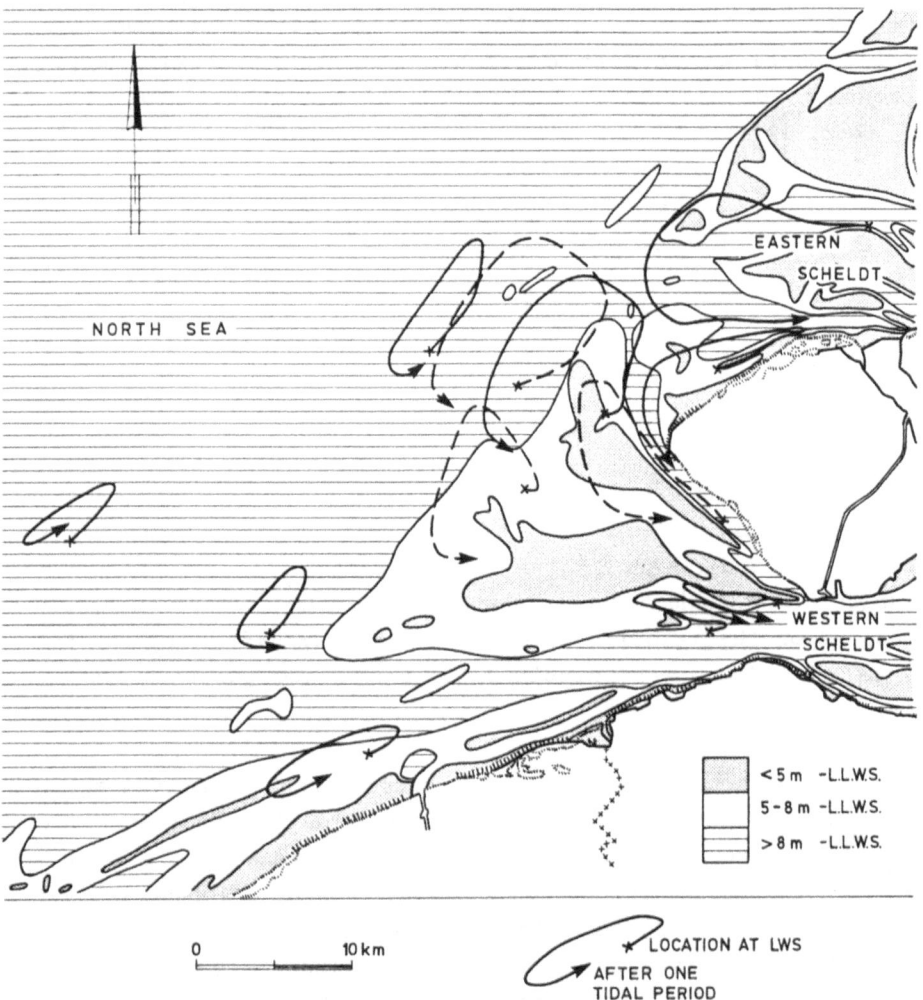

Fig. 10. Particle tracks in the southern Rhine-Meuse-Scheldt ebb tidal delta computed from velocity data (after Terwindt 1965).

In tidal lagoons with more than one inlet flushing can be induced by differences in tidal stress. The term "tidal stress" designates the tidally averaged contribution of the non-linear terms $\rho(u^2 + g\,^2/H)$ in the one dimensional momentum equation.

Suppose that semidiurnal components of velocity and tidal level, u_1 and ζ_1 dominate all other tidal components and that $\zeta_1 \ll$ depth H. Then a difference between the amplitudes of tidal level and current velocity at two inlets x_a and x_b will drive a residual flow from x_a to x_b or inversely. In the case that the inlets x_a and x_b are connected by a tidal channel of uniform cross-section, the following approximate expression for the residual discharge q_o can be derived (v.d. Kreeke and Dean, 1974).

$$q_o = \frac{g H^2 T}{4 \pi \ell} \left\{ A \frac{|\zeta_1^a||\zeta_1^b|}{H^2} \sin(\varphi_1^a - \varphi_1^b) + B \frac{|\zeta_1^a|^2 - |\zeta_1^b|^2}{4 H^2} \right.$$
$$\left. + C \frac{\rho^a - \rho^b}{\rho^a + \rho^b} + C \frac{\zeta_o^a - \zeta_o^b}{H} \right\} \tag{1}$$

In this expression ζ_1^a and ζ_1^b are the tidal amplitudes at the inlet x_a and x_b, φ_1^a and φ_1^b the tidal phase lags, ζ_o^a and ζ_o^b the mean tidal levels and ρ^a and ρ^b the mean densities. For the bottom stress a quadratic law $C_B |u_1| u$ has been assumed. The coefficients A, B, and C are functions of the friction parameter $\beta = (2 C_B / 3 \pi^2) |u_1| T/H$. For usual values of β the coefficients A, B and C are positive and of order unity.

Equation (1) shows that a tidally induced circulation from x_a to x_b is favoured in situations where the tidal amplitude decreases along the coast from x_a to x_b or the tidal phase lag increases from x_a to x_b. The influence exerted on the residual circulation by a longshore mean level gradient (caused, for instance, by wind stress) or by a longshore density gradient is displayed by the two last terms in equation (1). Examples of this flushing mechanism have been reported for the lagoons along the coast of Florida (v.d. Kreeke and Dean, 1974).

2.3. The offshore coastal zone

As mentioned in the previous section, tidal currents are stronger in partially enclosed seas than on open shelves. In these semi-enclosed shelf seas the tide may contribute substantially to exchange processes. Okubo (1971, 1974) has shown that the characteristics of turbulent diffusion in different shelf areas around the world are similar and compatible with the theory of locally isotropic turbulence. The radial variance of a patch of dye is proportional to the rate of energy transfer and proportional to the third power of time, t. In partially enclosed seas the tidal motion provides the main energy input for that part of the turbulent energy spectrum corresponding to time scales of a tidal period or less. The standard deviation of an initial "point" discharge (diameter 10-100 m) after a tidal period is at most 1 km. This corresponds to values of the dispersion coefficient less than 10 $m^2 s^{-1}$, an order of magnitude smaller than in lagoons or estuaries. The dispersion increases with the length scale which is considered. Dispersion coefficients of an order of magnitude similar to those observed in coastal basins apply in the coastal sea to dye patches with a diameter of approximately 100 km. Some tidal influences on flushing and mixing of a coastal sea will be mentioned below. Most processes are similar to those occurring in inshore basins.

Lateral gradients in current velocity generate shear diffusion. However, the spatial variability of the velocity field is smaller offshore than inside coastal basins or near inlets. For this reason shear diffusion is relatively less effective.

Coastline irregularities contribute to longitudinal dispersion due to temporary retention of water masses. Behind cliffs or headlands flow separation may occur, creating "dead zones" which exchange with the main flow by a headland eddy. Also the tendency of the main flow to follow isobaths (conservation of potential vorticity) may lead to the formation of pockets along the coast (Unlüata et al., 1983).

Probably the most important effect is the so-called topographic rectification of tidal currents (Zimmerman, 1981). Due to the presence of sand ridges, dunes and other bottom structures the tidal current is locally accelerated or decelerated. In addition to small variations in current velocity, a meandering behaviour of the tidal flow results. Associated ebb and flood dominated channels have been observed experimentally (Fig. 11; Robinson, 1966).

A topography with elongated bottom structures making an angle with the main tidal flow has been considered by Zimmerman (1981). Figure 12 illustrates the way in which a residual Eulerian eddy cell arises. Bottom friction slightly deviates the flow along the ridge. At the top the velocity is increased due to shallower depth. In response to this acceleration earth rotation forces the flow in a clockwise direction (northern hemisphere). This change in flow direction counteracts flow deceleration downstream of the bottom ridge if the ridge is inclined to the right with respect to the main flow (Fig. 12a). Therefore residual eddy cells are formed particularly around bottom structures inclined in

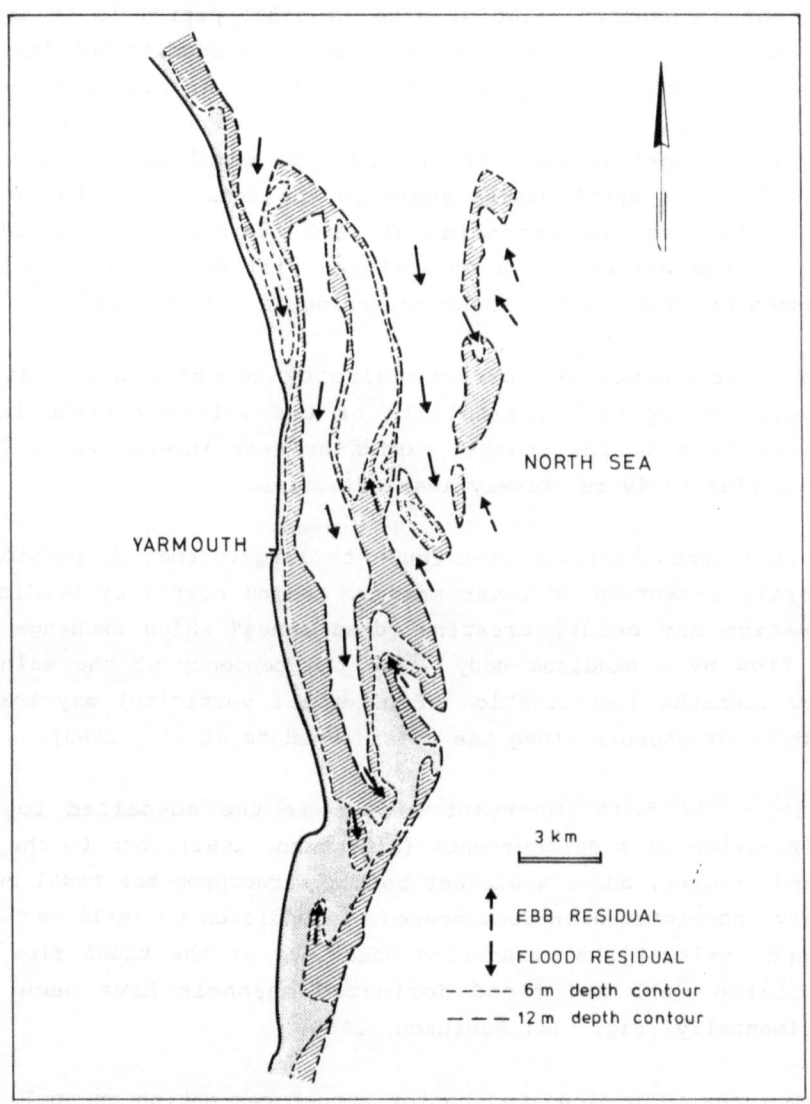

Fig. 11. Flood and ebb dominated flow around sandbanks along the East Anglian coast (after Robinson, 1966).

anticlockwise direction with respect to the main flow (Fig. 12b).
Following experimental evidence cited by Zimmerman (1981) this
coincides with the preferred orientation of sand ridges in the
northern hemisphere. Numerical models demonstrate that bottom
topography in a shelf sea is entirely reflected in the residual eddy
structure (Abraham et al., 1987). Zimmerman (1978) shows that the
strength of the residual eddy field is greatest if the characteristic
length scale of the topographic variability is in the same order as
the tidal excursion.

The interaction of tidal flow with a pattern of bottom ridges results
in a two dimensional velocity field with spatial fluctuations in phase
and amplitude. This yields in general non-cyclic Lagrangian
displacements, even if Eulerian residuals are absent. This is
illustrated, for example, in Figure 9. The residual Lagrangian
displacements strongly depend on the initial phase and position.
Residual pathways thus display a random character, described as
chaotic diffusion by Pasmanter 1986. In this way the large scale
dispersion in coastal seas may be explained. It is not yet clear
whether this mechanism or whether shear dispersion in the field of
residual eddies is the dominant process (Zimmermann, 1986b).

Komen and Riepma (1981) compare tidally induced residual vorticity with
vorticity induced by wind stress. They conclude that the latter
dominates in the southern bight of the North Sea for moderate to high
windspeeds. Diffusion of tracer material proceeds from tidal or wind
driven transport through the field of residual eddies.

Finally it should be mentioned that the tide also contributes to long
shore transport of water and dissolved constituents. This is due to the
progressive character of the tidal wave. The associated mass transport
is called "Stokes drift". The direction of propagation is to the left
of a vector pointing to the coast (northern hemisphere).

3.   NON-TIDAL EXCHANGES
3.1. Inshore coastal basins (type IV-VII)

The term "non-tidal exchanges" stands for the flushing and mixing processes caused by density gradients (buoyancy) and wind. Currents driven
by buoyancy and wind interact with the tidally driven water motion, and

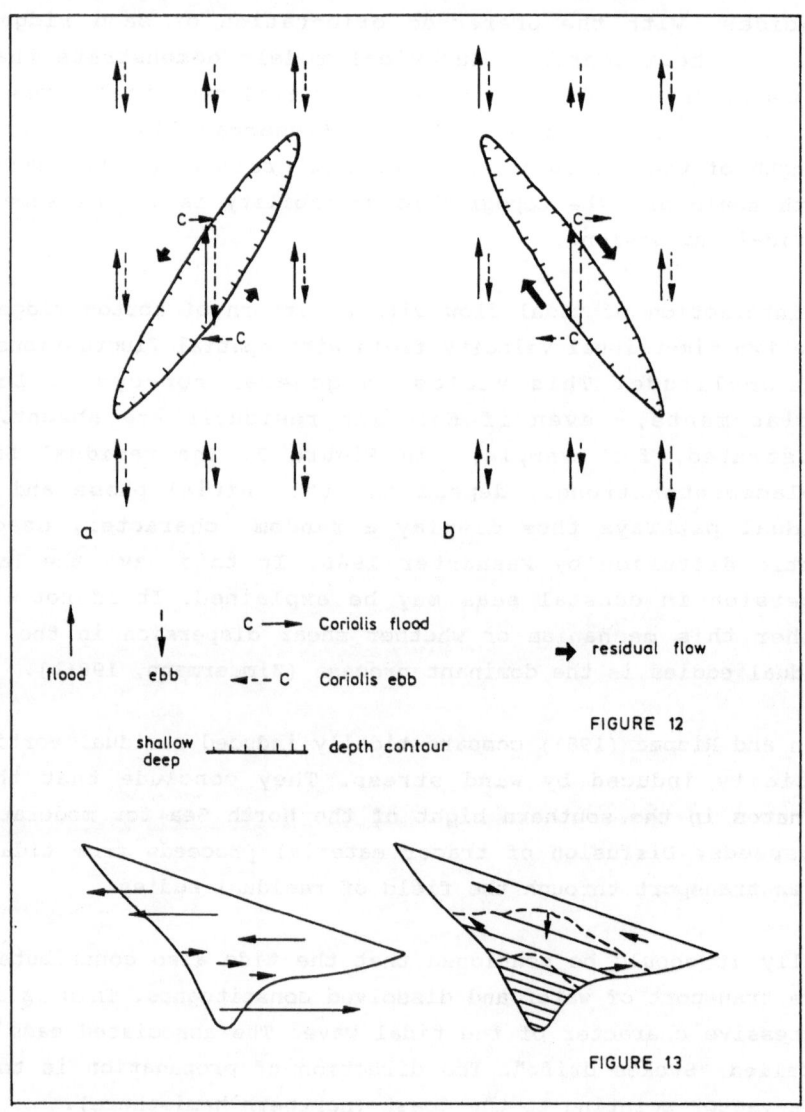

Fig. 12. Schematic representation of ebb and flood flow around sand ridges inclined to the right (a) and to the left (b) on the northern hemisphere.

Fig. 13. The three-dimensional buoyancy driven residual flow field in an estuary.

therefore display fluctuations on the tidal time scale. As will be
discussed, wind and density effects also interact.

Only shallow inshore systems are considered, with a depth H which is
much smaller than the Ekman depth of frictional influence $\sqrt{2N_z/f}$. The
effects of earth rotation can be ignored in that case. Flow
accelerations are balanced mainly by bottom stress. This balance is
reached on a relatively short time scale; it often permits the
establishment of quasi-equilibrium conditions.

3.1.1. Buoyancy

In most inshore basins a density gradient is present, caused by
admixture of fresh inland waters with marine waters. In some cases the
density gradient is reversed due to an excess of evaporation. Only the
first situation is considered.

A longitudinal increase of density induces a "baroclinic" pressure
gradient which increases with depth. In a channel with a rectangular
cross-section a vertical "estuarine" circulation is superimposed on,
and modulated by, the tidal flow. This circulation is directed seaward
near the surface and landward near the bottom. The strength of the
circulation is proportional to the longitudinal density gradient; it
strongly increases with depth H (approximately as the third power)
and decreases inversely proportional to vertical momentum exchange
coefficient N_z (Hansen and Rattray, 1965).

In practice, the topography of coastal basins is more complex,
characterized by channels and shoals. The longitudinal density gradient
then induces a three-dimensional residual flow pattern. The vertically
averaged baroclinic pressure gradient increases with the local depth,
causing a landward depth-averaged residual flow in the channel,
compensated by a seaward residual counter flow in the shallow portions.
The residual seaward flow takes place in the entire upper layer; the
residual landward is confined to the channel except the near surface
region (Fig. 13; Fischer 1972).

This density-induced circulation pattern influences the salinity
distribution. On the average, salinities will be higher in the channel
than on the shoals. During tidal phases with sufficiently strong
vertical mixing a lateral density gradient will appear, near surface

salinity in the channel being higher than at the channel boundary and on the shoals. The lateral baroclinic pressure gradient drives a transverse circulation, which along the bottom is directed to the channel boundary and the shoals and along the surface towards the channel axis. Near the channel axis flow convergence takes place with downwelling of surface water (Fig. 13; Nunes and Simpson, 1985). Surface flow convergence and fronts are commonly observed in estuaries; examples are reported by Bowman and Iverson (1977), Klemas and Polis (1977) and Dronkers and Van de Kreeke (1986). Observations of axial convergence in the nearly homogeneous seaward range of the eastern Scheldt estuary suggest that transverse circulation induced by channel bends constitutes another frontogenenis mechanism (L. Kohsiek, pers. comm.).

Turbulent mixing in non-homogeneous flows is a process which is still not well understood. Accurate mathematical formulations suitable for a broad field of applications are lacking. A recent account has been given by Kullenberg (1983). This is an important obstacle in quantifying the influence of density gradients on exchange processes in the mixing zones of inland and offshore waters.

In a very qualitative manner it can be stated that density differences tend to inhibit locally the mixing of inland and marine waters, but at the same time tend to enhance the relative displacement (advection) of these water masses. Large-scale coastal-offshore water exchange may ultimately be favoured by the presence of density differences. Buoyancy tends to increase the correlation time scale of successive particle displacements and a random walk description of the mixing process inside the coastal basins is less likely to apply than in the homogeneous case (Dronkers, 1982). For example, the dispersal of dissolved constituents from point sources cannot be described accurately in the vicinity of the source by using dispersion coefficients derived from the salinity distribution.

3.1.2. Wind

Wind-shear stress exerted on the water surface induces a surface drift, made up partly by a surface current and partly by wave-generated Stokes drift. In the shallow basins considered here, the wind-induced surface drift follows the direction of the wind vector. If the basin has a closed boundary in the wind direction a surface inclination is built up and a counter flow is established.

In a rectangular channel of finite length directed along the wind vector the counter flow extends over the major part of the vertical below the relatively thin wind-drift layer. For more complex geometries the equilibrium flow pattern is three dimensional and bears some resemblance to buoyancy-induced flow. The depth integrated barotropic pressure gradient opposing the wind stress is proportional to depth. Therefore wind stress dominates in the shallow parts of the cross-section where the current essentially follows the wind direction. In the deeper parts of the cross-section counter flow dominates, except in a thin surface layer (Simons, 1980).

The presence of density differences modifies the above flow pattern. In a stratified two-layer basin an inclination of the interface will oppose the surface slope. As a result the vertically integrated transport in the upper layer will be in the direction of the wind in the nearshore zone and against the wind in deep water, similar to the homogeneous case. In the lower layer, however, the transport is in the direction of the wind at the center and against the wind in the shallower portions (Lee and Ligget, 1970).

Wind-driven currents interact with tide and density-driven flow. Observations show that estuarine circulation is disturbed. It can be enhanced but also reversed; temporarily, three-layer flow may occur (Elliot, 1978; Vieira, 1985).

Wind and density influences cause seasonal variations in the flushing time scale of inshore coastal basins, corresponding to seasonal variations in storm activity and river discharge. Examples have been reported, _inter alia_, for the Ems-Dollard by Helder and Ruardy (1983) and for San Francisco Bay by Walters et al. (1985). For the Ems-Dollard it is shown that the observed seasonal variation of the dispersion coefficient correlates with run-off. Nevertheless, it is more likely that the seasonally varying wind activity is mainly responsible.

3.2. The inlet region of coastal basins type V-VII

Meterological forcing in the coastal shelf region can have an important impact on water levels and consequently on the water volumes exchanged between coastal basins and the shelf. Generally the tidal velocities are not strongly influenced by these subtidal water motions because of the relatively long time scale (typically 3 days and longer at mid-latitude locations). However, the excursion of water parcels is in some cases much larger than the excursion due to the tidal motion (Smith, 1977, 1985; Swenson and Chuang, 1983).

The subtidal motion decreases the time scale for renewal of basin waters. This occurs in particular if the tidal excursion is smaller than the average residual displacement of individual water parcels during one tidal period. Examples are lagoons with internal wind or density driven circulations and small tidal exchanges through the inlet, which are found, for instance, along the coast of Florida (Smith, 1985).

Longshore winds play an important role in the flushing of coastal basins. As shown by Smith (1977), they can produce a set-up or set-down of water levels generating non-tidal exchanges through the inlet. More important, however, is the longshore advection of the outflowing water. The tidal and non-tidal prisms which have left the basin will not return if this longshore transport is sufficiently strong. New coastal waters are advected before the inlet, where a density front may form. If the coastal waters have a higher salinity, the tidal inflow will be enhanced by a density driven inflow along the bottom In Figure 14 this is shown for the eastern Scheldt basin, as an example.

3.3. Non-tidal exchange in the coastal zone

For large rivers the mixing zone of inland and marine water is mainly situated on the coastal shelf. This holds in particular for rivers with a small tidal penetration and restricted channel cross-section due to sediment deposition. Only shallow coastal areas will be considered here, with a depth typically in the order of 20 to 50 meters. The shelf area will be assumed so large that upwelling of deep ocean waters does not directly influence the coastal mixing process.

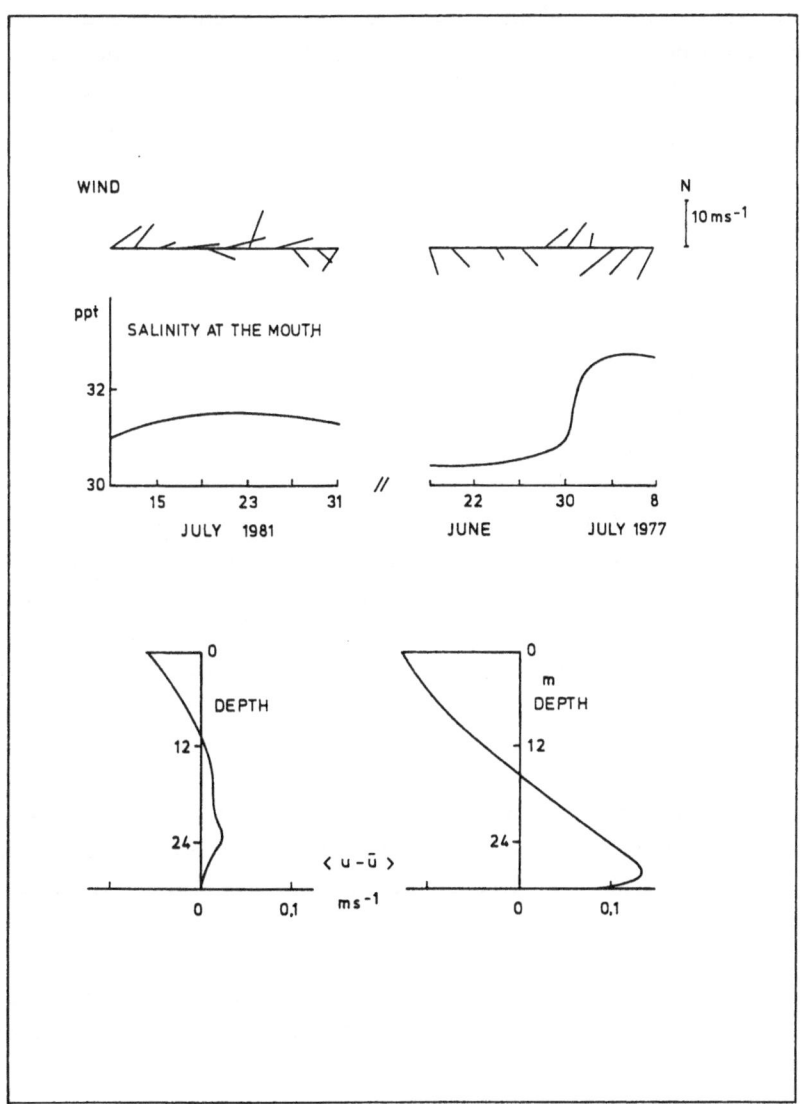

Fig. 14. Normally the longitudinal density gradient at the inlet of the Eastern Scheldt is small and estuarine circulation is nearly absent (situation on 31-7-81, location A, figure 18). Longshore advection of high salinity seawater is caused by a change of wind direction from north to south on 29-6-1977. An important residual vertical circulation results, measured on 8-7-1877.

The complexity of mixing processes in shallow coastal areas is largely due to the impact of bottom friction and topography in addition to influences of tide, wind, buoyancy and earth rotation. The tidally induced transport and mixing processes have been discussed in section 2.3.; here the general features of wind and density-induced processes will be reviewed. Some topics will be illustrated with experimental data collected in the Dutch North Sea coastal zone.

3.3.1. Buoyancy

A commonly observed feature in coastal zones is the presence of fronts separating coastal and offshore waters. Sharp small-scale fronts with strong discontinuities in physical and chemical properties occur when the river discharge is larger than, or in the same order as the tidal discharge at the outlet (rivers of type I-III and some rivers of type IV at high run-off). In the latter case the tidally pulsating river outflow causes a new front to appear at each ebb tide. Large-scale fronts often exist at a greater distance offshore even in the absence of distinct small scale fronts. They form the separation between coastal and ocean water; in general the discontinuity across the front is less important.

Small-scale fronts have been described for several rivers, for instance, the Connecticut River (Garvine and Monk, 1974; Garvine, 1977), the Hudson River (Bowman and Iverson, 1978), the Tees (Lewis, 1984) and the Rhine (Van Alphen et al., 1986). The most characteristic features (Fig. 15) are as follows:

- The low salinity surface plume extends offshore over a distance in the order of 10 km. The thickness h of the layer is a few meters. Near the surface outcropping the slope of isopycnals strongly increases to values exceeding 10^{-2}

-A well defined stable mixing layer with substantial internal wave motion is present beneath the surface front.

-A strong flow convergence exists at the surface towards the front from both sides, the seawater diving under the brackish upper layer. Velocities may reach supercritical values ($u \geq \sqrt{gh\delta\ \rho/\rho}$).

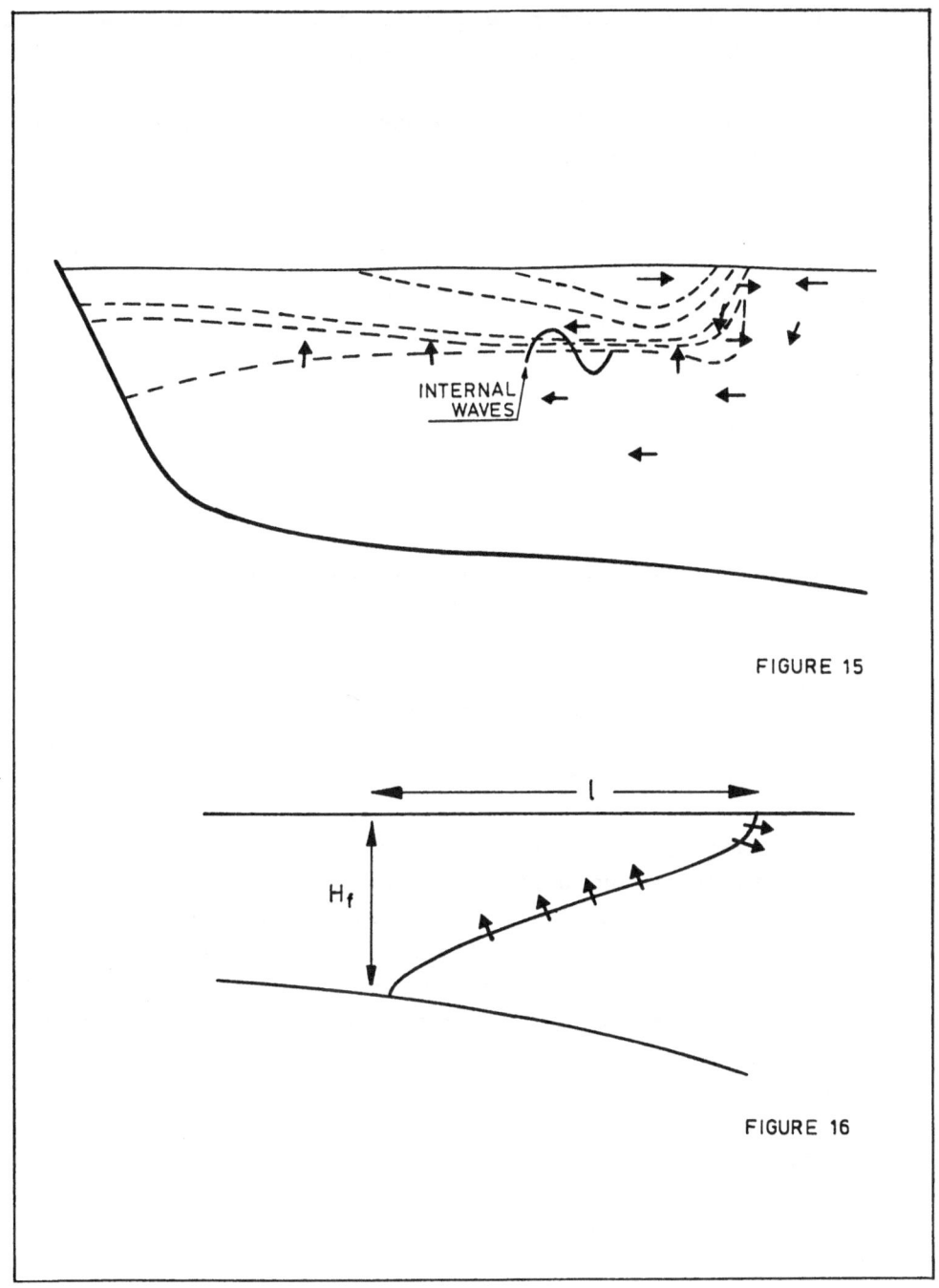

INTERNAL
WAVES

FIGURE 15

H_f

l

FIGURE 16

Fig. 15. Schematic representation of a small scale front.

Fig. 16. Schematic representation of a large scale front.

-Along the front a geostrophic current v(z) exists with a vertical shear related to the horizontal density gradient.

-A cross-frontal circulation u(z) exists in the brackish upper layer, caused by the increase of the longitudinal pressure gradient with depth ($\delta^2 p/\delta x \delta z = -g\delta \varsigma/\delta x$). If the front moves with constant velocity then this circulation is approximately in equilibrium either with vertical exchange of longitudinal momentum ($\delta(N_z \delta u/\delta z)/\delta z = \varsigma^{-1} \delta p/\delta x$, cf estuarine circulation), or with shear induced by the along-front current ($u = f^{-1}\delta(N_z \delta v/\delta z)/\delta z$), or with both. In any case it is strongly conditioned by internal friction (Garrett and Loder, 1981). Due to this circulation, buoyant material and organisms living near the surface are collected at the front.

 -The width of the surface outcropping zone being shorter than the internal Rossby radius of deformation Ro = $f^{-1}\sqrt{\frac{1}{2}gh\delta \varsigma/\varsigma}$, the approximation of geostrophic equilibrium locally fails.

-Detrainment occurs in the frontal zone. Brackish water is mixed downward in the marine bottom layer, causing depletion and finally disappearance of the plume (timescale in the order of a few hours). This detrainment in the frontal zone contrasts with entrainment occurring through the interface in the region behind the front.

Two-layer models yield a more quantitative description of the plume dynamics (Beardsley and Hart, 1978; Brown and Iverson, 1978). The motion of the plume depends not only on density gradients, but also on tidal currents, wind stress and earth rotation. The latter factor causes a deflection of the plume to the right in the northern hemisphere. This feature is also predicted by an analytical model for density-driven flow in a schematised coastal region which has been discussed by Heaps (1972) and applied to Liverpool Bay. The near-bottom flow is friction-dominated and therefore directed towards the coast in the absence of tide and wind.

Large-scale fronts separating coastal and oceanic waters are often found at greater depth and greater distance offshore. They are fed by coastal run-off and they respond to variations in river discharge on time scales much larger than the tidal period. The cross-frontal density contrast is not very large (order 1 kg m^{-3} or less). Both an along-front geostrophic current and a cross-frontal circulation are present. The cross-shore momentum transport integrated over the depth of the upper layer is approximately in equilibrium with longshore wind stress and longshore surface inclination (geostrophic equilibrium).

On the basis of this assumption a qualitative model of the frontal characteristics has been established by Csanady (1984a). From a mass balance of the coastal front, taking into account the entrainment of oceanic water, an expression results for the slope of the pycnocline. From the balance of buoyancy flux and turbulent energy dissipation a relation is obtained for the depth at which the front is formed (anchor depth H_f). The different relationships have the following form:

slope pycnocline	H_f/l	qfresh	$u*^{-3}$
depth of front formation	Hf	qfresh	$u*^4 \cos\theta$
cross frontal density contrast	$\delta\varrho$	qfresh	$u*^{-2} (\cos\theta)^{-1}$

Here

$\quad$ qfresh = fresh water discharge per unit coastal length,

$\quad u_*\quad$ = wind induced friction velocity

$\quad \theta\quad$ = wind direction with respect to the coast

are the parameters which determine entirely the structure of the front in this model (Fig. 16). From the above relationships it can be seen than an increase of fresh water discharge produces a sharper front closer to the shore; wind stress does the opposite.

Cross frontal admixture of coastal waters in oceanic waters can take place by several processes:

- Detrainment at the front
- Incorporation in gulf stream meanders, or in vorticity waves at the shelf edge
- Baroclinic instability

The latter process corresponds to explosive growth of wave-like disturbances at the pycnocline. This can occur when the width of the upper layer l becomes an order of magnitude larger than the Rossby radius of deformation Ro (Pedlosky, 1979; Griffiths and Linden, 1981).

3.3.2. Wind

Wind is the major cause of longshore transport in the coastal zone. For cross-shore transport a close competition between buoyancy and wind stress may exist. Based on a simple analytical model of coastal shelf circulation, Stommel and Leetma (1972) found that besides wind stress (especially the longshore component) and cross-shore density gradient, the Ekman number $E = N_z \cdot f^{-1} \cdot H^{-2}$ plays an important role. For large values of E (corresponding to shallow coastal regions with a high eddy viscosity N_z) buoyancy tends to be the major effect. In Figure 17 this is illustrated for the Dutch North Sea coast. The cross-shore component of the near-surface flow mainly depends on the wind (magnitude and direction); in spite of strong wind fluctuations, the near-bottom flow remains nearly all the time directed to the shore as a consequence of the cross-shore density gradient.

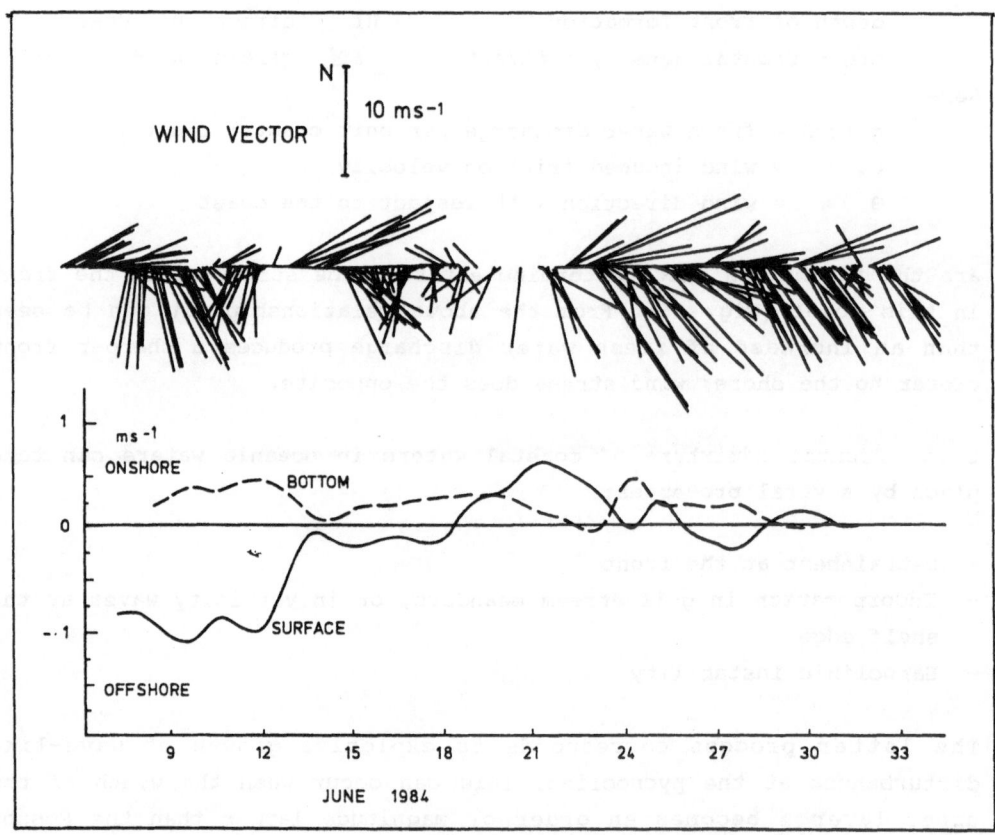

Fig. 17. The history of the wind vector and of the subtidal variations of the cross-shore velocity component near the surface and near the bottom, at location B, figure 18, along the Dutch North Sea coast.

Wind-driven flow in shallow coastal areas where both bottom friction and earth rotation have to be considered is discussed in the textbook of Csanady (1984b). Cross-shore wind drives a cross-shore circulation, and a longshore current to the right of the wind direction (in the northern hemisphere). The magnitude of the current velocity decreases strongly with depth. Longshore wind drives a longshore flow in the wind direction and a cross-shore circulation, with surface flow to the right of the wind direction. The magnitude of this cross-shore circulation is typically in the order of 10 cm s^{-1} for average wind conditions and a factor 2-4 larger for strong winds. Wind directed offshore or directed longshore with the coast to the left (northern hemisphere) contributes to the lateral spreading and mixing of coastal water on the shelf. The inverse occurs for wind directed onshore or directed longshore with the coast to the right.

The latter situation is most characteristic for the Dutch North Sea coast. This may provide an explanation for the generally observed coast parallel salinity distribution (Fig. 18). The average longshore current velocity (mainly wind-driven) amounts to 5-10 cm s^{-1} and is predominantly directed northward (Otto, 1983). It may also turn southward, however, for periods of approximately a week. Consequently a large coastal area is influenced by the Rhine outflow of the Rotterdam Waterway and Haringvliet. As an example of this influence the salinity fluctuation and the mouth of the Eastern Scheldt estuary is shown in Figure 19. The fresh water inflow in the Eastern Scheldt is low and constant and therefore cannot explain the salinity fluctuations. No correlation is observed with variations of the Rhine outflow, but a strong correlation with periods of NW-wind shows up. A westerly on-shore wind component is necessary because off-shore Ekman transport in the upper layer prevents the occurrence of low coastal salinities for northerly longshore wind.

4. TRANSPORT OF PARTICULATE MATTER

A substantial part of the inshore/offshore exchange of constituents - nutrients, pollutants - takes place by particulate transport (Darnell and Soniat, 1979). Fine cohesive sediments can easily bind many chemical substances which are initially dissolved. As long as the settling velocity of these particles is smaller than the upward component of turbulent water motion under all flow conditions, then inshore/offshore exchanges proceed in the same way as for dissolved

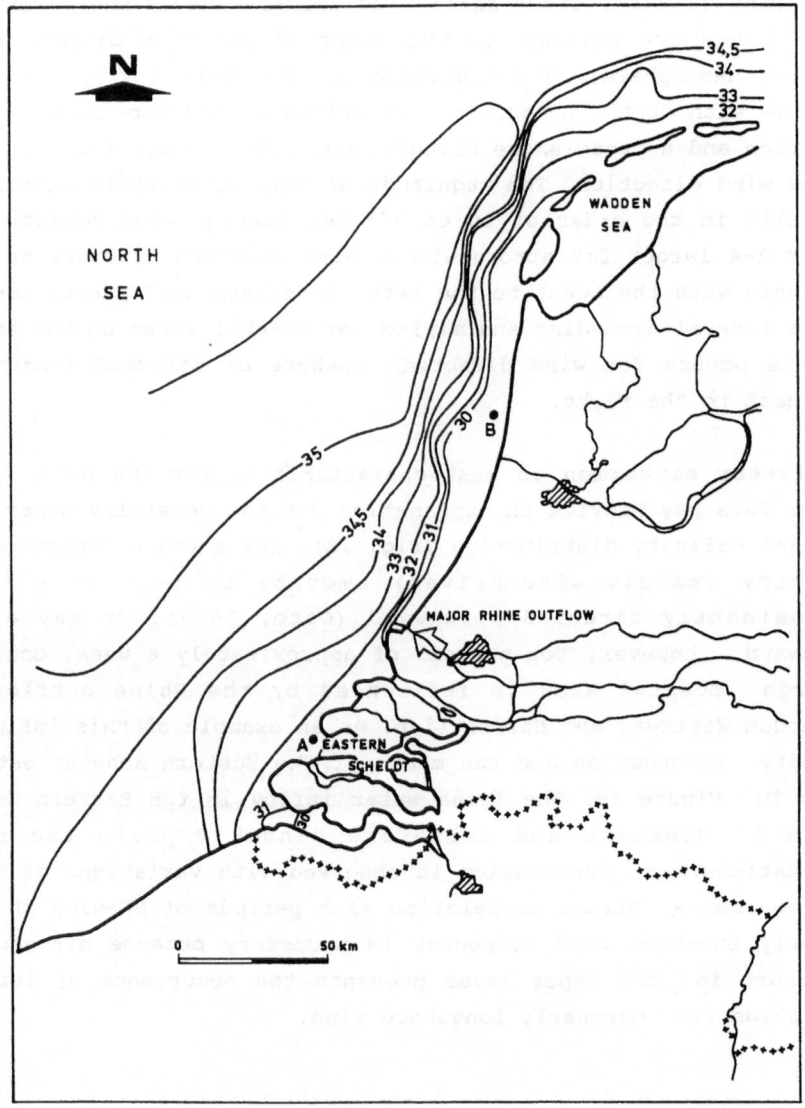

Fig. 18. Salinity distribution along the Dutch North Sea coast, measured in a period of wind and runoff close to mean conditions.

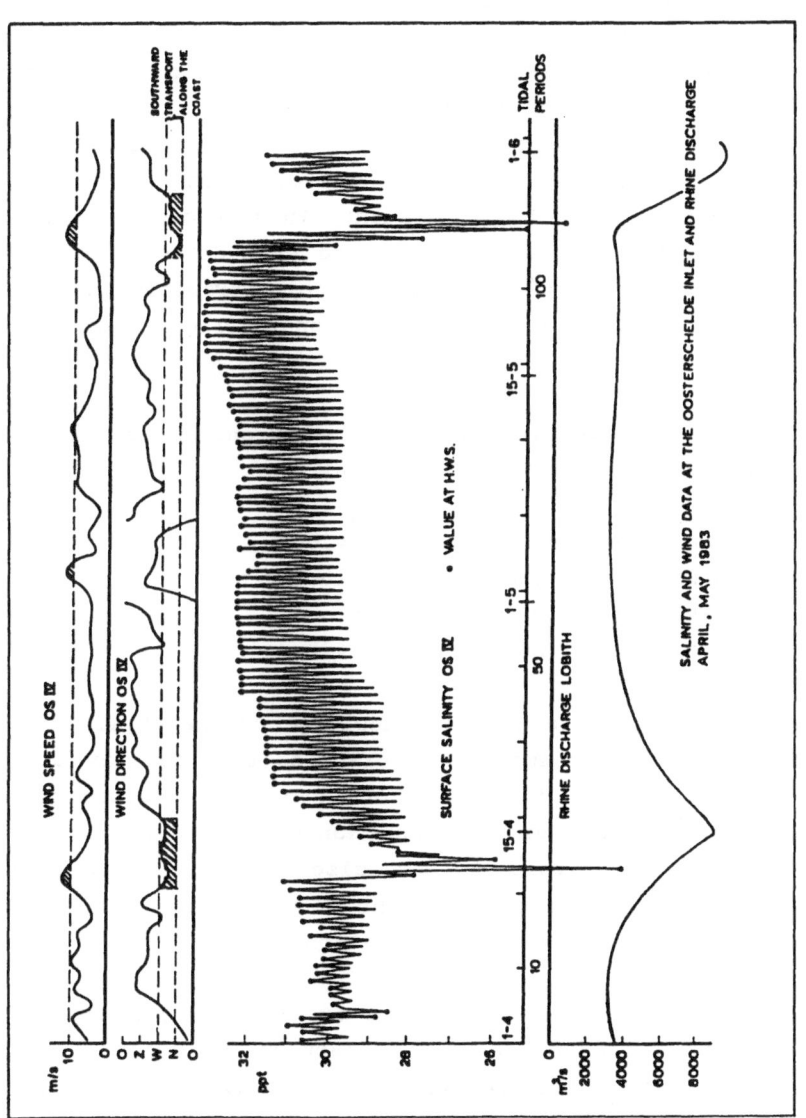

Fig. 19. Salinity fluctuations at the Eastern Scheldt inlet (location A, figure 18) caused by wind driven southward transport of Rhine water along the coast.

substances. However, due to the process of flocculation even the smallest particles may become part of flocs which easily settle under quiet flow conditions. Particles which are settled on the bottom during certain time intervals - for instance, during periods of slack tide or periods of neap tide or periods of calm weather - possess transport properties which are different from the transport properties of passive dissolved substances. The settling mechanism implies that only certain flow conditions are selected during which transport takes place, for instance, only when tidal currents exceed a certain critical value, or only during spring tide or only during storms. This will generally affect the long-term residual displacement. In some cases the residual displacement of particulate material can be even directed opposite to the residual current. For this reason particle transport deserves as much attention as water transport when considering inshore/offshore exchange of constituents.

In order to obtain insight into the residual transport of particulate matter one has to concentrate especially on:

(a) the flocculation and settling behaviour
(b) the consolidation versus erosion processes
(c) the near-bottom velocity distribution
(d) the time history of current velocities

Aspects (a) and (b) require a solid knowledge of the physico-chemical properties of the fine sediments. In practice this is very difficult due to spatial and seasonal variations of the sediment composition. The importance of (c) is rather obvious. Not only settling and erosion depend on the near bottom velocity, but also a substantial part of the transport, because near bottom sediment concentrations are generally much higher than depth-averages concentrations.

Item (d) is related to time lag effects in settling and erosion. For instance, the proportion of depositing particulate matter depends on the duration of low turbulent activity. If this duration is short, the concentration remaining in suspension is high and subsequent transport when the velocity has increased is large. In this way an asymmetry in the time history of currents between ebb and flood will occasion a residual tidal transport (Postma, 1967; Dronkers, 1986a). The role of tidal asymmetry for residual sediment transport in different types of coastal systems has been investigated by Dronkers (1986b). In this study also the important influence of wind-waves on sediment transport is discussed.

The present knowledge of sediment transport processes - especially aspects (a) and (b) - is insufficient for the development of reliable mathematical simulation models. Not only the magnitude of the transport of fine sediment is uncertain, but even the direction often cannot be predicted with certainty. An important obstacle in the current research is the lack of experimental instruments suitable for reliable long-term and large-scale surveys in nature.

Finally, the adsorption and desorption of soluble substances should also be mentioned. An important joint research effort of chemists and physicists on these processes is still needed in order to provide the knowledge necessary for the development of general rules and mathematical models for inshore/offshore nutrient or pollutant exchanges.

5. DISCUSSION

The water motion and exchange processes in shallow coastal systems are conditioned by tidal motion, density effects and wind stress. Bottom topography plays an important role and - to a lesser degree - earth rotation. A detailed simulation of inshore/offshore exchange processes requires the inclusion of all these aspects in a mathematical model. Such a model is extremely complex. Different types of coastal systems can be distinguished and different compartments (inshore basin, inlet and coastal shelf) can be considered separately. For each situation certain simplifications can be justified. The most important processes which should be represented in each model are discussed in the previous sections.

The present knowledge and computer facilities permit, in principle, simulations of water motion and exchange processes for vertically homogeneous waters (no density differences) which are sufficiently reliable for many purposes. Certain compartments of certain coastal systems can therefore be modelled with a satisfactory accuracy. These are in particular the regions with strong tidal currents and important bottom topography effects.

Generally, however, the mixing zones of inland and marine waters are characterized by some degree of vertical density stratification. In that case turbulence and eddy viscosity are suppressed and fronts are formed. Mixing may proceed on a small scale from breaking internal waves and on a large scale from baroclinic instability. The present knowledge of these processes is insufficient to obtain a reliable mathematical formulation.Only some crude parametrisations exist of the processes of turbulence suppression and entrainment (Kullenberg, 1983). Higher order turbulence closure models form an improvement, but are not yet firmly established. Simulations with a three-dimensional numerical shelf-sea from model by James (1984) gave rise to the following conclusions:

- many qualitative features of frontal dynamics are independent of the details of eddy mixing coefficients and basic density structure (for instance, the presence of cross-frontal circulation, surface convergence and eddy pairs developing at the frontal outcrop).

-the amount of frontal sharpening and the pattern of cross-frontal circulation strongly depend on eddy viscosity. Fronts developing in shallow water where tidal mixing is strong are most affected. This result has also been demonstrated by Garrett and Loder (1981).

Further fundamental research on turbulence in stratified flows should be stimulated. Much information on frontal structures has been obtained from remote sensing, and in particular from infrared satellite images (see, for instance, Klemas and Polis, 1977; Pingree and Griffiths, 1978). The development of new techniques such as the microwave radar enables the detection of the sea surface current distribution. Research in this direction will contribute to reveal an overall picture of the dynamics of coastal mixing processes on different space and time scales.

The inshore/offshore exchange of nutrients and pollutants does not depend uniquely on the exchange of water masses. Many substances are transported adsorbed to fine sedimentary particles. Knowledge of adsorption and desorption processes is scarce up to present. The description of the residual transport of fine cohesive sediment still meets serious difficulties due to the complexity of flocculation, consolidation and erosion processes. Joint physico-chemical research in these processes should be strongly recommended.

REFERENCES

Abraham, G., H. Gerritsen & G.J.H. Linijer, 1987. Subgrid tidally induced residual circulations. - Cont. Shelf Res. 7: 285-305.

Abraham, G., P. Jong de & E. van Kruiningen, 1986. Large scale processes in a partly mixed estuary. - In J.v.d. Kreeke (ed.): Proc. Symp. on the physics of shallow bays and estuaries, Miami, 1984. Springer Verlag.

Van Alphen, J.S.L.J., W.P.H. de Ruyter & J.C. Borst, 1986. Outflow and spreading of river Rhine water in The Netherlands coastal zone. - In Dronkers, J. & W. Van Leussen (eds.): Physical Processes in Estuaries. Springer Verlag, in press.

Awaji, T., N. Imasato & H. Kunishi, 1980. Tidal exchange through a strait: A numerical experiment using a simple model basin. - J. Phys. Oceanogr. 10: 1499-1508.

Beardsley, R.C. & J. Hart, 1978. A simple theoretical model for the flow of an estuary onto a continental shelf. - J. Geophys. Res. 83: 873-883.

Bowden, K.F., 1965. Horizontal mixing in the sea due to a shearing current. - J. Fluid Mechanics 21: 83-95.

Bowman, H.J. & R.L. Iverson, 1977. Estuarine and plume fronts. - In Oceanic fronts in coastal processes. Springer Verlag, Berlin.

Csanady, G.T., 1984a. The influence of wind stress and river runoff on a shelf-sea front. - J. Phys. Oceanogr. 14: 1383-1392.

Csanady, G.T., 1984b. Circulation in the coastal zone. D. Reidel Publ. Co., Dordrecht/Boston/Lancaster.

Darnell, R.M. & T.M. Soniat, 1979. The estuary/continental shelf as an interactive system. - In R.J. Livingston (ed.): Ecological processes in coastal and marine systems, pp. 487-525. Plenum Press, N.Y.

Dronkers, J., 1978. Longitudinal dispersion in shallow well mixed estuaries. Proc. 16th Conf. Coastal Engineering 3: 2761-2777.

Dronkers, J., 1982. Conditions for gradient-type dispersive transport in one-dimensional tidally averaged transport models. - Estuarine, Coastal and Shelf Sci. 14: 599-621.

Dronkers, J. & J.T.F. Zimmerman, 1982. Some principles of mixing in tidal lagoons. - Oceanologica Acta No SP, pp. 107-118.

Dronkers, J. & J. Van de Kreeke, 1986. Experimental determination of salt intrusion mechanisms in the Volkerak estuary. - Neth. J. Sea Res. 20:1-19.

Dronkers, J. 1986a. Tide induced residual transport of fine sediment. - In J. V.d. Kreeke (ed.): Proc. Symp. Physics of Shallow Bays and Estuaries, Miami, 1984. Springer Verlag.

Dronkers, J. 1986b. Tidal asymmetry and estuarine morphology. - Neth. J. Sea Res. 20: 117-131.

Elliot, A.J., 1978. Observations of meteorologically induced circulation in the Potomac estuary. - Estuar. Coastal Mar. Sci. 6: 285-299.

Fischer, H.B., 1972. Mass transport mechanisms in partially stratified estuaries. - J. Fluid Mechanics 53: 671-687.

Garrett, C.J.R. & J.W. Loder, 1981. Dynamical aspects of shallow sea fronts. - Philosoph. Trans. Royal Soc. Lond. A-302: 563-581.

Garvine, R.W. & J.D. Monk, 1974. Frontal structure of a river plume. - J. Geophys. Res. 79: 2251-2259.

Garvine, W., 1977. Observations of the motion field of the Connecticut river plume. - J. Geophys. Res. 82: 441-454.

Griffiths, R.W. & P.F. Linden, 1981. The stability of buoyancy-driven coastal currents. - Dynamics Atmos. Oceans - 5: 281-306.

Hansen, D.V. & M. Rattray, 1965. Gravitational circulation in straits and estuaries. - J. Mar. Res. 23: 319-326.

Hansen, D.V. & M. Rattray, 1965. Gravitational circulation in straits and estuaries. - J. Mar. Res. 23: 104-122.

Hansen, D.V. & M. Rattray, 1966. New dimensions in estuary classification. - Limnol. Oceanogr. 11: 319-325.

Heaps, N.S., 1972. Estimation of density currents in the Liverpool bay area of the Irish Sea. - Geophys. J. Royal Astronom. Soc. 30: 415-432.

Heathershaw, A.D. & F.D.C. Hammond, 1980. Secondary circulations near sand banks and in coastal embayments. - Deutsche Hydrogr. Zeitschr. 33: 135-151.

Helder, W. & O. Ruardij, 1983. A one-dimensional mixing and flushing model of the Ems-Dollard estuary: Calculation of time scales at different river discharges. - Neth. J. Sea Res. 17: 293-312.

James, I.D., 1984. A three-dimensional numerical shelf-sea front model with variable eddy viscosity and diffusivity. - Cont. Shelf Res. 3: 69-98.

Klemas, V. & D.F. Polis, 1977. Remote sensing of estuarine fronts and their effects on pollutants. - Photogramm. Eng. Remote Sensing 43: 599-612.

Komen, G.J. & H.W. Riepma, 1981. The generation of residual vorticity by the combined action of wind and bottom topography in a shallow sea. - Oceanologica Acta 4: 267-277.

Kullenberg, G., 1983. Mixing processes in the North Sea and aspects of their modelling. - In J. Sünderman & W. Lenz (eds.): North Sea dynamics, pp. 349-369. Springer Verlag.

Lee, K.K. & J.A. Ligget, 1970. Computation for circulation in stratified lakes. - J. Hydraul. Div. A.S.C.E. 96: 2089-2115.

Lewis, R.E., 1984. Circulation and mixing in estuary outflows. - Cont. Shelf Res. 3: 201-214.

Nunes, R.A. & J.H. Simpson, 1985. Axial convergence in a well-mixed estuary. - Estuar. Coastal Shelf Science 20: 637-649.

Okubo, A., 1967. The effect of shear in an oscillatory current on horizontal diffusion from an instantaneous source. - Int. J. Oceanol. Limnol. 1: 194-204.

Okubo, A., 1971. Oceanic diffusion diagrams. - Deep-Sea Res. 18: 789-802.

Okubo, A., 1973. Effect of shoreline irregularities on streamwise dispersion in estuaries and other embayments. - Neth. J. Sea Res. 6: 213-224.

Okubo, A., 1974. Some speculations on oceanic diffusion diagrams. - Rapp. Proc.-v. Cons. Int. Explor. Mer 167:77-85.

Otto,L., 1983. Currents and water balance in the North Sea. - In J. Sünderman & W. Lenz (eds.): North Sea dynamics, pp. 26-43. Springer Verlag.

Pasmanter, R., 1986. Dynamical systems, deterministic chaos and dispersion in shallow tidal flow. - In J. Dronkers & W Van Leussen (eds.): Physical Processes in Estuaries. Springer Verlag, in press.

Pedlosky, J., 1979. Geophysical fluid dynamics. Springer Verlag.

Pingree, R.D. & D.K. Griffiths, 1978. The tidal physics of headland flows and offshore tidal bank formation. - Mar. Geol. 32:269-289.

Postma, H., 1961. Transport and accumulation of suspended matter in the Dutch Wadden Sea. - Neth. J. Sea Res. 1: 148-190.

Postma, H., 1980. Sediment transport and sedimentation. - In E. Olausson & I. Cato (eds.): Chemistry and biochemistry of estuaries, pp. 153-186. John Wiley.

Robinson, A.H.W., 1966. Residual currents in relation to shoreline evolution of the East Anglian coast. - Mar. Geol. 4: 57-84.

Schubel, J.R., 1971. Estuarine circulation and sedimentation. - Lecture Notes, Am. Geol. Inst., Washington D.C.

Simons, T.J., 1980. Circulation models of lakes and inland seas. - Can. Bull. Fish. Aquat. Sci. 203, Ottawa 1980.

Smith, N.P., 1977. Meteorological and tidal exchanges between Corpus Christi Bay, Texas, and the northwestern Gulf of Mexico. Estuar. Coastal Shelf Science 5: 511-520.

Smith, N.P., 1985. The decomposition and simulation of the longitudinal circulation in a coastal lagoon. - Estuar. Coastal Shelf Science 21: 623-632.

Stommel, H. & H.G. Farmer, 1952. On the nature of estuarine circulation. - Woods Hole Oceanogr. Inst. Ref. no. 52-51.

Stommel, H. & A. Leetmaa, 1972. The circulation on the continental shelf. - Proc. Nat. Acad. Sci. U.S. 69: 3380-3384.

Swenson, E.M. & W.-S. Chuang, 1983. Tidal and subtidal water volume exchange in an estuarine system. - Estuar. Coastal Shelf Science 16: 229-240.

Swift, D.J.P., 1976. Coastal sedimentation. - In D.J.P. Swift & D.J. Stanley (eds.): Marine sediment transport and environmental management, pp. 255-310. John Wiley.

Taylor III, R.B., 1974. Dispersive mass transport in oscillatory and unidirectional flows. - Coastal Oceanogr. Eng. Lab., College Eng., Univ. Florida, Techn. Rep. 24.

Terwindt, J.H.J., 1965. Orienterend onderzoek naar de herkomst en beweging van slib in het Deltagebied bij de huidige toestand. - Rep. K-239, Rijkswaterstaat (in Dutch).

Ünlüata, Ü., T. Oguz & E. Özsoy, 1983. Blocking of steady circulation by coastal geometry. - J. Phys. Oceanogr. 13: 1055-1062.

Van de Kreeke, J. & R.G. Dean, 1975. Tide induced mass transport in lagoons. - Jour. Waterways, Harbors Coastal Eng. Div. 101: 393-403.

Vieira, M.E.C., 1985. Estimates of subtidal volume flux in mid-Chesapeake Bay. - Estuar. Coastal and Shelf Science 21: 411-427.

Walters, R.A., R.T. Cheng & T.J. Conomos, 1985. Time scales of circulation and mixing processes of San Fransisco Bay waters. - Hydrobiologia 129: 13-36.

Wanless, H.R., 1976. Intracoastal sedimentation. - In D.J.P. Swift & D.J. Stanley (eds.): Marine sediment transport and environmental management, pp. 221-239. John Wiley.

Zimmerman, J.T.F., 1976. Mixing and flushing of tidal embayments in the western Dutch Wadden Sea II: Analysis of mixing processes. - Neth. J. Sea Res. 10: 397-439.

Zimmerman, J.T.F., 1978. Topographic generation of residual circulation by oscillatory (tidal) currents. - Geophys. Astrophys. Fluid Dynamics 11: 35-47.

Zimmerman, J.T.F., 1981. Dynamics, diffusion and geomorphological significance of tidal residual eddies. - Nature 290: 549-555.

Zimmerman, J.T.F., 1986a. A comparison of 3D and 2D models of topographically rectified tidal currents. - In J.v.d. Kreeke (ed.): Proceedings of the symposium on the physics of shallow estuaries and bays, in press. Springer Verlag.

Zimmerman, J.T.F., 1986b. The tidal whirlpool: a review of horizontal dispersion by tidal and residual currents. - Neth. J. Sea Res. 20:133-154.

COASTAL/OFFSHORE WATER EXCHANGE IN NARROW, DEEP SHELF AREAS

T.A. McClimans
Norwegian Hydrotechnical Laboratory
and Division of Port and Ocean Engineering, NTH
Trondheim, Norway

1. INTRODUCTION

To define coastal and offshore waters as separate entities is to set up a dividing line either on geographical, biological, chemical or physical basis, or a combination of these. In my opinion, the density front that often exists between coastal and offshore waters is as good a dividing line as one could wish, especially for areas with a deep, narrow shelf. Large gradients of temperature, salt, biomass and current velocity occur at these fronts. For deep, narrow shelves, geographical (topographical) constraints are very weak. In the following we will be concerned primarily with the motions of these fronts and the processes occurring in frontal regions. Most of the presentation is based on my studies of the Norwegian Coastal Current (NCC), the results of which are congruent with other deep, narrow shelf areas.

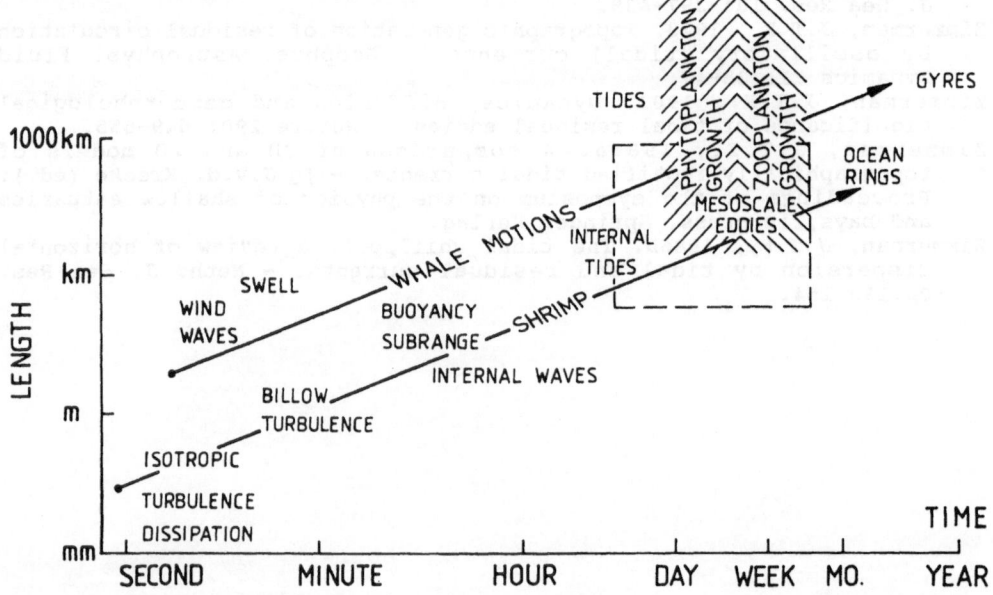

Fig. 1. Time and space scales of some physical features together with some biological scales. The dashed box indicates important scales for coastal current exchange.

Lecture Notes on Coastal and Estuarine Studies, Vol. 22
B.-O. Jansson (Ed.), Coastal-Offshore Ecosystem Interactions.
© Springer-Verlag Berlin Heidelberg 1988

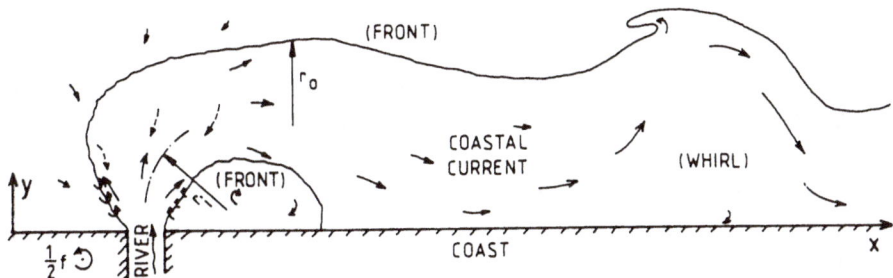

Fig. 2. Schematic of a river plume and coastal current (northern hemishpere) (McClimans, 1983).

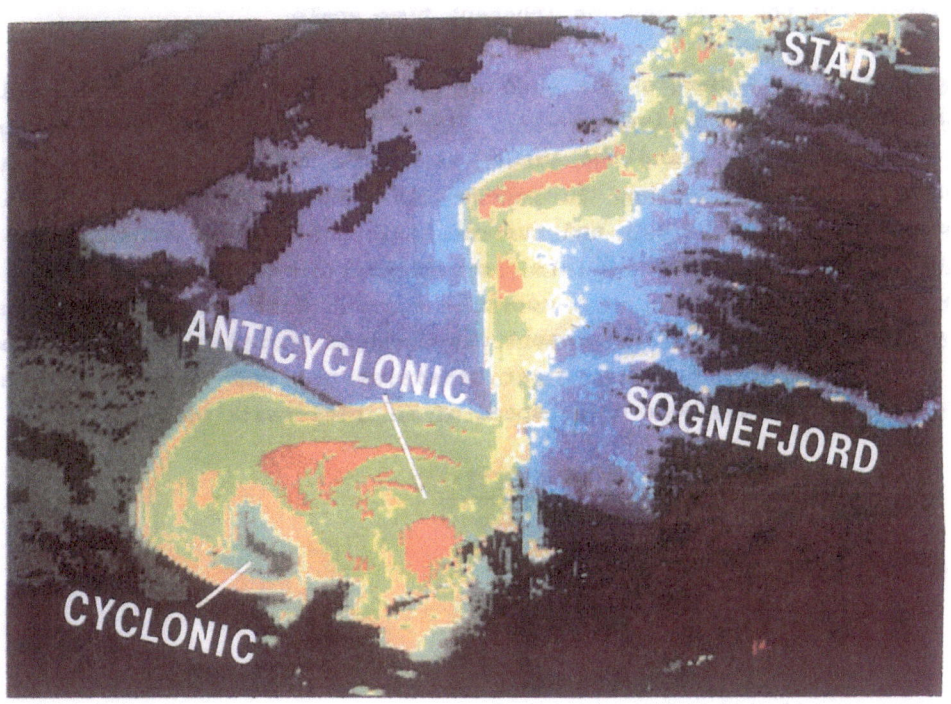

Fig. 3. NOAA satellite thermal image of the NCC (Courtesy Tromso Telemetry Station).

2. SCALES

Exchange processes occur within a large range of time and space scales. Their biological importance varies from the survival of the individual (respiration, nutrient supply, violent shears) to the survival of their species (migration, predation, recruitment, food supply). The development of an ecologic system depends on the types and stability of the water exchange processes. Figure 1 gives a general overview of the various physical and biological scales of interest for the present discussion. Scale factors are essential to modellers for interpreting results from one experiment to the conditions of another. The dynamics of coastal current flows and exchange depend on the scales of various boundary conditions like fresh (or brackish) water outflow, mixing (wind, tides, plume, artificial) and stability. Topography and the earth's rotation steer the flow in a variety of ways depending on the stability of the flow and its inherent time scale.

Figure 2 gives a bird's eye view of the development of a densimetric coastal current from river runoff. Figure 3 shows a NOAA thermal image from the NCC revealing the large, mesoscale eddies that are typical for most coastal regions throughout the world, e.g. British Columbia (Emery and Mysak, 1980) and Australia (Griffiths and Pearce, 1985). Long, interfacial waves regulate flow changes along the coast. Figure 4 shows a cross section of a coastal current with longshore speed u, greatly simplified for the present discussion. Rotation ($f/2$) produces a sloping interface. With layer densities p_i and acceleration of gravity g, the longshore phase speed of an interfacial wave is

$$c_i = \frac{h_1 h_2 \ g(\varsigma_1 - \varsigma_1)}{(h_1 + h_2) \ \varsigma_2}^{\frac{1}{2}}$$

and the ratio

$$\frac{u}{c_i} = F$$

is the so-called densimetric Froude number. Many large scale flows in nature, including the buoyant coastal currents have $F \approx 1$. For such large-scale flows it is often convenient to look at the densimetric Froude number as a ratio of length scales

$$F = \frac{r_i}{r_o}$$

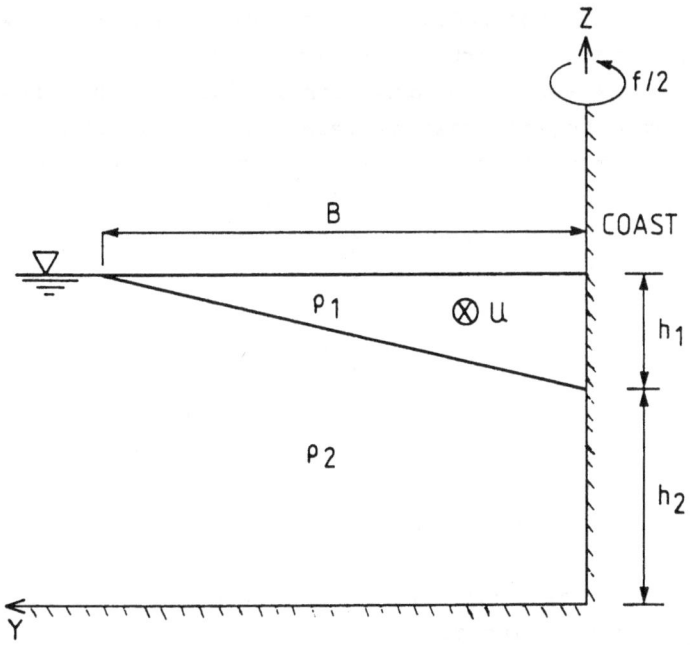

Fig. 4. Cross section of a two layer (Margules) coastal current.

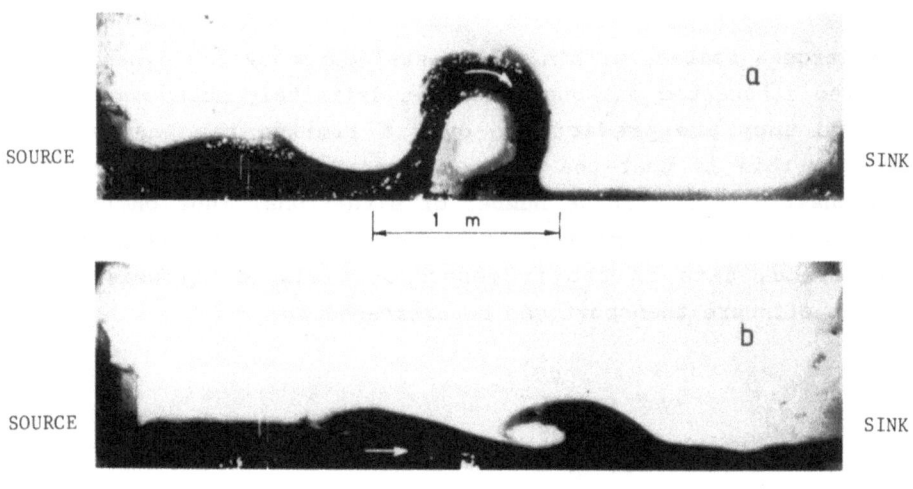

Fig. 5. Laboratory experiments of densimetric coastal currents.
a) F <1; b) F>1.

where $r_i = u/f$ is the radius of curvature of a free jet with velocity u moving on a rotating earth with local angular rotation speed $f/2$ and $r_o = c_i/f$ is the Rossby deformation radius, or the offshore spreading achieved by a buoyant mass of water before the effect of the earth's rotation thwarts it to the right (in the northern hemisphere). These length scales are on the order of 10 km. The parameter $f = 2\omega\sin\theta$ is called the Coriolis parameter. Here ω is the angular rotation speed of the earth and θ is the latitude.

In the laboratory experiments shown in Figure 5 the general character of the flow can be seen to change from F<1 (subcritical) to F>1 (supercritical). In the subcritical case the cyclonic (counter-clockwise) eddies are accompanied by a large growth and separation of anticyclonic (clockwise) eddies within the coastal current. The pools may form large lenses which carry large amounts of coastal water from the coast to the offshore waters. The lifetime of these isolated pools may be several weeks (or months), depending on the size of the outflow. The large-scale mixing to the coastal current occurs primarily in the cyclonic eddies. There is, however, mixing on a multitude of scales along the convolved front. For the largest scales, the tides make a minor contribution to mixing. Wind effects will be discussed later.

3. TURBULENCE

On the largest scales, within the coastal current, the dynamics may be considered stochastic although there is definitely much order in these flows and they are predictable over a limited time and space. The reason for this is that the boundary conditions forcing these motions are stochastic over a large number of situations. Thus on time scales of months (seasonal averages), it has been common in the past to look at historical data in the framework of a simple diffusion model in which the offshore transport can be expressed as

$$Q_y = A_T \frac{dC}{dy}$$

where Q_y is the offshore (y) flux of concentration C and A_T is the turbulent exchange coefficient. Advanced diffusion models depend on a good algorithm for A_T. With our present day knowledge of the dynamics of buoyant coastal currents we can estimate the exchange coefficient to be proportional to ur_0. The factor of proportionality must be determined and will undoubtedly depend strongly on whether F is greater than or less than 1 for the given area. Wind and tidal mixing will of course increase A_T.

The large scale flows discussed above are essentially two-dimensional in the horizontal plane. At scales for which the random motions are smaller than the thickness of the spreading flow, the turbulence becomes more isotropic, that is three dimensional. This is the scale for which wind and wind-wave induced turbulence acts to mix the flow both within each layer and across the fronts. For scales much less than the thickness of the transition zone the random motions are nearly isotropic. It is in this region that unusually large current velocity strains may tear apart small plankton that are not adapted to the environment. The threshold below which molecular viscosity reduces the turbulent velocity shears is the so-called Kolmogorov length

$$= \frac{\nu^3}{\varepsilon}^{1/4}$$

where ε is the energy dissipation (= energy supply) and ν is the kinematic molecular viscosity. λ decreases weakly with increased turbulence and is usually on the order of 1 mm.

4. WIND

Wind is an important agent for the transport and mixing of water masses at sea. Concepts like "wind-mixed layer" and "coastal upwelling" are well known. In coastal regions the stability of the water masses modify the effects of wind. For example the densimetric coastal current acts as a wave guide to transmit wind effects (downstream) along the coast at the speed of the interfacial waves c_i. This is on the order of 1 m/s. In this manner, some of the remote effects of wind can be predicted several days in advance.

In view of the large momentum connected with mesoscale eddies, they may often produce effects which resemble coastal upwelling and/or downwelling. Confusion in the interpretation of several data sets has arisen from a lack of knowledge of these events. In cases where there are counter currents shoreward of anticyclonic eddies it is tempting to assign the process to the wind (if it blows in the right direction).

The actual amount of wind forcing necessary to reverse a coastal current the size of the NCC is enormous. McClimans and Nilsen (1982) concluded that it requires a sustained northerly wind of 30 m/s. At best, normal wind stresses can move the thin surface layer shoreward or seaward a few kilometers, during which time wind-induced mixing may be important. The deeper mixing is presumably induced indirectly by wind-induced motions. This process was explained briefly in three paradoxes of tail winds over coastal currents (McClimans and Eidnes, 1983).

Wind may at times have strange effects in densimetric coastal currents. This arises from the fact that a spreading flow contains a large amount of potential energy. A measure of the transfer of this energy to turbulence is provided by the Richardson number

$$ Ri = \frac{g}{\varrho} \frac{d\varrho}{dz} \left(\frac{du}{dz} \right)^{-2} $$

in which the stabilizing effect of the vertical density gradient $d\varrho/dz$ is compared to the destabilizing effect of the vertical current shear du/dz. For $Ri < 1/4$ the spreading flow is unstable, producing billow turbulence and rapid vertical mixing across the pycnocline (density gradient). Thus, a wind that produces and increase of du/dz in a flow that is close to $Ri = 1/4$ may cause a breakdown of the density field driving the flow.

The interplay between local and remote wind effects is very complicated and it is tempting to return to the large space and time scales to reduce the problem to a stochastic diffusion model. Unfortunately, the data collection programs are limited in space and time and, without more detailed knowledge of the situation at hand, the seasonal estimates may be quite misleading.

5. TRANSPORT

Wind drift in coastal regions, like tidal currents, give often smaller residual transports than the densimetric flows. However, they account for much of the variability. Thus the time scale of interest should determine where the efforts are to be directed. Cases in point are spawning, larva incubation, food transport, species migrations and perhaps catastrophic events like cold fronts and red tides. Many of these processes have a time scale of weeks, the timing of successive

events being important for optimum (maximum) production. A very simple
formula for estimating the longshore transport of densimetric
(baroclinic) coastal currents in regions with narrow, deep shelves is
the so-called hydrographers equation

$$Q = \frac{g(\varrho_2 - \varrho_1)h_1^2}{2f\varrho_2}$$

Although this calculation may be affected by the variability caused by
wind and tidally induced interfacial waves, it is based on the same
geostrophic balance as the celebrated Margules current and appears to
be quite robust for rough estimates. The most convenient aspect with
this formula is that the transport is computed (within the conditions
of the theory) on the basis of only one vertical density profile, or
one "cast" in oceanographic terminology.

Beneath the coastal flow there is also a current, although often
assumed weak. In the case of the California current, the underflow is
in the opposite direction most of the time (Ikeda et al., 1984). At
times the flow is so strong that it transports the densimetric coastal
current "upstream". Large mesoscale eddies in the external flow manage
to scavenge large portions of water from the coastal region off the
coast of California (Simpson, 1984).

6. ENTRAINMENT

The process of a coastal current transport of the admixture of seawater
to river runoff is known as entrainment. Garvine (1974) was perhaps the
first to point out that the river water was rather detrained to the
larger ocean. His ideas were first formulated for the Connecticut River
plume which is mixed downward due to tidal action, but he has proven
that most pools of lighter water are entrained by the much more
voluminous external sea (Garvine, 1979). There may of course be local
entrainment to the coastal flow, but large losses occur when a ring of
coastal water migrates seaward. A rough estimate based on the
laboratory results of McClimans and Green (1982) indicated that, in the
absence of an external flow and renewed fresh water supply, most of the
coastal current will peel off within about 30 r_0 from its major source.
The NCC is, however, contained by a residual onshore current that
reduces this loss. The Alaskan Coastal Current is contained often by
wind-induced downwelling (Royer, 1982).

7. NEWER DEVELOPMENTS

Recent developments in coastal current studies are relevant to water transport and offshore exchange. Patchiness has for several years been a problem for quantifying the production of the sea. Numerous satellite images and laboratory experiments show that there are multiple fronts separating coastal waters from offshore waters. Filaments of highly productive regions are deformed in large spirals and mixed through wind action. On smaller scales, winds and waves produce long filaments called windrows which appear to be a highly anisotropic, near surface turbulence. The action of waves propagating through regions of high current shears is being studied in Trondheim together with the wind paradoxes.

In Norway there has been some concern for the effects of man-made changes in the hydrologic cycle (anthropogenic changes), in particular, the extensive regulation of fresh water sources for the production of hydroelectric energy. Although the natural variability of the Norwegian Coastal Current greatly exceeds the amount of water which is diverted to hydroelectric production, the ecology may be highly sensitive to systematic change - especially in the seasonal timing of the various biological processes. This topic has been the subject of two workshops in Norway (Skreslet et al., 1976; Skreslet, 1986) and the questions are difficult to formulate due in part to the complexity of the physical exchange processes.

ACKNOWLEDGEMENTS

This work has been supported in part by the Fund of Licence Fees.

REFERENCES

Emery, W.J. & L.A. Mysak, 1980. Dynamical interpretation of satellite-sensed thermal features off Vancouver Island. - J. Phys. Ocean. 10: 961-970.

Garvine, R.W., 1974. Dynamics of small scale oceanic fronts. - J. Phys Ocean. 4: 557-569.

Garvine, R.W., 1979. An integral hydrodynamic model of upper ocean frontal dynamics: Part II. Physical characteristics and comparison with observations. - J. Phys. Ocean. 9: 19-36.

Griffiths, R.W. & A.F. Pearce, 1985. Satellite images of an unstable warm eddy derived from the Leeuwin Current. - Deep Sea Res. 32: 1371-1380.

Ikeda, M., W.J. Emery & L.A. Mysak, 1984. Seasonal variability in meanders of the California Current System off Vancouver Island. - J. Geophys. Res. 89: 3487-3505.

McClimans, T.A., 1983. Laboratory simulation of river plumes and coastal currents. ASME Symposium on modeling of environmental flow systems, Boston. - FED 8: 3-9.

McClimans, T.A. & G. Eidnes, 1983. Three paradoxes of tail winds over baroclinic coastal currents. - Ocean Modelling 50: 12-13.

McClimans, T.A. & T. Green, 1982. Phase speed and growth of whirls in a baroclinic coastal current. - River and Harbour Lab. Rep STF 60 A82108.

McClimans, T.A. & J.H. Nilsen, 1982. Whirls in the Norwegian Coastal Current. - In H.G. Gade, A. Edwards & H. Svendsen (eds.): Coastal oceanography, pp. 311-320. Plenum Press.

Royer, T.C., 1982. Coastal fresh water discharge in the northeast Pacific. - J. Geophys. Res. 87: 2017-2021.

Simpson, J.J., 1984. Lateral entrainment of non-local waters by offshore mesoscale eddies in the California Current System. - Paper S5.03 at the 10th EGS annual meeting, Louvain, Belgium.

Skreslet, S., 1986. The role of freshwater outflow in coastal marine ecosystems. - NATO ASI Series G. Springer Verlag.

Skreslet, S., R. Leinebo, J.B.L. Matthews & E. Sakshaug, 1976. Fresh water on the sea. - Ass. Norweg. Oceanogr.

SATELLITE REMOTE SENSING FOR ESTIMATING COASTAL OFFSHORE TRANSPORTS

U. Horstmann
Institut für Meereskunde an der Universität Kiel
Düsternbrooker Weg 20
2300 Kiel, Federal Republic of Germany

1. INTRODUCTION

The present capability of remote sensing in biological oceanography makes it a logical tool in the study of coastal/offshore ecosystem coupling. Sea surface temperature images from the NOAA satellite systems give indications of water flow and concomitant transport of nutrients. The Coastal Zone Color Scanner (CZCS) can trace the transport of suspended matter and show changes in spatial distribution of chlorophyll concentrations. LANDSAT and NIMBUS G images depict increased light attenuation near inlets which can be attributed to eutrophication and other anthropogenic disturbances.

A review of the application of remote sensing in the study of coastal-offshore transports should refer to various examples around the globe. Such examples do exist, especially from areas where cloudfree skies permit frequent recordings of images, like in the Mediterranean (Sturm, 1983) or along the subtropical and tropical Atlantic or Pacific coasts (NASA 1982; Zion and Abbot, 1984) However, by choosing all the examples from the Baltic Sea, where frequent cloud cover limits the amount of sea-surface images tremendously, a more convincing demonstration of the capability of remote sensing can be achieved. The intent of this paper is to demonstrate to what extent coastal-offshore processes can be traced with the present state of the art of remote sensing in oceanography.

2. MATERIALS AND METHODS

Since 1978 daily passes of the TIROS NOAA satellites with their infrared scanners and NASA's NIMBUS 7 satellite carrying amongst other instruments the Coastal Zone Color Scanner (CZCS) have accumulated a tremendous amount of data from coastal and offshore regions around the world. Such data are available from receiving stations in different regions through the national points of contact or directly from NASA. The data for Europe, especially the northwestern part, as semi-processed images can be

obtained from the University of Dundee's receiving station in Scotland
(Baylis, 1981). The data used in this presentation originate mainly from
Dundee. For background information of the Coastal Zone Color Scanner, see
Table 1.

```
    Swath width . . . . . . . . . . . . . . . . . . . 1550 km
    Resolution(pixel size at nadir). . . . . . . . . . . 825 m
    Equator crossing . . . . . . . . . . . . . . . . . noon
    Coverage . . . . . . . . . . . . . . . . . . . . . (daily above 55°N)


    Spectral bands

    Center:        Half width:      Characteristics:

    443 nm         20 nm            Peak of chlorophyll-a absorption (blue)
    520 nm         20 nm            Wing of chlorophyll-a absorption (blue)
    550 nm         20 nm            Minimum of chlorophyll-a absorption
    670 nm         20 nm            Atmospheric reference
    750 nm         100 nm           Land/cloud versus water separation
    11.5 μm        2 μm             Temperature (non-operational since 1981)
```

Table 1. Nimbus 7/CZCS specifications

The sea surface temperature images presented here are mainly contrast-
enhanced raw data of the infrared channel no. 4 of the Advanced Very High
Resolution Radiometer (AVHRR) of the TIROS and NOAA satellites. A few
AVHRR scenes have been processed for absolute temperatures using the
algorithms of McClain (Llewllyn-Jones et al., 1984). The Coastal Zone
Color Scanner data originate partly from the receiving station in Dundee
and partly from NASA's Goddard Space Flight Center, USA. The data are
processed using the digital interactive image processing system (DIBIAS)
at the DFVLR, the German Aerospace Research Center where a software
package consisting of 12 different programs has been developed for this
purpose (van der Piepen et al., 1985).

For all images presented here the following data processing was
undertaken:

1) Geometric corrections

A panorama correction was applied to all scenes so as to compensate for the distortions originating from the scanning process.

In order to be able to compare individual pixels of the satellite image with ground truth collected by ships, geographic reference points were used to transform some sub-scenes more accurately into a Lambert projection.

2) Radiometric corrections

Radiometric corrections, were done to determine the Rayleigh radiance (scattering by air molecules) and the Mie radiance (scattering by aerosols) including a pixel to pixel processing taking the changes of illumination and the scan angles into account. For these computations Violliers (1982) algorithms were used.

Pigment concentrations (chlorophyll and phaeophytin) were estimated using radiance ratios of wave lengths of maximum chlorophyll absorption (443 nm) to minimum chlorophyll absorption (550 nm).

We applied the following algorithms of Gordon and Morel (1983) with their respective constants to obtain the images shown here.

$$Chl = 1.92 \, \frac{R(\lambda_1)}{R(\lambda_3)}^{-1.8} \qquad \lambda_1 = 443 \text{ nm} \\ \lambda_3 = 550 \text{ nm}$$

Where R (1) : R(3) is the reflectance ratio of CZSC channels one and three. The algorithms used have been developed and partly verified for waters of the western Atlantic and the Mediterranean Sea. They were not specifically produced for waters where inorganic and/or organic suspended matter as well as Gelbstoff make a dominant contribution to the optical properties, in nearshore waters, especially near river outlets (so called "Case 2" waters). In these waters specific algorithms are needed especially for quantitative estimates of phytoplankton pigments.

3. RESULTS

3.1. <u>The detection of coastal-offshore transport through sea surface temperature anomalies</u>

By observing the images obtained from the infrared bands of the AVHRR it is possible to resolve sea surface temperature anomalies from which

Fig. 1. Seasurface temperature on two consecutive days of the Central Baltic Sea in winter demonstrating heat loss of nearshore waters (NOAA 6, AVHRR Infrared of 22 and 23 Feb. 1981, light = cold, dark = warmer waters).

Fig. 2. Coastal upwelling in the Southern Baltic Sea in August 1982. (NOAA 7, AVHRR Infrared Channel 4).

conclusions on processes prior to the image can be drawn. Processes can directly be observed on sequences of images on consecutive days or scenes received at greater intervals. Even in areas with frequent cloud cover such series of images can be obtained, although at less frequent intervals.

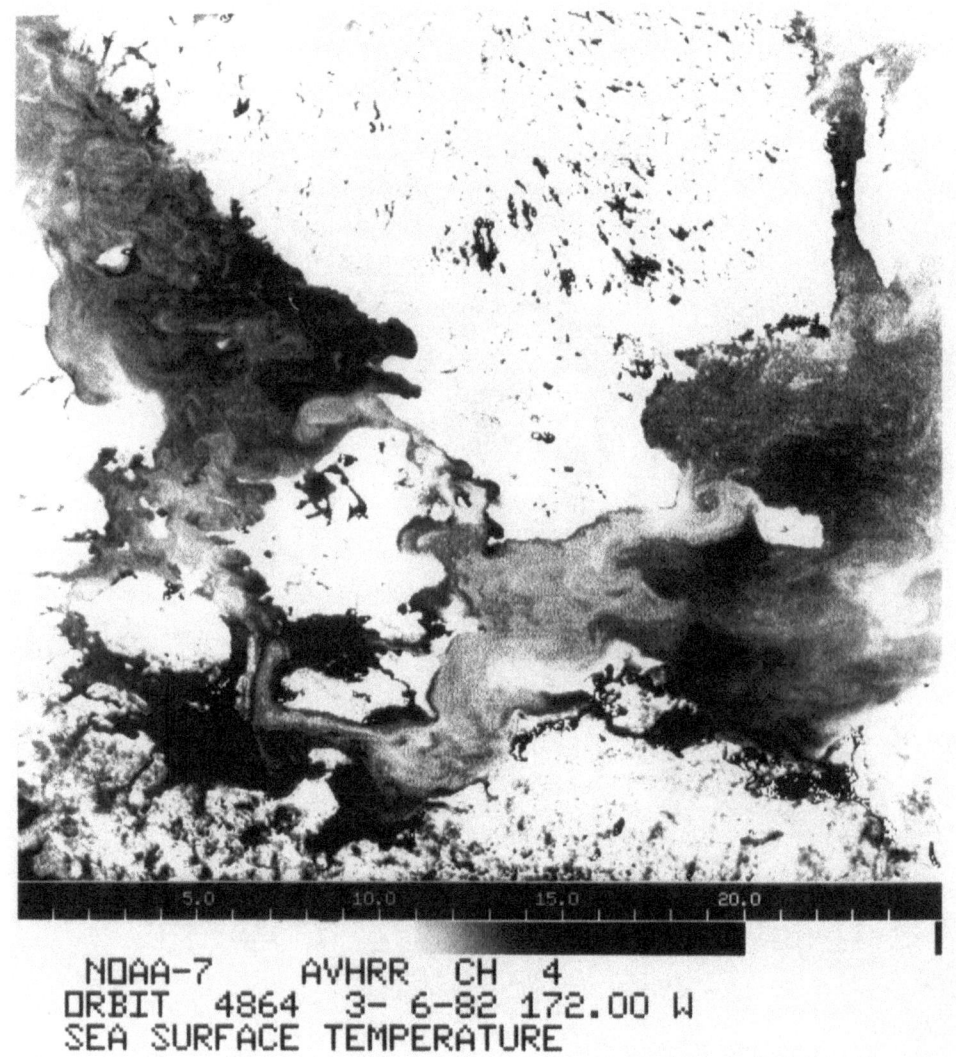

Fig. 3. Upwelling of cold water (11°C) at the Darss Sill drifting into the Belt Sea. Image processed for absolute temperatures, see scale. (NOAA 7, AVHRR 3, June 1982).

AHVRR infrared images on two consecutive days of the central Baltic Sea demonstrates the heat loss of the nearshore waters in winter (Fig. 1). Aside from the information on the extent of colder coastal waters the images can be evaluated for spatial and temporal changes of surface temperature patterns which include estimations on nearshore-offshore processes. Fig. 2 shows an example of coastal upwelling from the

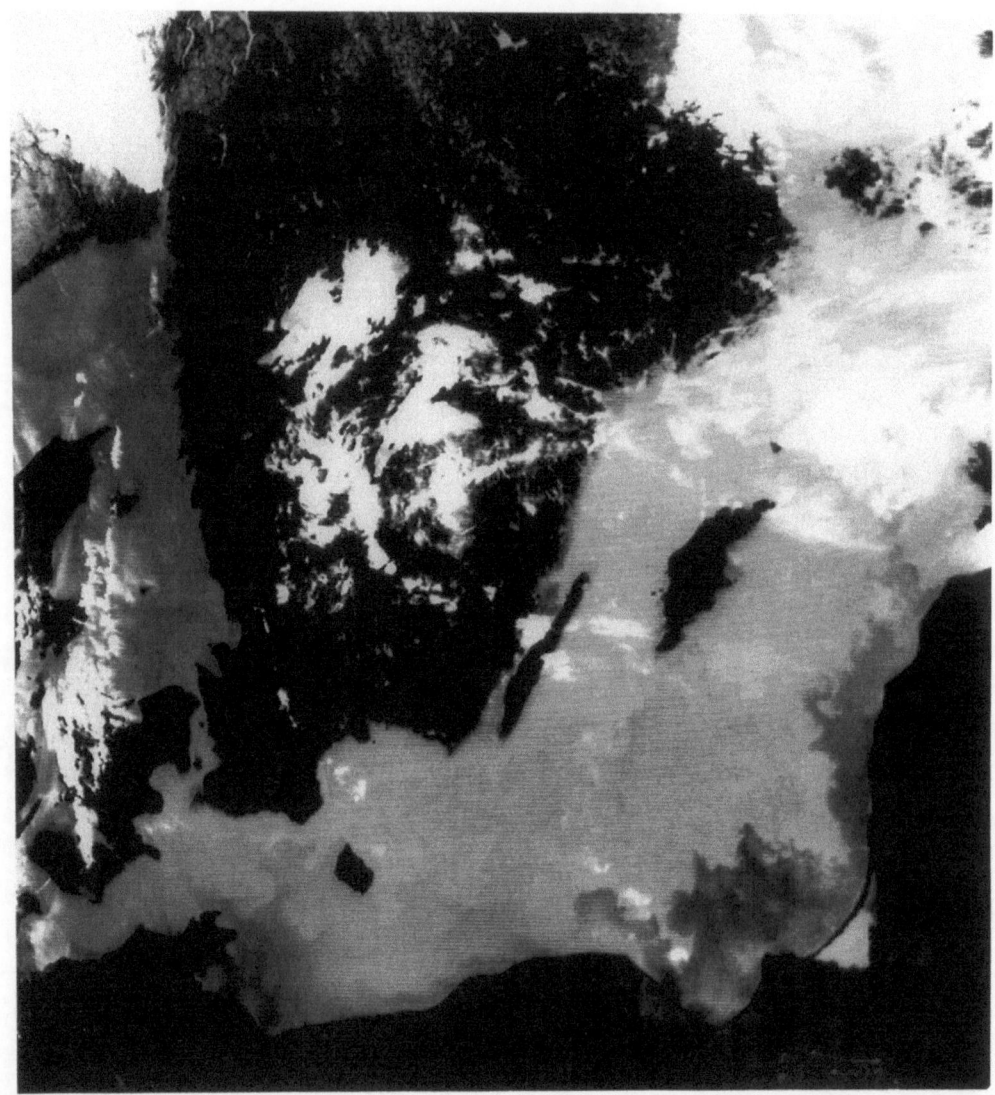

Fig. 4. Pattern of warm river water in the Southern and Southeastern Baltic extending from the river mouth far into the Central Baltic. (TIROS-N, AVHRR Infrared Channel 4, 12, April 1979).

southern Baltic Sea from August 1982. Due to easterly winds parallel to the coast cold water masses appear within a period of 6 days (three consecutive days shown here) along the Polish coast as a consequence of Ekman upwelling. Upwelling phenomena along the coastlines can be observed quite often on infrared satellite images of the Baltic Sea. Temperature anomalies of surface water are only of limited relevance for biological processes. However, upwelled water depending on its depth of origin contains increased mineral nutrient concentrations. From absolute temperature calculations obtained from satellite data using the McClain algorithms, we can conclude nutrient concentrations of upwelled surface water if we have ground truth nutrient depth profiles from the area. From the western Baltic Sea (Fig. 3) cold water of $11^{\circ}C$ from a depth of 45 m with increased phosphate and nitrogen values, appears at the sea surface near the Darss Sill and drifts into the Belt Sea where it may affect growth of nutrient-limited phytoplankton.

Sea surface temperature images clearly define the boundaries of water bodies. Aside from upwelling phenomena sea surface images can inform on current systems in coastal and offshore waters, can point out fronts and eddies, and can also trace the discharge of river water which usually differs in temperature from the surrounding sea. In temperate latitudes during springtime warmer river water can be traced in estuarine areas extending sometimes far into the open sea. Fig. 4 shows the Baltic Sea in April 1979, with warm river water from the rivers Odra, Wistula and Nemunas extending far into offshore areas.

3.2 Multispectral water color images for detecting coastal-offshore ecosystem relations

While sea surface temperature images only give indirect evidence of the major factors influencing coastal-offshore ecosystems, data from color scanners show the appearance of organisms in the upper water layers. This can be accumulations of small phytoplankton as well as zooplankton and sometimes schools of fishes.

Fig. 5 is a pigment map of the southeastern Baltic on 7 July 1981 representing mass accumulation of cyanobacteria (bluegreen algae), (Horstmann, 1983). An image of suspended matter processed by means of the pigment algorithm shows a similar pattern (Fig. 6). This is caused by the fact that cyanobacteria with Nodularia spumigena as the dominating species have lost their ability to disperse in the euphotic zone and have accumulated near the surface, resulting in an unusually high back-scattering signal which must not be misinterpreted with regard to

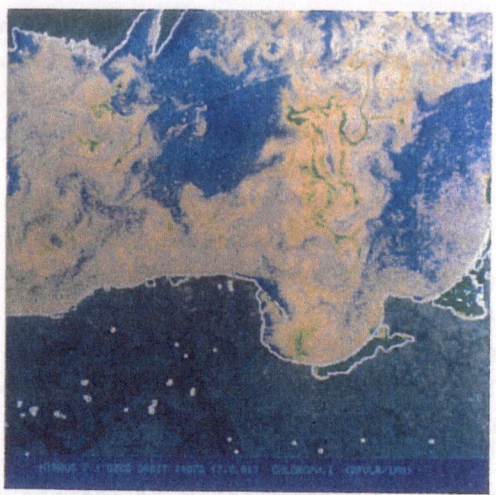

Fig. 5. Pigment concentrations in the surface waters of the Central Baltic on 7 July 1981, exhibiting distribution pattern of bluegreen algae accumulations. (NIMBUS-7, CZCS-processed, green = high, Yellow = medium and blue = low values.

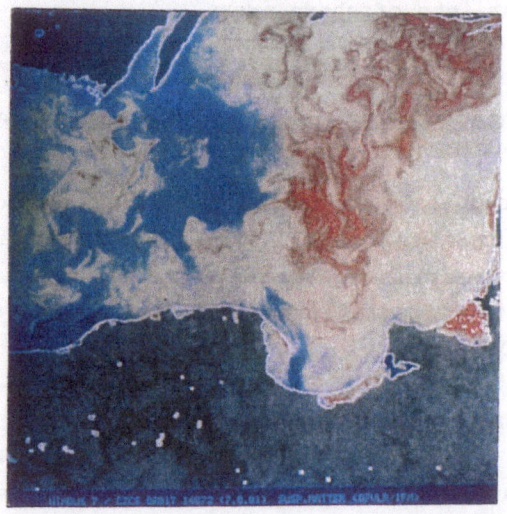

Fig. 6. Suspended matter in the surface waters of Central Baltic on 7 July 1981. (NIMBUS-7, CZCS enhanced Channel 3, red = high, light grey = medium and blue = low contrencations).

Fig. 7. LANDSAT-2 (MSS4) image of the Central Baltic, south of Bornholm Island showing high-resolution pattern of surface-drifting bluegreen algae on 9 Sep 1977.

Fig. 8. NIMBUS-7 (CZCS processed). Distribution pattern of pigments in the South-eastern Baltic Sea on 30 Mar 1983. Bay of Gdansk is the U-shaped bay at lower left, outside the Wistula lagoon. To the right of this the Kurland lagoon.

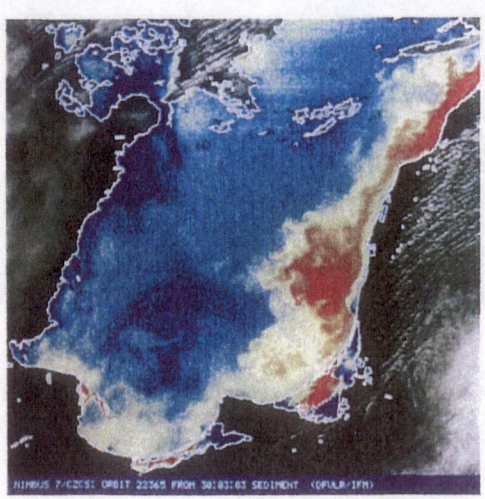

Fig. 9. NIMBUS-7 (CZCS processed). Distribution pattern of suspended matter on 30 Mar 1983.

pigment concentration. Wind-induced waves can redistribute the
cyanobacteria within the upper layers. However, the evaluation of series
of satellite images of bluegreen algae blooms from the Baltic exhibit a
number of interesting details: the general distribution pattern in fronts
and eddies, or the fact, that in nearshore areas, where nutrients are
expected to be relatively high in summer, bluegreen algae rarely appear
floating near the surface, but are distributed in the euphotic zone,
exhibiting high chlorophyll signals on satellite images. Fig. 7 shows the
same phenomenon south of Bornholm Island on a LANDSAT satellite image
with its higher areal resolution (75 m per pixel).

Much more apparent and clearly visible on satellite images are
coastal-offshore processes in estuaries. Examples are taken from the
eastern Baltic Sea since this area, due to the more continental climate,
is more often cloudfree than the western Baltic Sea or the North Sea.
Atmospheric corrections have been applied to all images shown here.

A considerable amount of suspended matter appears along the Lithuanian
and Latvian coast (Fig. 8). The patterns extend far into the central
Baltic, especially in the area influenced by the Nemunas river. Pigment
distribution patterns (Fig. 9) partly coincide with the suspended matter
and do not extend into other areas. At this time of the year the euphotic
zone is limited to the very surface, and phytoplankton growth in these
latitudes is still light-limited. Therefore, eutrophication effects cannot
be expected.

Suspended matter is also observed in the Bay of Gdansk off the mouth of
the Wistula river, and especially in the Wistula and the Courland
lagoons, matching the high values in the pigment image. Due to the
limited water depth in the lagoons, the entire water column is euphotic,
enabling an early start of the spring plankton bloom. The western part of
the central Baltic is covered by clouds.

Figs. 10 and 11 show the southeastern Baltic Sea on 12 April 1979. The
distribution of suspended matter shows distinct features extending from
the mouths of the rivers (from left to right) Wistula, Nemunas and Venta
(Fig. 10). The pigment distribution (Fig. 11), however, extends much
further into offshore waters, especially in the Bay of Gdansk and its
northern extension. The pigment concentration appears also to be high in
the western central Baltic and along the Polish north coast. In these
areas the plankton growth is apparently stimulated by river water
containing considerable amounts of nutrients.

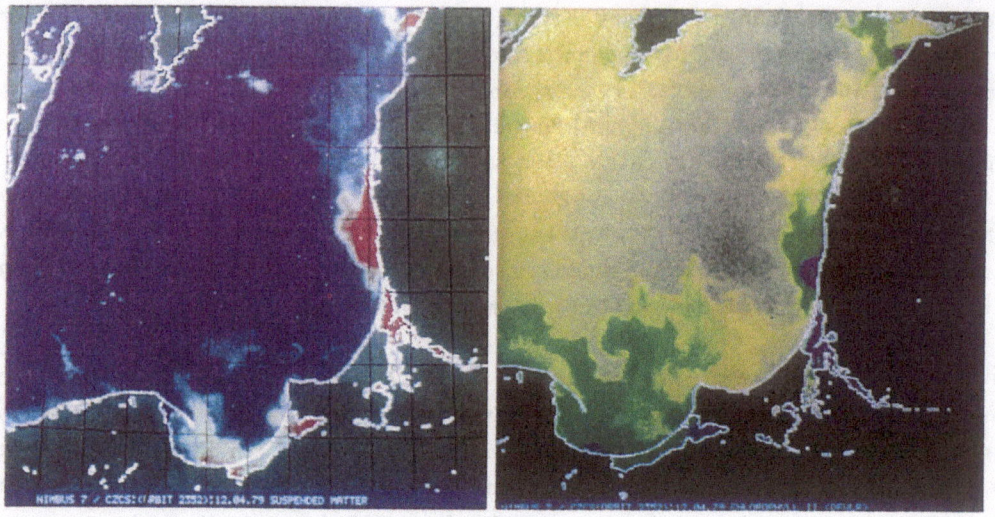

Figs. 10-11. NIMBUS-7 (CZCS processed). Fig.10. Concentration of suspended matter on 12 Apr. 1978 near the mouths of the rivers (red areas): Wistula (bottom), Nemunas (middle right) and Venta (top right). Fig. 11. Distribution of pigments in the Southeastern Baltic Sea on 12 Apr. 1978.

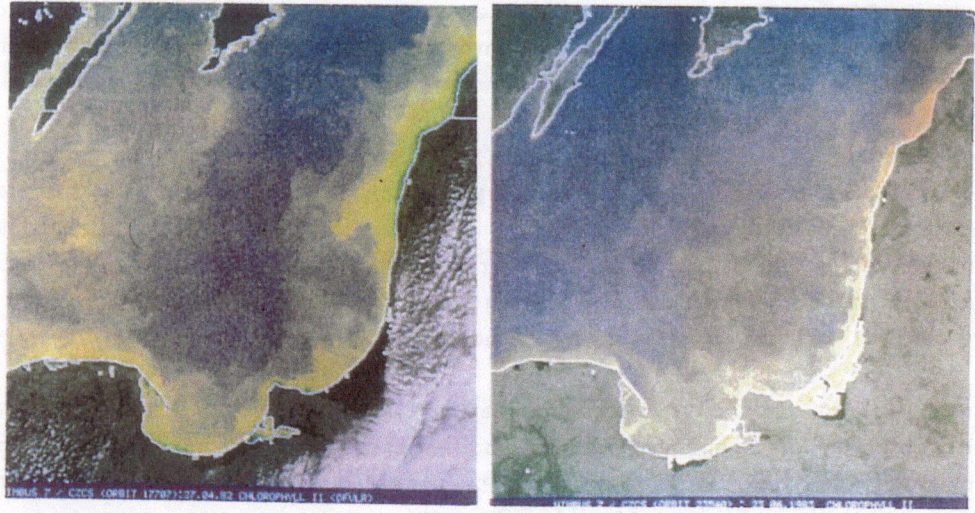

Figs. 12-13. NIMBUS-7 (CZCS processed). Pigment distribution in the Southeastern Baltic Sea on 27 Apr. 1982 (Fig. 12) and on 23 Jun. 1983 (Fig. 13).

Remarkable details in these scenes, show strips of water parallel to the Hela peninsula with low pigment concentration as evidence of upwelling which is frequently observed during southeasterly winds (see also Fig. 13). The pigment and temperature distribution in the Wistula and Courland lagoons indicates good algal growth possibly starting below and in the ice, as supported by sea surface temperature images at the same time. The blue spots close to the river outlets in Fig. 11 are artifacts due to the present limitations in the atmospheric corrections for this area.

The image of 27 April 1982) (Fig. 12) represents another pattern of pigment distribution in the southern Baltic when northwesterly winds cause the increased pigment concentrations to occur mainly along the Latvian, Lithuanian and Polish coasts.

On 23 June 1983 (Fig. 13) the pigment concentration in the central and southeastern Baltic seems to be slightly lower. The concentration along the Lithuanian and Latvian coast is moderate in comparison to the spring situation, partly caused by the reduced water loads in summer, partly by the weather (northwesterly winds from 18 to 23 June 1983).

In temperate latitudes the initiation of the phytoplankton spring bloom can be considered as one of the most puzzling consequences of coastal-offshore couplings. Up till now it has not been possible to acquire sequences of images of the spring bloom development from the Baltic. A LANDSAT (MSS4) image (Fig. 14) on 3 March 1976 shows the beginning of the spring bloom in the western Baltic. As known from ground truth data, the white pattern in the water coincides with high chlorophyll data caused by diatom blooms. The image shows that phytoplankton development starts in shallow waters

LANDSAT images demonstrate that for near-coastal processes, scanners with better resolution (in this image 75 m) are required. This can be confirmed by the first products of the multispectral "LANDSAT Thematic Mapper" scanner with a resolution of 30 m on which ship traces can be recognized.

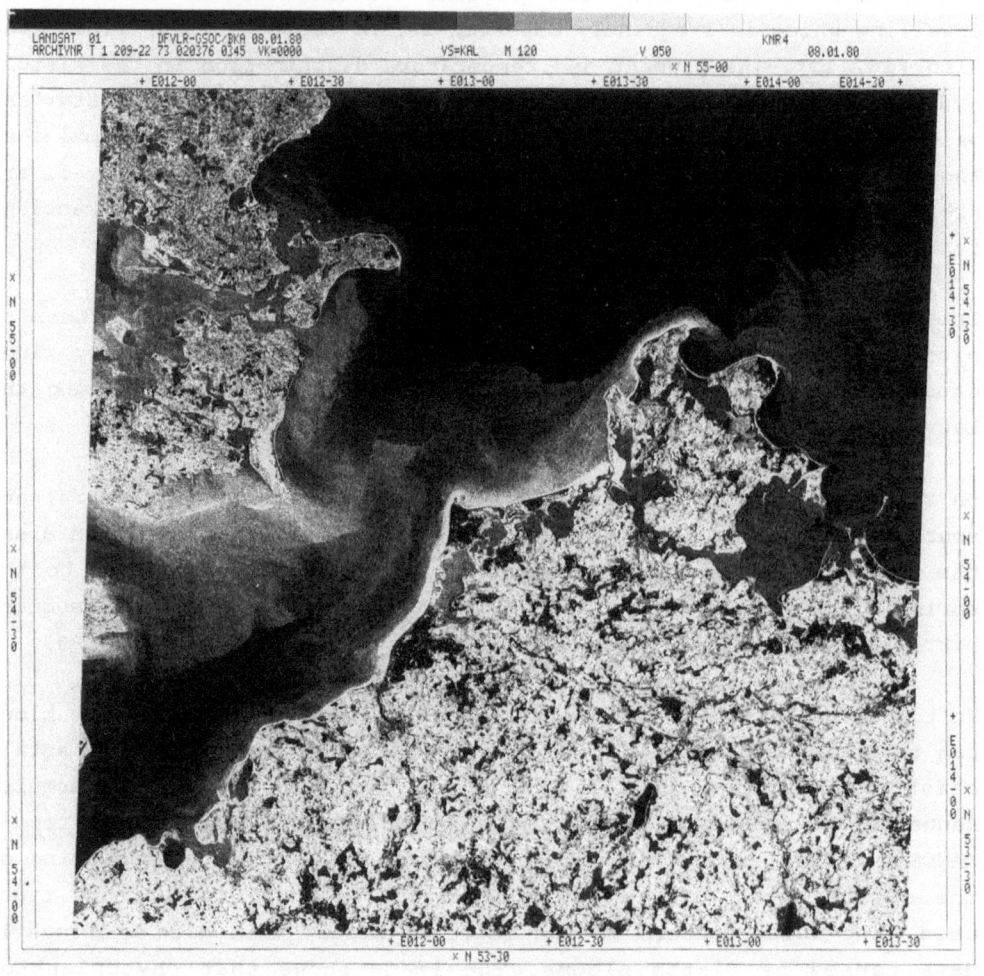

Fig. 14. LANDSAT (MSS-4) image of 3 Mar. 1976 demonstrating the beginning of the spring phytoplankton bloom in the Western Baltic Sea.

4. DISCUSSION

The examples presented here are taken from a region which, due to frequent cloud cover, sea fogs and high aerosol concentration, is least suited for remote sensing of the sea surface. Brilliant images of coastal offshore processes exists from the relatively more cloud-free Mediterranean Sea, coasts of Florida and San Francisco Bay. The fact, however, that coastal offshore processes can be monitored even in the Baltic Sea demonstrates the capability of remote sensing. Satellite images exhibit sea surface temperature patterns and give rough information on concentrations of suspended matter and pigments and especially chlorophyll-a. Sequences of images of two or more consecutive days can demonstrate transport processes in the surface layer. Ocean color data and infrared sea surface temperature images, relate phytoplankton growth to coastal upwelling, divergence or convergence. Satellite data of the Coastal Zone Colour Scanner in the open ocean where well known phytoplankton communities exist have led to realistic estimates of primary productivity in those areas (Eppley et al., 1984). Conditions are more variable in the Baltic Sea and the interpretation of satellite data is more difficult. Light penetration is limited by the amount of particles and dissolved substances in the Baltic. Consequently, the back-scattered radiance received by the satellite's scanner originates from the upper water layers. In the Baltic Sea the appearance of bottom structures on satellite images in shallow areas in summer indicates that at this time the radiometric information originates from the upper ten to twelve meters. During phytoplankton growth in eutrophication blooms, however, satellite images represent conditions in the uppermost surface layer only. The satellite-received back-scattered radiance from the sea gives almost no information on the vertical distribution of phytoplankton in the water column. This is why ground truth information should always include a vertical resolution of pigment data.

For calibration of satellite obtained data, sea truth data from ships are urgently needed. The coincidence of ship's investigations and cloudfree satellite recordings is still low all over the world. The existing satellite data from nearshore and offshore regions are capable of telling us in what areas and at what time relevant ground truth data should be taken.

As for quantitative estimates of pigment or particle concentrations from satellite images, a number of facts should be considered: the high characteristics of different parts of the Baltic Sea require specific algorithms. They also require improved color scanning systems which will be launched in the first half of the next decade.

At present a tremendous amount of satellite data lies dormant in the libraries of the receiving station or the respective national point of contact. The examples of data from the Baltic Sea should encourage more utilization of remote sensing for studies of coastal processes.

REFERENCES

Baylis, P.E., 1981. University of Dundee satellite image data acquisition and archiving facility, matching remote sensing technologies and their applications. - Proc. 9th Ann. Conf. Remote Sens. Soc., London. 517 pp.

Eppley, R.W., E. Stewart, M.R. Abbott & U. Heyman. 1985. Estimating ocean primary production and statistics for the Southern California Bight. - J. Plankt. Res. 1: 57-70.

Gordon, H.R. & A.Y. Morel, 1983. Remote assessment of ocean color for interpretation of satellite visible imagery. - Lecture notes on Coastal and Estuarine studies. Springer Verlag, New York, Berlin, Heidelberg, Tokyo.

Horstmann, U., 1983. Distribution patterns of temperature and water colour in the Baltic Sea as recorded in satellite images: indicators of plankton growth. - Bericht Inst. Meereskunde Univ. Kiel 1: 1-147.

Llewllyn-Jones, P.J., R.W. Minet, R.W. Saunders & A.M. Zavody, 1984. Satellite multichannel infrared measurements of seasurface temperature of the North-east Atlantic using AVHRR/2. - Quart. J. Royal Meteorol. Soc. 110: 613-632.

NASA, 1982. The Marine Resources Experiment Program (MAREX). - Report of the ocean color science working group NASA Goddard Space Flight Center.

Piepen van der, H., V. Amann & R. Fiedler, 1985. Erkundung und Interpretation der Meeresfarben. - DFVLR Nathrichten 46: 21-26.

Viollier, M., 1982. Radiometric calibration of the Coastal Zone Color Scanner on Nimbus 7 : a proposed adjustment. - Applied Optics 21: 1142-1145.

Zion, P. & M. Abbott, 1984. Oceanography from space. A research strategy for the decade 1985-1995. - JOJ Executive Summary, July 1984.

II. MASS BALANCE STUDIES

THE USE OF STABLE ISOTOPE RATIOS FOR TRACING THE NEARSHORE-OFFSHORE EXCHANGE OF ORGANIC MATTER

J.N. Gearing
Department of Fisheries and Oceans
Maurice Lamontagne Institute
Mont-Joli, Quebec G5H 3Z4, Canada

Stable isotopes particularly of carbon, nitrogen, sulphur, and hydrogen, can be very useful for tracing and quantifying the amounts of anthropogenic and natural terrigenous organic matter in the marine environment. In many circumstances, they provide the only way of deter mining the origins of this material. However, the method has definite limitations which must be taken into account. In particular, one must always measure the possible sources of the organic matter in individual locations.

Results indicate that there is little transport of terrestrially-derived organics more than a few kilometers from shore. The terrestrial material which is present in offshore sediments appears to be relatively refractory to biological assimilation. Studies of anthropogenic organic matter confirm that such pollution is primarily confined to local areas.

Different sorts of information can be obtained from isotope ratios depending on the type of sample examined. Ratios of particulate organic matter (POC) reflect not only conservative transport of material but also _in situ_ biological processes altering organic matter. Sediments provide information on the integrated result of organic matter deposition throughout the year. The ratios in organisms result from the organic matter present in the environment as well as its bioavailability. Dissolved organic matter can also be measured and its ratio does differ depending on the relative amounts of terrestrial- to marine-derived constituents. However, little work has been yet done in this area. Individual chemical compounds or classes of compounds can also be examined isotopically, providing information on particular chemicals rather than organic matter as a whole.

Stable isotope ratios should be used in conjunction with physical and biological information as well as C/N ratios, lignin oxidation products.

Lecture Notes on Coastal and Estuarine Studies, Vol. 22
B.-O. Jansson (Ed.), Coastal-Offshore Ecosystem Interactions.
© Springer-Verlag Berlin Heidelberg 1988

1. INTRODUCTION

Stable isotope ratios act as naturally-occurring tracers for organic matter, making possible, under certain conditions, the quantification of coastal-offshore exchanges. In general, organic matter has isotope ratios characteristic of its origin (e.g. plants with different modes of photosynthesis and different growth conditions, anthropogenic compounds). These ratios are maintained as the organic matter moves through the biosphere and geosphere. A mixture of organic matter from two sources has isotope ratios intermediate between those of the two sources, in proportion to the fraction of material from each source.

Isotope ratios are one of the few methods which can trace organic matter as it moves through natural ecosystems. Ratios can be measured on both the total organic matter and on particular chemical fractions or compounds. When used on organisms, isotope ratios provide information of organic matter actually assimilated into body tissues, not just material ingested.

As with all tools, this method has certain limitations which must be borne in mind when interpreting its results. Firstly, specific environmental conditions must be met. This generally means an ecosystem with a limited and known number of sources of organic matter having different isotope ratios. Two sources with different isotope ratios are ideal; additional sources with other isotope ratios complicate interpretation. Secondly, the difference in isotope ratios of the two sources should be large compared with analytical variability. Thirdly, the ratios within each source should vary as little as possible. And finally, the ratios of the organic matter should not change, or should change in a predictable manner, as it moves through the food web or during diagenesis.

Most of the major elements of organic matter have stable isotopes which can be used as tracers (Table 1). The methodology is essentially the same for all these elements. Samples are converted completely to an appropriate gas which can be analyzed using a Nier-type mass spectrometer with dual detectors. Organic matter is usually combusted and analyzed as CO_2, N_2 and SO_2. Hydrogen is analyzed as H_2 formed from the water produced by combustion. Standards used and analytical variability for these analyses are given in Table 1.

Table 1. Parameters used for stable isotopes of organic matter. Del (δ), expressed in $^o/oo$ or per mil, is defined as $[(R_{sample}/R_{standard})-1] \times 1000$, where R = ratio of isotopes.

ELEMENT	R	NAME	AR[a]	STANDARD
HYDROGEN	$^2H/^1H$	δD of $\delta^2 H$	± 3.0	SMOW (standard mean ocean water)
CARBON	$^{13}C/^{12}C$	$\delta^{13}C$	± 0.3	PDB (Pee Dee belemnite)
NITROGEN	$^{15}N/^{14}N$	$\delta^{15}N$	± 0.5	Atmospheric N_2
SULFUR	$^{34}S/^{32}S$	$\delta^{34}S$	± 0.5	CDT (Cañon Diablo troilite)

[a] Analytical reproducibility

By far the most work on organic matter has been concerned with carbon. Coastal-offshore changes in carbon isotopes will be the main focus of this paper. Several review articles on $\delta^{13}C$ give details of other uses of this ratio (Degens, 1969; Schwarcz, 1969; Parker and Calder, 1970; Parker, 1971; Smith, 1972; Erlenkeuser, 1978; Deines, 1980; van der Marwe, 1982; Fry and Sherr, 1984; Rounick and Winterbourn, 1986). Nitrogen and sulphur have a great potential for natural isotopic variations because they undergo many chemical reactions between several valence states. General reviews of their variations include those of Kaplan (1975 and 1983), Krouse (1980), Létolle (1980), and Wada (1980). Hydrogen isotopes are also potentially useful for organic compounds. However, only a few studies relevant to coastal-offshore exchanges of organic matter have been conducted with these three isotopes. Hydrogen isotopes have been primarily used for tracing inorganic compounds such as water (see Anderson and Arthur, 1983: Kaplan, 1983).

2.   MAJOR TRACERS FOR ORGANIC MATTER - A CRITICAL REVIEW

2.1. Carbon

The average $\delta^{13}C$ values for some of the main reservoirs of organic carbon are shown in Fig. 1. These cover a wide range and include many pairs of sources whose $\delta^{13}C$ values are different enough compared with analytical variability ($\pm$ 0.2 to 0.3$^o/oo$) to allow them to be distinguished isotopically. The isotopic variability which is normally found in each of these reservoirs, however, means that isotopic results often have large standard errors. Deviations from the final criterion, that the isotope ratio is conserved during diagenesis and movement through the food web, also introduces some errors.

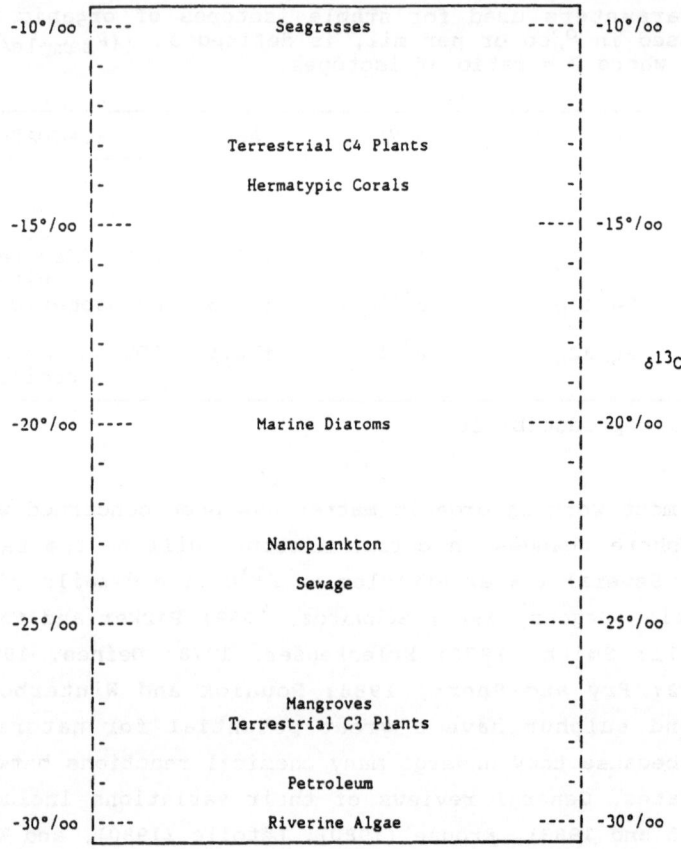

Fig. 1. Average $\delta^{13}C$ of some reservoirs of organic matter

Variability within a source (criterion 3) depends on the environmental conditions present when the material was being formed. For example, plants using the C_3 photosynthetic pathway have, on average, more negative $\delta^{13}C$ values than those using the C_4 pathway, yet certain plants can use both types of systems and thus have intermediate isotope ratios. Smith and Epstein (1971) and O'Leary (1981) present more complete explanations of these differences in plants; Gearing et al. (1984a) summarize the current findings on the variations within marine phytoplankton. There are also differences of 1 to 3°/oo between species of plants in each group. McMillan et al. (1980) and Benedict et al. (1980) report surveys of various species of seagrasses; Wong and Sackett (1978) report isotope ratios for different species of phytoplankton cultured in the laboratory. Within the same species, variability can also be significant due to different growth conditions and biochemical composition. Stephenson et al. (1984) reported differences as great as 8°/oo within an individual macroalgae; 2°/oo is a more usual intraspecies range. Other material such as sewage and petroleum can also vary depending on the origin of individual samples (Sofer, 1984).

Thus while Fig. 1 provides a good estimate of which sources are potentially distinguishable by carbon isotope ratios, it is absolutely necessary to measure the end members of sources at each study location. For example, phytoplankton in Narragansett Bay average -21.3 ± 1.1°/oo over a year (Gearing et al., 1984a), whereas phytoplankton worldwide range between -12 and -31°/oo (Fry and Sherr, 1984). The runoff from an individual river depends on the vegetation and other organic inputs to its particular drainage area; the $\delta^{13}C$ of its POC may differ significantly from the average of all rivers studied and will, in general, have a more restricted range (see later section for examples). Measurements at particular locales increase the accuracy and reduce the uncertainty of isotopic results and should not be omitted.

Carbon isotope ratios do not appear to change during short-term diagenesis although loss of the more soluble and labile proteins and sugars has been postulated to result in more negative $\delta^{13}C$ values with increasing diagenesis. Studies of phytoplankton and macrophytes held in the laboratory for periods of up to two years have shown little or no change in $\delta^{13}C$ (see summary in Gearing et al., 1984a). Seagrasses and mangrove leaves allowed to decompose in litter bags in Florida bays were isotopically unchanged after six weeks (Zieman et al., 1984), but roots of seagrasses buried for 18 months in a salt-marsh showed a preferential loss of C-13 (Benner et al., 1987). Mangrove wood from the Straits of Malaca dated at around a thousand years old (-27.3 and -27.6°/oo, Geyh et al., 1979) is within the range of living mangroves from the same area (-24.5 to -28.5, average -27.1°/oo, Rodelli et al., 1984). Spiker and Hatcher (1984) have postulated that diagenesis over 4000 years caused a shift of about 4°/oo in sedimentary organic matter, but it is uncertain how much of this change may have resulted from the change in inorganic carbon isotope ratios which they found in the same core. Dean et al. (1986) conclude from a review of many papers that diagenesis has little or no effect on $\delta^{13}C$ of organic matter over geologic time. For organisms and sediments over tens or a few hundreds of years, the possibility of diagenetic changes in $\delta^{13}C$ of organic matter appear to be minimal under most environmental conditions.

Finally, carbon isotope ratios can change as organic matter moves through a food web. DeNiro and Epstein (1978) found animals to be an average of $0.8 \pm 1.1^{o}/oo$ more positive than their food (13 experiments). Forty-four animals collected in the field averaged $0.6 \pm 1.4^{o}/oo$ more positive than their inferred food (Sackett al., 1965; Degens et al., 1968a; Deuser, 1970; Minson et al., 1975; Fry, 1977; Land et al., 1977; Fry et al., 1978; McConnaughey and McRoy 1979b; Petelle et al., 1979). Gearing et al. (1984a) found zooplankton to be $0.6 \pm 1.0^{o}/oo$ more positive than concurrently collected phytoplankton (35 sets). Fry et al. (1984) reported zooplankton to be an average of $1.8^{o}/oo$ more positive than POC. Rau et al. (1983) calculated changes of 0.7 and $1.4^{o}/oo$ per trophic level. Thus there is on average an enrichment in $\delta^{13}C$ of approximately $1^{o}/oo$ per trophic level, and a correction should be made when attempting to quantify carbon sources in animals.

Fig. 2 illustrates how important can be both variations within sources at different locations (different environmental conditions) and variations with trophic level. The values of similar organisms are not greatly different worldwide, allowing general conclusions to be drawn, but the accuracy of which the method is capable can only be achieved by measuring several individual organisms over a year (to reduce individual variability) whose biology is relatively well-known (to reduce errors due to trophic differences) at the particular location of interest (to reduce spatial variations).

Working within these limitations, stable carbon isotope ratios have proven to be very useful for a variety of studies in ecosystems where there are only two dominant, isotopically-distinct sources of carbon. Many of these elucidate coastal-offshore exchanges of organic matter by studying the mixing of organics derived from marine phytoplankton with terrestrial-derived material, either natural riverine and estuarine runoff or pollutants such as sewage or petroleum. Later sections of this paper examine each of these in detail. In addition, isotopes have been used to examine the sources of carbon within estuarine ecosystems, including marsh plants and seagrasses (see, for example, Parker, 1964; Smith and Epstein, 1970; Johnson and Calder, 1973; Haines, 1976a, b; Fry et al., 1977; Thayer et al., 1978; Fry and Parker, 1979; Haines and Montague, 1979; McConnaughey and McRoy, 1979a; Hackney and Haines, 1980; Fry et al., 1982a; Sherr, 1982; Fry et al., 1983; Hughes and Sherr, 1983; Peterson and Howarth, 1983; Schwinghamer et al., 1983; Fry, 1984; Kitting et al., 1984; Zieman et al., 1984; Peterson et al.,

1985; Simenstad and Wissmar, 1985). Unfortunately, estuarine marshes and seagrass meadows have several isotopically-distinct sources of carbon, including seagrasses (-10°/oo), seagrass epiphytes (-15°/oo), benthic diatoms (-19°/oo), phytoplankton (-21°/oo), and C_3 marsh grasses (-25°/oo). Carbon isotopes alone cannot separate these sources. In other ecosystems, investigators have attempted to characterize and trace organic matter derived from macrophytes (Dunton and Schell, 1982; Stephenson et al., 1984), peat (Schell, 1983; LeBlanc and Risk, 1985), and mangroves (Gearing et al., 1984b; Rodelli et al., 1984; Zieman et al., 1984; Torgersen and Chivas, 1985).

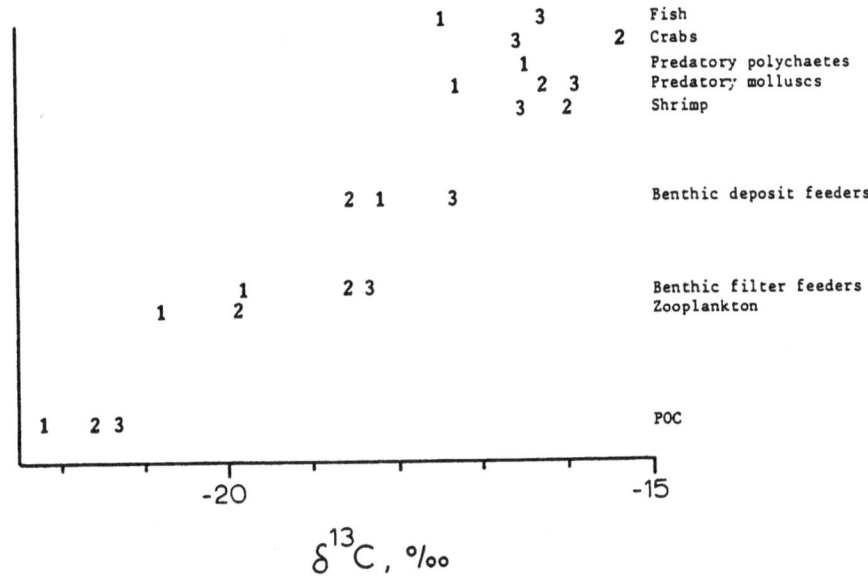

Fig. 2. Variations in carbon isotope ratios with location and trophic level. Data are from: 1) Narragansett Bay (Gearing et al.1 1984a), 2) northern Gulf of Mexico (Fry and Parker, 1979; Fry et al., 1984), and 3) offshore Malaysia (Rodelli et al., 1984)

2.2. Nitrogen

Fig. 3 shows the average $\delta^{15}N$ values of some of the reservoirs of organic matter. The naturally-occurring range of stable nitrogen isotope ratios is smaller than that of carbon. Because there are fewer pairs of isotopically-distinct nitrogen sources, these isotopes tend to be useful in fewer environments than $\delta^{13}C$. Terrestrial organic matter, however, is more positive than marine organic matter.

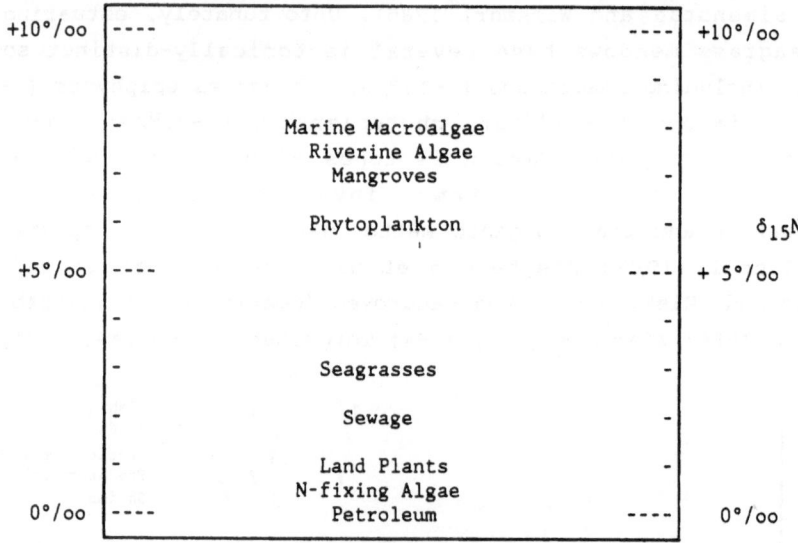

Fig. 3. Average $\delta^{15}N$ of some reservoirs of organic matter

Nitrogen isotopes, like those of carbon, show considerable variation within sources. For example marine phytoplankton have been reported to range between -2 and +10°/oo (Macko, 1981). N_2-fixing and non-N_2-fixing plants have different nitrogen isotope ratios (Virginia and Delwiche, 1982). Soils vary widely in the $\delta^{15}N$ of their organic matter depending on the type of soil and on the presence of fertilizers (Cheng et al., 1964). Unfortunately, not enough data have been collected to fully understand the reasons for such variability, thus corrections are not yet possible. Nitrogen isotopes can also change significantly during short-term diagenesis due to microbial fixation of nitrogen (Zieman et al., 1984). Finally, nitrogen isotope ratios, like those of carbon, increase at higher trophic levels. DeNiro and Epstein (1981a) reported an average difference of +3.0 $\pm$ 2.6°/oo (n = 13) between animals in the laboratory and the food on which they were raised. Macko et al. (1982), Rau (1982), and Schoeninger and DeNiro (1984) also found differences in $\delta^{15}N$ according to trophic level.

In summary, nitrogen isotope ratios are potentially useful in several situations. These include distinguishing terrigenous and anthropogenic organics from marine-derived organic matter and in studies using multiple isotopic tracers. Until more work has been done on the extent and causes of its variations in nature, nitrogen will remain less useful than carbon for individual isotope studies.

2.3. Sulphur

The stable isotope ratios of sulphur have been widely used to trace the origin of inorganic sulphur compounds (see, for example, Kaplan 1975 and 1983 and Krouse, 1980). Their principle application for organic matter has been as a tracer of microbial degradation over geologic time scales (see, for example, Thode and Reese, 1970). In the present day, rooted plants (-10 to + 5°/oo) differ in δ^{34}S from marine phytoplankton (+20°/oo) (Fry et al., 1982b). There is also evidence that marsh plants growing in anoxic sediments have more negative δ^{34}S values than other terrestrial vascular plants growing in oxic sediments (Peterson et al., 1985). Thus δ^{34}S may be useful for distinguishing organic sulphur arising from these sources. It can also distinguish sewage-derived material in the marine environment (Sweeney and Kaplan, 1980; Sweeney et al., 1980). However, these types of studies are still in their infancy and intra-source variability and differences due to short-term diagenesis and trophic level have yet to be fully examined.

2.4. Hydrogen

In addition to their utility for tracing water masses and indicating paleoclimate, the stable isotopes of hydrogen appear to be useful for distinguishing organic matter of vascular plants from that of algae (Nissenbaum, 1974). Algae are, on average, 100°/oo more negative than the water from which they derive their hydrogen (Estep and Hoering, 1980) whereas the difference for woods is -30°/oo (Schiegl and Vogel, 1970) and for salt marsh plants -55°/oo (Smith and Epstein, 1970). Although these differences are greater than for the other isotopes discussed, the analytical variability for hydrogen is correspondingly great. Moreover, the intra-source variability is large because the δD of source water varies with location and climate.

A problem unique to hydrogen is the ease with which organically-bound hydrogen may exchange with that of surrounding water. DeNiro and Epstein (1981b) showed that up to 25% of the organic hydrogen in dead tissue could exchange with water. On the other hand, Estep and Dabrowski (1980) and Macko et al. (1983) have reported the δD of animals to reflect that of their food rather than that of their water. It is not known whether shifts with trophic level occur for hydrogen. Obviously more work needs to be done to determine whether hydrogen isotopes, are in truth, capable of tracing terrestrial and marine organic matter in the oceans.

3. DISTINGUISHING TERRESTRIAL FROM MARINE ORGANIC MATTER

Perhaps the most useful attribute of stable isotope ratios for understanding coastal-offshore exchanges is their ability to distinguish and quantify the amount of terrestrially-derived organic matter present in the marine environment. There have been a number of such studies conducted in many different environments throughout the world, particularly measuring $\delta^{13}C$ in sediments. Investigations with other isotopes, usually nitrogen, in both sediments and organisms are now beginning to appear in the literature. In general, these studies have found little evidence for transport of particulate organic matter far offshore (beyond one to tens of kilometers). In fact, the organic matter in many estuaries appears to be derived from phytoplankton production rather than from higher plants. The following sections summarize this work.

3.1. Particulates (Suspended and Sedimented)

The use of stable carbon isotope ratios for determining the origin of organic particulates was begun in the 1960's. Sackett and Thompson (1963) demonstrated that the stable carbon isotopes of sedimentary organic matter became less negative in a regular manner going from rivers to beyond the barrier islands of the Mississippi Sound; they postulated that this change could be used to quantify the extent of transport of terrestrial organic carbon offshore. Hunt (1966) surveyed sediments along the Atlantic coast of the United States and found that $\delta^{13}C$ values indicating terrestrially-derived organics often did not extend even to the mouth of estuaries. In contrast, a study of sediment cores concluded that terrestrial organic carbon had been deposited on the abyssal plain of the Gulf of Mexico during the last glacial period when sea level was lower and rivers were flowing directly onto the outer edge of the present-day continental shelf (Newman et al., 1973). Such studies established the capacity of carbon to trace terrestrial material, but had to accept many assumptions since little was then known of the extent of isotopic variation within sources or during diagenesis. Their conclusions, however, are in agreement with more recent findings.

Studies of sedimentary organic carbon have now been conducted worldwide in almost all types of environments. In North America, Hunt's survey has been augmented with studies of the St. Lawrence Estuary (Rashid and Reinson, 1979; Tan and Strain, 1979a, b; Pocklington and Tan, 1983; Tan and Strain, 1983), small rivers in Nova Scotia (LeBlanc and Risk, 1985) and Maine (Macko, 1981; Mayer et al., 1981; Incze et al., 1982), Narragansett Bay (Gearing et al., 1984a), and Chesapeake Bay (Spiker and Kendall, 1983; Sigleo and Macko, 1985) along the Atlantic coast. The Beaufort Sea along the north coast of Alaska (Gearing et al., 1977) and San Francisco Bay on the Pacific shore (Spiker and Schemel, 1979) have also been examined. The Gulf of Mexico is probably the most well known area in the world for stable isotopes. Shultz and Calder (1976) reported values for sediments taken along transects of two rivers in northern Florida and off two passes of the Mississippi River. Gearing et al. (1977) examined the southern and western passes of the Mississippi and 16 transects perpendicular to the coast around the Gulf to Veracruz, Mexico. Fry et al. (1977) and Gearing et al. (1977) made detailed studies of the sediments of the Laguna Madre, noting the influence of organic matter from seagrasses in the restricted areas behind a barrier island. In the Caribbean, one study has been done off the north shore of Jamaica (Gearing, 1975) and in South America, only the Orinoco River, Venezuela (Kennicutt et al., 1987) has been extensively investigated. In Europe, isotopic transects have been reported for the North Sea (Salomons and Mook, 1981), in the Ems estuary of the Netherlands (Létolle and Martin, 1970; and the Loire (Letolle et al., 1986) and Gironde (Létolle and Martin, 1970; Fontugne, 1983) rivers. In Africa, the Nile (Kennicutt et al., 1987) and the Niger (Gearing et al., 1977) have been investigated. In Asia, sediments of the Bay of Bengal have been examined (Gearing et al., 1977; Fontugne, 1983) as well as transects off the Changjiang (Kennicutt et al., 1987) and two Malaysian rivers (Gearing et al., 1982). Australia is represented by several studies in Queensland (Torgersen et al., 1983; Torgersen and Chivas, 1985).

These data make possible a good understanding of the extent of carbon isotopic variability worldwide. Average $\delta^{13}C$ values for the two end members (riverine and marine) are summarized in Tables 2-3 and Fig. 4.

Table 2. Average organic carbon and nitrogen isotope ratios of riverine sediments (*) and suspended particulates (no symbol) from the published literature.

RIVER	LAT[a]	$\delta^{13}C$[b]	$\delta^{15}N$[b]	REFERENCE
Colville (Alaska, USA)	71	-27.0 (1)		Schell 1983
Ems (Germany-Netherlands)	54	-27.3 (5)		Létolle and Martin 1970
Rhine (Netherlands)	52	-27.0 (4)		Salomons and Mook 1981
St. Lawrence (Quebec, Canada)	47	-25.5 (9)		Pocklington and Tan 1983
Loire (France)	47	-25.1 (5)	+6.0 (2)	Létolle et al. 1987
Dordogne (France)	45	-26.4 (3)		Fontugne 1983
Gironde (France)	45	-25.8 (3)		Fontugne 1983
Garonne (France)	45	-26.1 (3)		Fontugne 1983
Garonne (France)	45	-23.7 (3)		Létolle and Martin 1970
Sheepscot (Maine, USA)	44	-25.0 (2)*		Mayer et al. 1981
Salmon Falls (Maine, USA)	44	-27.0 (1)*		Hunt 1966
Merrimack (Massachusetts, USA)	43	-29.2 (1)*		Hunt 1966
Klamath (California, USA)	42	-24.9 (1)*	+6.9 (1)*	Peters et al. 1978
Redwood (California, USA)	42	-24.9 (1)*	+5.4 (1)*	Peters et al. 1978
Eel (California, USA)	41	-25.7 (1)*	+2.9 (1)*	Peters et al. 1978
Thames (Connecticut , USA)	41	-24.2 (1)*		Hunt 1966
Connecticut (Connecticut, USA)	41	-24.4 (1)*		Hunt 1966
Patuxent (Maryland, USA)	38	-24.3 (4)	+6.6 (4)	Sigleo and Macko 1985
Sacramento (California, USA)	38	-28.2 (1)		Spiker and Schemel 1979
San Joaquin (California, USA)	37	-26.3 (2)*	+6.0 (2)*	Peters et al. 1978
Pamlico (North Carolina, USA)	36	-27.5		Brinson and Matson 1983
Kern (California, USA)	36	-25.6 (1)*	+5.3 (1)*	Peters et al. 1978
Savannah (Georgia, USA)	33	-28.1 (1)*		Hunt 1966
Ogeechee (Georgia, USA)	32	-26.7 (1)*		Hunt 1966
Altamaha (Georgia, USA)	32	-26.1 (1)*		Hunt 1966
Altamaha (Georgia, USA)	32	-28.0 (8)		Sherr 1982
Saltilla (Georgia, USA)	31	-26.1 (1)*		Hunt 1966
St. Mary's (Florida, USA)	31	-25.6 (1)*		Hunt 1966
St. John's (Florida, USA)	30	-26.5 (1)*		Hunt 1966
Fenholloway (Florida, USA)	30	-26.3 (4)		Shultz and Calder 1976
Ochlockonee (Florida, USA)	30	-25.8 (1)		Shultz and Calder 1976
Biloxi (Mississippi, USA)	30	-26.8 (4)		Sackett and Thompson 1963
Houma (Louisiana, USA)	30	-25.5 (1)*		Gearing 1975
Mississippi (Louisiana, USA)	29	-25.5 (7)		Shultz and Calder 1976
Orinoco (Venezuala)	9	-27.1 (3)*		Kennicutt et al. 1987
Sangga (Malaysia	3	-27.7 (1)		Gearing et al. 1982
Selangor (Malaysia)	3	-24.0 (1)		Gearing et al. 1982
Amazon (Brazil)	0	-29.4 (1)		Williams and Gordon 1970
Marrett (Queensland, Australia)	14S	-25.0 (5)*		Torgersen et al. 1983

AVERAGE: POC	-26.4 (20)	+6.3 (2)	
SEDIMENTS	-26.0 (19)*	+5.3 (5)*	
TOTAL	-26.2 (39)	+5.6 (7)	

[a] Approximate latitude in degrees north unless otherwise noted
[b] Mean (number of samples)

Table 3. Average organic carbon and nitrogen stable isotope ratios of marine sediments (* or ** for carbonat-rich sediments) and particulates (no symbol) from the published literature.

BODY OF WATER	LAT[a]	$\delta^{13}C$[b]		$\delta^{15}N$[b]		REFERENCE
Beaufort Sea	71	-22.8	(3)*			Gearing et al. 1977
Pacific (off Alaska)	59	-21.0	(2)			Rau et al. 1982
Eastern Bering Sea	57	-21.3	(4)*	+8.0	(4)*	Peters et al. 1978
Bering Sea	55	-20.8	(4)			McConnaughey 1978
Gulf of St. Lawrence	48	-22.4	(13)*			Tan and Strain 1979a
Gulf of St. Lawrence	48	-23.9	(11)			Tan and Strain 1983
Pacific (off Washington)	47	-22.4	(1)*			Hedges and VanGeen 1982
Pacific (off Japan)	45			+2.3	(1)	Saino and Hattori 1985
Atlantic (Gulf of Maine)	44	-21.6	(32)*	+6.4	(32)*	Macko 1981
Atlantic (Wilkinson Basin)	43	-21.6	(1)*			Hunt 1966
Narragansett Bay	42	-21.8	(8)*			Gearing et al. 1984a
Atlantic (Nantucket Shoals)	41	-20.9	(7)*			Hunt 1966
Atlantic (Hudson Shelf)	40	-22.1	(13)*			Burnett and Schaeffer 1980
Pacific (off San Francisco)	38	-21.2	(1)*			Spiker and Schemel 1979
Atlantic (off Cape Hatteras)	35	-21.5	(5)*			Hunt 1966
Pacific (off Japan)	34			+6.4	(1)	Wada 1980
Pacific (off California)	33	-20.3	(2)			Rau et al. 1982
Pacific (southern California)	33	-20.8	(5)*	+7.6	(5)*	Emery 1960
Pacific (Tanner Basin)	33	-21.3	(5)*			Peters et al. 1978
Mediterranean Sea (off Nile)	32	-19.6	(9)*			Kennicutt et al. 1987
Atlantic (Blake Plateau)	30	-18.8	(4)**			Hunt 1966
Pacific (off California)	30	-23.2	(1)			Williams and Gordon 1970
Pacific (off Shanghai)	30	-21.8	(1)*			Kennicutt et al. 1987
Pacific (off Tokyo)	26			+4.1	(1)*	Wada 1980
Gulf of Mexico (north shelf)	29	-22.1	(9)*			Sackett 1964
Gulf of Mexico (north shelf)	29	-21.2	(1)			Calder and Parker 1968
Gulf of Mexico (north shelf)	29	-21.0		-0.9		Macko et al. 1984
Gulf of Mexico (north shelf)	29	-20.6	*	+6.5	*	Macko et al. 1984
Gulf of Mexico (north shelf)	29	-21.5	(1)*			Sackett et al. 1973
Gulf of Mexico (east shelf)	28	-20.7	(12)*			Gearing et al. 1976
Gulf of Mexico (northwest s.)	28	-21.6	(8)			Fry et al. 1984
Gulf of Mexico (west shelf)	27	-20.7	(19)*			Gearing et al. 1977
Gulf of Mexico (north slope)	27	-21.4	(3)*			Northam et al. 1981
Gulf of Mexico (abyssal)	25	-21.0	(30)*			Newman et al. 1973
Gulf of Mexico (abyssal)	25	-21.2	(18)			Eadie and Jeffrey 1973
Gulf of Mexico (east shelf)	25	-18.5	**	+3.6	**	Macko et al. 1984
Gulf of Mexico (east shelf)	25	-19.4		+7.5		Macko et al. 1984
Atlantic (off West Africa)	21	-21.3	(1)*			Gaskell et al. 1975
Caribbean (off Jamaica)	18	-20.8	(6)**			Gearing 1975
Caribbean (off Jamaica)	18	-21.8	(12)			Land et al. 1977
Bay of Bengal	18	-20.2	(18)*			Gearing 1975
Gulf of Bengal å Andaman Sea	15	-20.5	(20)*			Fontugne 1983
Atlantic (off Venezuela)	11	-21.6	(9)*			Kennicutt et al. 1987
Atlantic (off Nigeria)	6	-21.8	(18)*			Gearing et al. 1977
Indian (off Malaysia)	3	-21.2	(1)			Rodelli et al. 1984
Indian (north of Australia)	10S	-21.8	(2)			Fry et al. 1983
Pacific (off Peru)	12S	-21.3	(1)*			Dean et al. 1986
Indian (north of Australia)	14S	-19.0	(3)**			Torgersen et al. 1983
Pacific (off Peru)	15S	-21.5	(2)*			Dean et al. 1986
Indian (south of Australia)	50S	-21.5	(8)*			Sackett et al. 1973
Indian (south of Australia)	60S	-25.8	(3)			Eadie and Jeffrey 1973

AVERAGES: POC	-21.7	(14)	+3.8	(4)
SEDIMENTS	-21.1	(34)	+6.0	(6)
TOTAL	-21.3	(48)	+5.2	(10)

[a] Approximate latitude in degrees north unless otherwise noted
[b] Mean (number of samples)

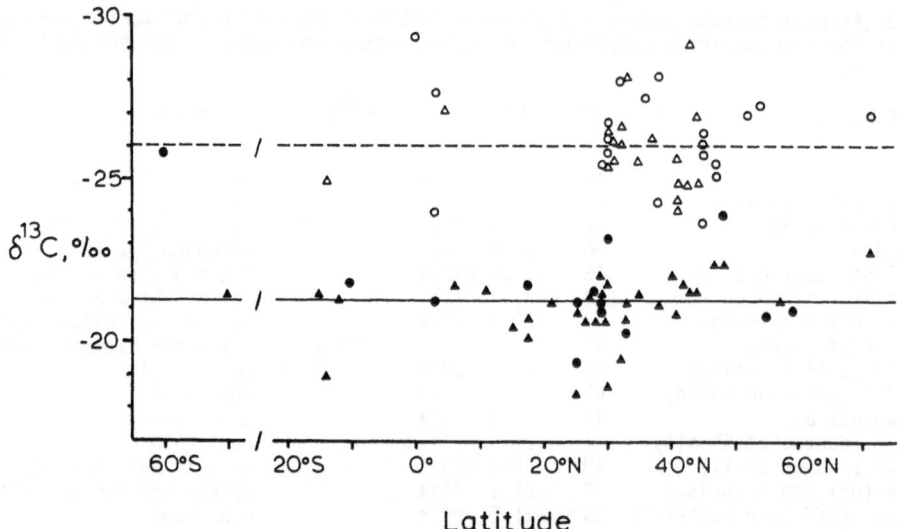

Fig. 4. Carbon isotope ratios of terrestrial sediments (open triangles) and particulates (open circles) and marine sediments (solid triangles) and particulates (solid circles) as a function of latitude. References are given in Tables 2 and 3.

The two types of organic matter are isotopically different; the values on Table 2-3 are separable with a confidence greater than 99.99% (two-tailed Mann-Whitney U-test). Particulate riverine organic carbon averages $-26.2^o/oo$ while marine particulates have a mean of $-21.3^o/oo$. This $4.9^o/oo$ difference is much greater than analytical variability ($\pm$ 0.2 to $0.3^o/oo$) and also larger than most differences between locales. Areal variability, however, is significant, a fact which reemphasizes the need to examine carbon sources at each locale.

These data also cast light on several hypothesized variations. There is no apparent trend with latitude, such as has been postulated to be due to differences in temperature or the relative proportion of different sorts of phytoplankton (Degens et al., 1968b; Fontugne and Duplessy, 1978, 1981; Wong and Sackett, 1978; Rau et al., 1982). There is, however, a trend to higher ratios in carbonate-rich sediments, possibly due to the relatively higher ratios in corals and related benthic algae (Land et al., 1975, 1977; Black and Bender, 1976) or to a high proportion of bound protein with relatively positive isotope ratios. A difference between POC and sediments has also been postulated, since on any given day at a given location POC-sediment differences of several per mille have been reported (Fry and Sherr, 1984). In Narragansett

Bay, the sediments were found to be within 0.5⁰/oo of the POC <u>averaged</u> over a year (Gearing et al., 1984a). Worldwide the values of riverine and marine POC are not separable from the associated sediments (two-sided Mann-Whitney U-test). Sedimentary organic matter is representative of inputs integrated over seasons and years, while POC values are more variable, changing with the tides, currents, and seasons.

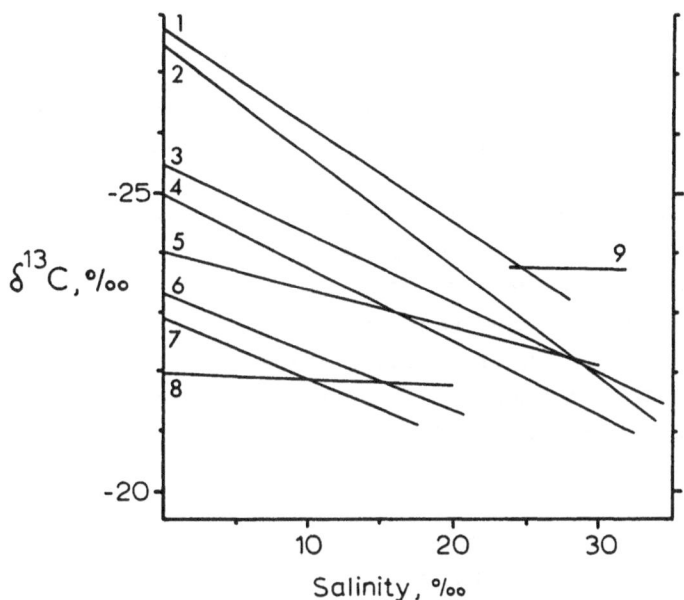

Fig. 5. Transects of del C-13 versus salinity reflect local conditions. Transect 1 (POC) is from the Sangga River, Malaysia (Gearing et al., 1982), 2(POC) is North Bay, San Francisco (Spiker and Schemel 1979), 3 (sediments) is Houma Navigational Canal, Louisiana (Gearing, 1975), 4 (sediments) is Southwest Pass, Mississippi River (Gearing, 1975),5 (POC) is Selangor River, Malaysia (Gearing et al., 1982), 6 (sediments) is South Pass, Mississippi River (Gearing, 1975), 7 (sediments) is Garden Bay, Mississippi River (Gearing, 1975), 8 (sediments) is East Bay, Mississippi River (Gearing, 1975) 9 (POC) is South Bay, San Francisco (Spiker and Schemel, 1979).

The extent of known variation with location can be obtained from Fig. 5 which shows nine transects from San Francisco Bay, the Gulf of Mexico and Malaysia. Different locations are, in general, difficult to compare because isotopic changes are usually compared with distance (from land, river mouth, estuary mouth, etc.) to facilitate calculating the areal extent of terrestrial influence. Where the data are available, I have compared the change in carbon isotope ratios with salinity as a rough approximation at normalizing the different locations. These lines are

the result of linear regressions of the data (usually around 5 to 10
samples) and incorporate considerable scatter. However, it can be seen
that the terrestrial end members vary over nearly 6°/oo, presumably
depending on the amount of terrestrial material present, the types of
plants which produced the organic matter, the current speed of the
rivers, etc. The marine end member is less variable, ranging over only
3°/oo, or 1°/oo if the values from San Francisco are omitted.

Comparisons of these nine transects (Fig. 5) provide insights into
local conditions. The transects from North and South Bays, San
Francisco differ greatly, North Bay (line 2) having a steep slope as
organic matter from the Sacramento and San Joachim rivers is lost via
deposition. South Bay (line 9) has no major riverine input; its
isotope ratios remain nearly constant at an isotope ratio indicative
of sewage carbon. Of the two Malaysian rivers, the Sangga (line 1) has
a terrestrial end member of -27.2°/oo, very similar to the average
ratio of plants from the mangrove forests through which it flows
(-27.1°/oo, Rodelli et al., 1984), while the Salangor (line 5) end
member is -24.0°/oo, more similar to an integrated value for human
wastes. The Selangor passes through settled agricultural areas rather
than the uninhabited mangrove preserves where the Sangga flows. When
extrapolated, the two transects intersect at -21.6°/oo, a ratio which
is analytically indistinguishable from the worldwide average for
particulate marine organic carbon (-21.3°/oo, Table 3). The five
transects off the Mississippi River (lines 4, 6, 7, 8) and Houma
Navigational Canal (line 3) have similar slopes, except for that taken
in East Bay which is relatively isolated from direct riverine runoff.

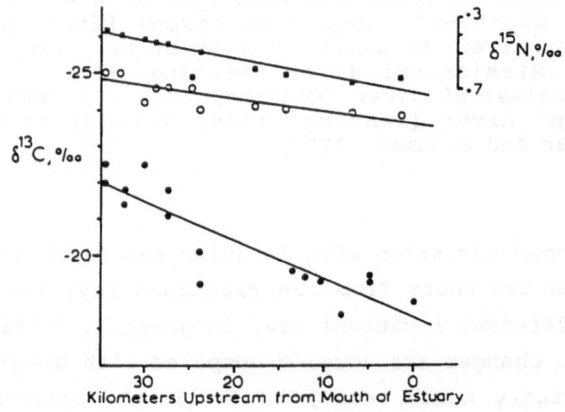

Fig. 6. Change in isotope ratios with distance in the Sheepscot
Estuary, USA (Macko, 1981, 1983; Incze et al. 1982). Solid squares are
values of del N-15 in sediment, open circles are del C-13 in sediment,
and solid circles are del C-13 in bivalves.

Nitrogen, like carbon, seems to be useful for tracing terrestrial organic matter. Peters et al. (1978) suggested its application and correlated $\delta^{13}C$ and $\delta^{15}N$ values of sediments from a variety of terrestrial and marine environments. They extrapolated the δ^{15} N of the two end members to be approximately +2.5°/oo for terrestrial material and +8°/oo for marine organic matter. Since their work nitrogen isotopes have been measured in transects of particulates in Nova Scotia (LeBlanc and Risk, 1985), two Maine estuaries (Macko, 1981), Chesapeake Bay (Spiker and Kendall, 1983) and the Loire River (Létolle et al., 1986). Fig. 6 shows the variation of both carbon and nitrogen isotopes in the sediments of the Sheepscot Estuary, Maine. The correlation between the two isotopes is good at this single location, however, a comparison of this correlation with that found by Peters et al. (1978) shows that the end members vary from place to place (Fig. 7). As for carbon alone, the greater part of the variability lies with the terrestrial end members. All of the limited amount of data available indicates that this variation is large for nitrogen isotopes as well as for carbon (Tables 2-3). More data need to be collected before the usefulness of nitrogen isotopes for organic particulates can be fully evaluated.

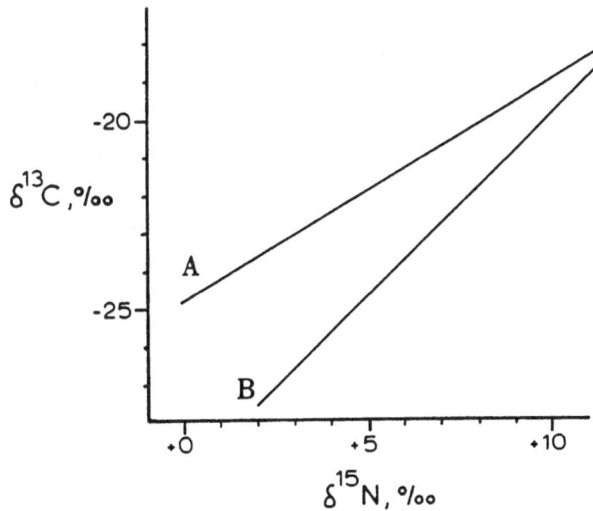

Fig. 7. Correlation of carbon and nitrogen isotope ratios in sediments from Maine (A; Macko 1981) and from the Pacific coast of North America (B; Peters et al., 1978)

3.2 Organisms

It is also possible to use the isotope ratios in organisms to trace the origin of carbon actually assimilated into body tissues. Isotope ratios, thus, provide complementary data to that obtained by analysis of gut contents. They give no information on individual organisms consumed, but rather data on the general type of organic matter assimilated over several days, weeks, or months.

Unfortunately, the isotopic variability between organisms of the same species is rather high and many individuals must be analyzed (either singly or as composite samples) in order to provide useful information. DeNiro and Epstein (1978, 1981a) reported individual laboratory animals raised on the same food to have carbon and nitrogen isotopes varying up to 1.8°/oo and 3°/oo respectively.

Trophic level must also be taken into account when calculating the origin of animal organic matter. As mentioned in section 2, the carbon isotope ratios of animals are an average of 1°/oo more positive than that of their food; their nitrogen isotope ratios are up to 3°/oo more positive. These differences are useful for determining the trophic level of an organism if the origin of the food is known (McConnaughey and McRoy, 1979b; Gearing et al., 1984a). In an area where one is trying to calculate the origin of the organic matter, an animal's trophic position must be known or presumed in order to correct for trophic effects. Fig. 2 shows the great extent to which trophic levels can change isotopic ratios in phytoplankton-based ecosystems.

Most of the studies of organisms have been concerned with understanding the origins of organic matter within a single ecosystem. Many of these are listed in the carbon section of the introduction. In addition, there have been a few comparative studies of animals in nearshore and adjacent offshore area in Alaska (McConnaghey and McRoy, 1979a, b), Texas (Fry and Parker, 1979), and Malaysia (Rodelli et al., 1984). Bivalves along onshore-offshore transects were analyzed in Nova Scotia (LeBlanc and Risk, 1985) and Maine (Incze et al., 1982) estuaries, and in New Zealand (Stephenson and Lyon, 1982).

Fig. 8 compares the carbon isotope ratios in Alaskan and Malaysian organisms. The offshore organisms differ slightly, but this may be due to different sampling techniques collecting organisms of higher or lower trophic levels. Species of omnivorous shrimp from the two areas

are virtually identical (Figs. 8A and 8B). Organisms from the mangrove
swamps of Malaysia are shifted to more negative values, mangrove carbon
being more negative than that of phytoplankton (-27.1°/oo versus -
21.1°/oo). However, the isotopic shift for either all animals (4.2°/oo)
or the two species of shrimp (3.2°/oo) is not as great as the
difference between the two carbon sources (6°/oo), indicating that
animals in the nearshore regions (within 2 km of shore) were utilizing
a significant amount of carbon derived from phytoplankton in addition
to the carbon from terrestrial plants. The same conclusion can be
drawn from the Alaskan data (Fig. 8B and 8D). The inshore-offshore
difference for two shrimp species was 3.4°/oo while the difference
between phytoplankton and eelgrass was estimated at 12°/oo. Moreover,
there were no animals (even higher consumers) in the eelgrass meadow
with isotope ratios higher than the average ratio of eelgrass in that
locale (-10.3°/oo). Higher ratios would be expected from animals
deriving their carbon exclusively from eelgrass.

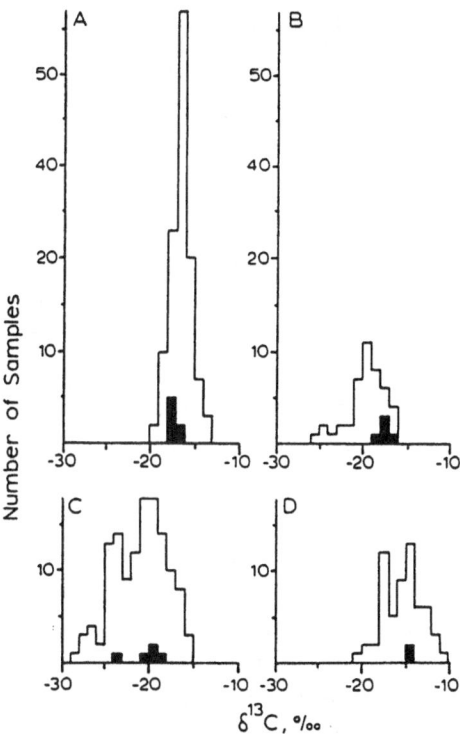

Fig. 8. Distribution of stable carbon isotope ratios in organisms from
offshore Malaysia (A) and the Bering Sea (B) as compared with nearshore
mangrove swamps, Malaysia (C) and an eelgrasslagoon, Alaska (D). Solid
areas show values for individual species of shrimp: Penaeus merguiensis
and Metapenaeus mutatus in Malaysia (A, C) and Crangon dalli and C.
septemspinosa in Alaska (B, D). Data are from McConnaughey and McRoy
(1979a, b) and Rodelli et al. (1984).

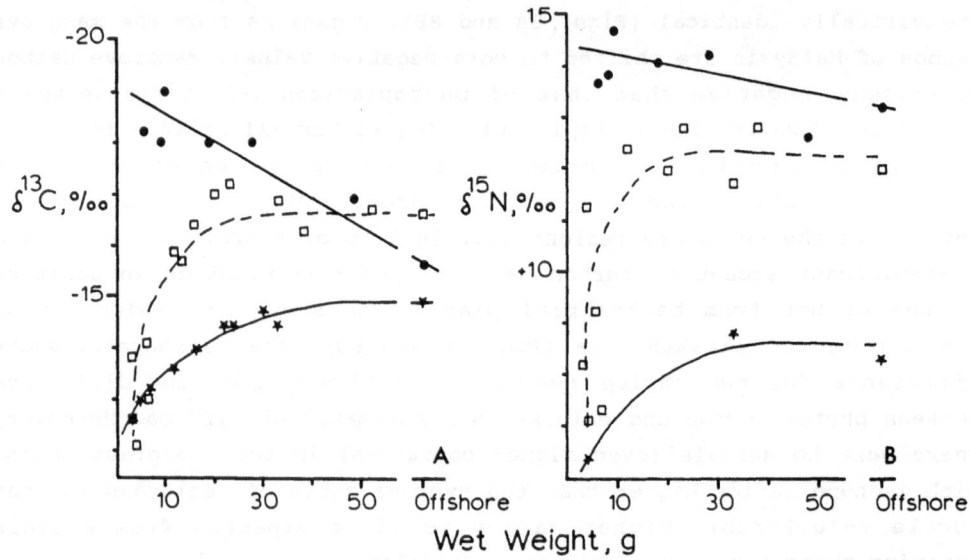

Fig. 9. Distribution of carbon and nitrogen in migratory shrimp (<u>Penaeus duorarum</u>) from south Florida seagrass meadows into coralline areas (stars), in <u>Penaeus aztecus</u> from south Texas seagrass meadows (open squares), and in <u>Penaeus aztecus</u> and <u>Penaeus setiferus</u> from the Texas-Louisiana coast (solid circles). Data from Fry (1983).

Another fact that must be borne in mind when examining data from organisms is that the results are influenced by the bioavailability of the organic matter. For example, the shrimp collected offshore in Malaysia showed no isotopic evidence of utilization of mangrove carbon, having the same ratios as similar shrimp collected worldwide in phytoplankton-based ecosystems. However, other samples of the same species from the same area contained significant amounts of mangrove detritus in their guts (Rodelli et al., 1984). Thus mangrove detritus may be present in the gut of animals without being assimilated into the body tissues. It is hypothesized that such detritus becomes more refractory to metabolism as it moves offshore. Such a trend may be indicated by the data from Maine (Fig. 6; Macko, 1981; Incze et al., 1982). The amount of terrestrial material in the sediments declines offshore, but the decline in terrestrial material present in bivalve tissues is much more rapid. Upstream where the terrestrial material is relatively fresh, bivalves incorporate relatively more terrestrial organic matter into their tissues than farther out in the estuary where the detritus may be more refractory.

Stable isotopes of carbon, nitrogen, and sulphur have been also used to trace the migration of fish and shrimp (Fry, 1981, 1983). Animals living in the lagoons have isotope ratios influenced by vascular plants, either very positive seagrasses and C_4 plants or very negative C_3 plants. As the animals move offshore and consume exclusively phytoplankton-derived organic matter, this signal was lost. Fig. 9 shows the results in three regions for carbon and nitrogen. It also illustrates the differing isotope ratios to be found in different regions for different species of animals having approximately the same trophic level.

3.3 Other Types of Samples

Different types of information is obtained from isotope ratios of different types of samples. For example as mentioned previously, sediments tend to give a more integrated picture of the total organic matter present over a year, while POC is more variable with time. Ratios in animal tissues are related not only to the presence of organic matter, but also its bioavailability and the trophic level of the organism. Likewise it is possible to obtain isotopic information on dissolved organic matter or on individual chemical compounds or classes of compounds. Unfortunately, there are few such studies.

Dissolved organic carbon does differ between terrestrial and marine environments. Terrestrial DOC is around $-25^o/oo$ (Parker and Calder, 1970; Williams and Gordon, 1970; Kerr and Quinn, 1980; Sigleo and Macko, 1985) while marine dissolved material is around $-22^o/oo$ (Calder and Parker, 1968; Parker and Calder, 1970; Williams and Gordon, 1970; Parker, 1971; Eadie et al., 1978; Kerr and Quinn, 1980). Other potentially useful DOC reservoirs are sewage (around $-23^o/oo$; Reimers 1968 (around -30 or $-31^o/oo$; Reimers, 1968; Kerr and Quinn, 1980), petroleum products (around -30 or $-31^o/oo$; Calder and Parker, 1968), and effluent from pulp and paper mills ($-28^o/oo$; Calder and Parker, 1968).

Examination of compounds or classes of compounds eliminates some of the variability due to biochemical differences between samples. The isotope ratios of compound classes such as humic and fulvic acids (for example, Nissenbaum and Kaplan, 1972; Nissenbaum, 1974), lipids (for example, Gearing et al., 1976; Shadskiy et al., 1982; Baturin et al., 1983), and bone collagen (for example, Chisholm et al., 1982; Schoeninger et al.,

1983; Schoeninger and DeNiro, 1984) are useful because they are relatively well preserved over time. Some individual chemical compounds have also been examined (for example, DesMarais et al., 1980; Nishimura and Baker, 1986).

4. TRACING ANTHROPOGENIC POLLUTANTS

Isotope ratios have proven useful in several cases for tracing the extent of local pollution by petroleum. Calder and Parker (1968) first used carbon isotopes of POC and DOC in the Houston and Corpus Christi (Texas) Ship Channels. They concluded that the $\delta^{13}C$ values of dissolved organic matter could be used to quantify petrochemical pollution. Macko (1981) used carbon and nitrogen ratios and Van Vleet et al. (1983) used carbon ratios to charterize the origin of tar balls. Spies and DesMarais (1983) measured carbon and sulphur isotope ratios in benthic organisms near a natural petroleum seep off California. Comparison with the values from organisms in a control area allowed the calculation that the animals near the seep contained approximately 15% extra carbon directly from petroleum ($\delta^{13}C$) and an additional 14% indirectly through sulfate reducers ($\delta^{34}S$).

Sewage can also be traced by isotope ratios, but it often has ratios not greatly different from those in uncontaminated areas. For example, sewage particulate $\delta^{13}C$ averages around -23.5°/oo only 2°/oo more negative than marine POC and almost the same as the ratios of nanoplankton. In order to improve the results under such circumstances, many samples and/or several isotopes should be used. In particular, sewage nitrogen is 4 to 8°/oo different from marine organic nitrogen, depending on the location (Sweeney et al., 1980).

Reimers (1968) first measured the $\delta^{13}C$ of inorganic carbon and dissolved organic carbon from sewage; he concluded that the first but not the second differed enough from natural values to be useful as a tracer. Myers (1974) measured the $\delta^{13}C$ of particulates from sewage and marine sediments around sewage outfalls in California. Sweeney et al. (1980) examined $\delta^{15}N$, $\delta^{34}S$, and uranium in sediments from the same area and used their results and those of Myers (1974) to model geochemical processes. Sweeney and Kaplan (1980) used similar measurements on flocculent suspended matter to trace sewage from outfalls. Burnett and Schaeffer (1980) measured $\delta^{13}C$ to quantify the amount of sewage carbon in sediment transects across the dumping area in the New York Bight. Rau et al., (1981) measured $\delta^{13}C$, $\delta^{15}N$, and δD in sole and prawns from a control and a sewage-contaminated site. They found significant inter-site differences for all three isotopes and suggested refinements in sampling to improve precision.

One possible complication to be kept in mind with sewage studies is that the nutrients released from sewage may act to change the isotope ratios of phytoplankton locally (Parker and Calder, 1970; Gearing and Gearing, 1982). In the MERL ecosystems, nutrient additions resulted in more positive $\delta^{13}C$ values for diatoms. Carbon isotope ratios of phytoplankton and benthic organisms have been measured in control tanks, tanks receiving nutrients alone, and tanks receiving sewage sludge (Gearing et al., 1987).

Finally, carbon isotopes may be useful in tracing the effluents from pulp and paper mills. Calder (1969) measured these effluents to be -27.4^{o}/oo at one mill. Rashid and Reinson (1979) concluded that pulp mills were a major source of organic carbon to the sediments of the Miramichi Estuary, New Brunswick, on the basis of $\delta^{13}C$ and other measurements.

5. COMBINATIONS OF TECHNIQUES

It is preferable, where possible, to use more than one technique in a study so that the weaknesses of each method may be balanced against the strengths of others. Thus for sewage, several isotopes were measured (Sweeney and Kaplan, 1980; Sweeney et al., 1980; Rau et al., 1981). Fry (1983) used carbon, nitrogen, and sulphur to trace fish and shrimp migrations in the Gulf of Mexico. Peterson et al. (1985) presented a good rationale for the use of multiple isotopic tracers in a New England salt marsh. Because of the variety of potential sources of organic matter, they measured $\delta^{13}C$, $\delta^{15}N$, and $\delta^{34}S$ in plants and several species of clams at various locations in the marsh. They were able to conclude that upland plants contributed little or no carbon to the bivalves and that the relative proportion of carbon from phytoplankton and _Spartina_ depended on the location of the animal in the marsh. They stated that the use of multiple tracers eliminated many of the ambiguities associated with the use of single isotopic tracers in salt marshes.

For the same reasons, the results of isotopic tracers have also been compared with other types of measurements made in the same location. Table 4 summarizes the results of comparisons of carbon isotope and C/N ratios. These measurements have also been closely compared in a freshwater ecosystem by LaZerte (1983). They generally correlate at a single location. However, there seems to be less environmental scatter with carbon isotope ratios than with C/N ratios, a fact particularly evident in the freshwater study where direct comparisons were made on over 30 samples from Lake Memphremagog.

Table 4. Comparison of carbon isotope and C/N ratios from various locations

$\delta^{13}C$	C/N RATIOS						
	A	B	C	D	E	F	G
-29							9-15
-28						15-25	
-27		8.5				20-28	
-26	15-22	9-10				18-28	
-25	10-12	10-11	11-13				
-24	9-11		10-14		6-15	10-30	
-23	7-9			10.4	8	15-18	
-22	6-7		5-8			8-14	
-21						10-12	

A St. Lawrence River: Tan and Strain, 1979a.
B St. Lawrence Estuary: Pocklington and Tan, 1983.
C Maine Estuary: Macko, 1981.
D Washington State, USA: Hedges and VanGeen, 1982.
E Changjiang River, China: Kennicutt et al., 1987.
F Orinoco River, Venezuela: Kennicutt et al., 1987.
G Amazon River, Brazil: Williams, 1968; Williams and Gordon, 1970.

Another widely used measure of the amounts of terrestrial and marine organic carbon is lignin. Pocklington and Leonard (1979) compared results from the two techniques in the St. Lawrence. Hedges measured both carbon isotope ratios and a parameter derived from the ratios of terrestrial and marine lignin oxidation products (Hedges and Parker, 1976; Hedges and Mann, 1979; Hedges and vanGeen, 1982). A comparison of

these data is shown in Fig. 10. The two methods correlate
significantly, even though the data are taken from several different
locations. Within a single locale, the correlation is even better (see,
for example, Hedges and Mann, 1979, Fig. 4). Due to differing mixtures
of plants contributing to different drainage areas, phenol yields and
lignin mixtures can vary greatly (Hedges et al., 1984). As with isotpoe
ratios, thorough examinations of local conditions are necessary for
accurate results. Overall, the two methods give comparable results for
tracing terrestrial organic matter.

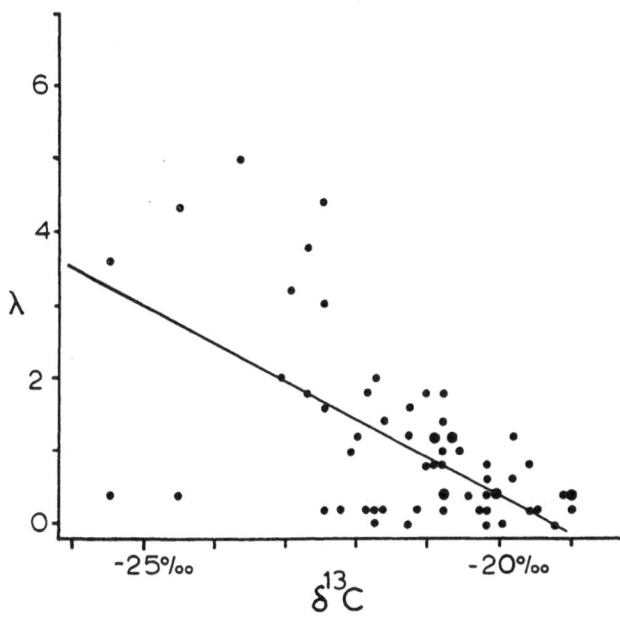

Fig. 10. Correlation of data from lignin oxidation products λ and $\delta^{13}C$
in marine sediments (Hedges and Parker, 1976; Hedges and van Geen,
1982)

Other measurements that have been compared with isotope ratios,
especially for quantification of terrestrially-derived organic matter
in the marine environment, include organic carbon and nitrogen
concentrations (Gearing, 1975; Newman et al., 1973; Sweeney et al.,
1980; and others), sediment type (Rashid and Reinson, 1979), ^{14}C
(Spiker and Schemel, 1979; Schell, 1983), fatty acids (Gaskell et al.,
1975; LeBlanc and Risk, 1985; Letolle et al., 1986; Nishimura and
Baker, 1986), and hydrocarbons (Gearing et al., 1976; VanVleet et al.,
1983; Kennicutt et al., 1987; Nishimura and Baker, 1986).

ACKNOWLEDGEMENTS

I would like to thank Drs. P. Gearing and J. Hedges for their helpful comments. Discussions with the other attendees at the SCOR Working Group 65 Workshop on Coastal-Offshore Exchanges have also been useful.

REFERENCES

Anderson, T.F. & M.A. Arthur, 1983. Stable isotopes of oxygen and carbon and their application to sedimentological and paleoenvironmental problems. - In M.A. Arthur, T.F. Anderson, I.R. Kaplan, J. Veizer & L.S. Land (eds.): Stable isotopes in sedimentary geology, pp. 1-151. Soc. Econ. Paleontologists Mineralogists, Tulsa, Okla.

Baturin, G.N., Y.A. Romankevich & I.P. Shadskiy, 1983. Carbon isotope composition of bone phosphate lipids in bottom sediments. - Oceanoglogy 23: 597-600.

Benedict, C.R., W.W.L. Wong & J.H.H. Wong, 1980. Fractionation of the stable isotopes of inorganic carbon by seagrasses. - Plant. Physiol. 65: 597-600.

Black, Jr., C.C. & M.M. Bender, 1976. δ13C values in marine organisms from the Great Barrier Reef. - Aust. J. Plant Physio. 3: 25-32.

Brinson, M.M. & E.A. Matson, 1983. Carbon isotope distribution in the Pamlico River estuary, North Carolina and tributaries. - Estuaries 6: 306 (Abstr).

Burnett, W.C. & O.A. Schaeffer, 1980. Effect of ocean dumping on 13C/12C ratios in marine sediments from the New York Bight. - Estuar. Coast. Mar. Sci. 11: 605-611.

Calder, J.A., 1969. Carbon isotope effects in biochemical and geochemical systems. Ph.D. Dissertation, Univ. Texas, Austin.

Calder, J.A. & P.L. Parker, 1968. Stable carbon isotope ratios as indices of petrochemical pollution of aquatic systems. - Environ. Sci. Technol. 2: 535-539.

Cheng, H.H., J.M. Bremner & A.P. Edwards, 1964. Variations of nitrogen-15 abundance in soils. - Science 146: 1574-1575.

Chisholm, B.S., D.E. Nelson & H.P. Schwarcz, 1982. Stable-carbon isotope ratios as a measure of marine versus terrestrial protein in ancient diets. - Science 216: 1131-1132.

Dean, W.E., M.A. Arthur & G.E. Claypool, 1986. Depletion of 13C in Cretaceous marine organic matter: source, diagenetic, or environmental signal? - Mar. Geol. 70: 119-157.

Degens, E.T., 1969. Biogeochemistry of stable carbon isotopes. - In G. Eglinton, & M.T.J. Murphy (ed.): Organic geochemistry, pp. 304-329. Springer-Verlag, New York.

Degens, E.T., M. Behrendt, B. Gotthardt & E. Reppmann, 1968a. Metabolic fractionation of carbon isotopes in marine plankton -II. Data on samples collected off the coasts of Peru and Ecuador. - Deep-Sea Res. 15: 11-20.

Degens, E.T., R.R.L. Guillard, W.M. Sackett & J.A. Hellebust, 1968b. Metabolic fractionation of carbon isotopes in marine plankton -1. Temperature and respiration experiments. - Deep-Sea Res. 15: 1-9.

Deines, P., 1980. The isotopic composition of reduced organic carbon. - In P. Fritz & J. Ch. Fontes (eds.): Handbook of environmental geochemistry, pp. 329-406. Elsevier Sci. Publ., New York.

DeNiro, M & S. Epstein, 1978. Influence of diet on the distribution fo carbon isotopes in animals. - Geochim. Cosmochim. Acta 42: 495-506.

DeNiro, M. & S. Epstein, 1981a. Influence of diet on the distribution of nitrogen isotopes in animals.- Geochim Cosmochim. Acta 45: 341-351.

DeNiro, M. & S. Epstein, 1981b. Hydrogen isotope ratios of mouse tissues are influenced by a variety of factors other than diet. - Science 214: 1374-1375.

DesMarais, D.J., J.M. Mitchell, W.G. Meinschein & J.M. Hayes, 1980. The carbon isotope biogeochemistry of the individual hydrocarbons in bat guano and the ecology of the insectivorous bats in the region of Carlbad, New Mexico. - Geochim. Cosmochim. Acta 44: 2075-2086.

Deuser, W.G., 1970. Isotopic evidence for diminishing supply of available carbon during diatom blooms in the Black Sea. - Nature 225: 1069-1071.

Dunton, K.H. & D.M. Schell, 1982. The use of 13C/12C ratios to determine the role of macrophyte carbon in an arctic kelp community. - Eos 63: 54 (abstr.).

Eadie, B.J. & L.M. Jeffrey, 1973. δ^{13}C analyses of oceanic particulate matter. - Mar. Chem. 1:199-209.

Emery, K.O., 1960. The sea off southern California. John Wiley and Sons, New York.

Erlenkeuser, H., 1978. Stable carbon isotpoe characteristics of organic sedimentary source materials entering the estuarine zone - In Biogeochemistry of estuarine sediments, pp. 199-206. UNESCO, Paris.

Estep, M.F. & H. Dabrowski, 1980. Tracing food webs with stable hydrogen isotopes. - Science 209: 1537-1538.

Estep, M.F. & T.C. Hoering, 1980. Biogeochemistry of the stable hydrogen isotopes. - Geochim. Cosmochim. Acta 44: 1197-1206.

Fontugne, M.R., 1983. Les isotopes stables du carbone organique dans l'ocean. Applications a la paleoclimatologie. These, Doc. d'Etat, Univ. Paris-Sud, Centre d'Orsay.

Fontugne, M.R. & J.-C. Duplessy, 1978. Carbon isotope ratio of marine plankton related to surface water masses. - Earth Plant, Sci. Letters 41: 365-371.

Fontugne, M.R. & J.-C. Duplessy, 1981. Organic carbon isotopic fractionation by marine plankton in the temperature range -1 to 31° C. - Oceanol. Acta 4: 85-90-

Fry, B., 1977. Stable carbon isotope ratios - a tool for tracing food chains. - MS. Thesis, Univ. Texas, Austin.

Fry, B., 1981. Natural stable carbon isotope tag traces Texas shrimp migrations. - Fish. Bull. 79: 337-345.

Fry, B., 1983. Fish and shrimp migrations in the northern Gulf of Mexico analyzed using stable C, N, and S isotope raios. - Fish. Bull. 81: 789-801.

Fry, B., 1984. 13C/12C ratios and the trophic importance of algae in Florida Syringodium filiforme seagrass meadows. - Mar. Biol. 79: 11-19.

Fry, B, & P.L. Parker, 1979. Animal diet in Texas seagrass meadows: 13C evidence for the importance of benthic plants. - Estuar. Coast. Mar. Sci. 8: 499-509.

Fry, B. & E. Sherr, 1984. δ13C measurements as indicators of carbon flow in marine and freshwater ecosystems. - Contrib. Mar. Sci. 27: 13-47.

Fry, B., R.S. Scalan & P.L. Parker, 1977. Stable carbon isotope evidence for two sources of organic matter in coastal sediments: seagrasses and plankton. - Geochim. Cosmochim. Acta 41: 1875-1877.

Fry, B., A. Joern & P.L. Parker, 1978. Grasshopper food web analysis; use of carbon isotope ratios to examine feeding relationships among terrestrial herbivores. - Ecology 59: 498-506.

Fry, B., R. Lutes, M. Northem, P.L. Parker & J. Ogden, 1982a. A 13C/12C comparison of food webs in Caribbean seagrass meadows and coral reefs. - Aquat. Bot. 14: 389-398.

Fry, B., R.S. Scalan, J.K. Winters & P.L. Parker, 1982b. Sulphur uptake by saltgrasses, mangroves, and seagrasses in anaerobic sediments. - Geochim. Cosmochim. Acta 46: 1121-1124.

Fry, B., R.S. Scalan & P.L. Parker, 1983. 13C/12C ratios in marine food webs of the Torres Strait, Queensland. - Aust. J. Mar. Freshwtr. Res. 34: 707-715.

Fry, B., R.K. Anderson, L. Entzeroth, J.L. Byrd & P.L. Parker, 1984. 13C enrichment and oceanic food web structure in the northwestern Gulf of Mexico. - Contrib. Mar. Sci. 27: 49-63.

Gaskell, S.J., R.J. Morris, G. Eglinton & S.E. Calvert, 1975. The geochemistry of a recent marine sediment off northwest Africa. An assessment of source of input and early diagenesis. - Deep-Sea Res. 22: 777-789.

Gearing, J.N., P.J. Gearing, D.T. Rudnick, S.G. Requejo & M.J. Hutchins, 1984a. Isotpoic variability of organic carbon in a phytoplankton-based, temperate estuary. - Geochim. Cosmochim. Acta 48: 1089-1098.

Gearing, J., P. Gearing, M. Rodelli & N. Marshall, 1984b. Initial findings from stable carbon isotope ratios in west coast mangrove areas of peninsular Malaysia, In E. Sepadmo, A.N. Rao & D.J. MacIntosh (eds.), Proc. Asian Symp. Mangrove Environ.: Res. & Manage., pp. 488-495. UNESCO, Kuala Lumpur.

Gearing, P.J., 1975. Organic carbon stable isotope ratios of continental margin sediments. Ph.D. Diss., Univ. Texas, Austin.

Gearing, P.J. & J.N. Gearing, 1982. Distribution of stable isotopes in control and eutrophied microcosms (MERL) and in Narragansett Bay. - EOS 63: 955 (Abstr.).

Gearing, P., F.E. Plucker & P.L. Parker, 1977. Organic carbon stable isotope ratios of continental margin sediments. - Mar. Chem. 5: 251- 266.

Gearing, P., J.N. Gearing, T.F. Lytle & J.S. Lytle, 1976. Hydrocarbons in 60 northeast Gulf of Mexico shelf sediments: a preliminary survey. - Geochim. Cosmochim. Acta 40: 1005-1017.

Gearing, P., J.N. Gearing & J. Maughan, 1987. Stable carbon isotopes trace sewage sludge in controlled ecosystems. - MS in prep.

Geyh, M.A., H.-R. Kudrass & H. Streif, 1979. Sea-level changes during the late Pleistocene and Holocene in the Strait of Malacca. - Nature 278: 441-443.

Hackney, C. & E.B. Haines, 1980. Stable carbon isotope composition of fauna and organic matter collected in a Mississippi estuary. - Estuar. Coast. Mar. Sci. 10: 703-708.

Haines, E.B., 1976a. Stable carbon isotopes in the biota, soils and tidal waters of a Georgia salt marsh. - Estuar. Coast. Mar. Sci. 4: 609-616.

Haines, E.B., 1976b. Relation between the stable carbon isotope composition of fiddler crabs, plants and soils in a salt marsh. - Limnol. Oceanogr. 21: 880-993

Haines, E.B. & C.L. Montague, 1979. Food sources of estuarine invertebrates analyzed using 13C/12C ratios. - Ecol. 60: 48-56.

Hedges, J.I. & P.L. Parker, 1976. Land-derived organic matter in surface sediments from the Gulf of Mexico. - Geochim. Cosmochim. Acta 40: 1019-1029.

Hedges, J.I. & A. van Geen, 1982. A comparison of lignin and stable carbon isotope compositions in Quaternary marine sediments. - Mar. Chew. 11: 43-54.

Hughes, E.H. & E.B. Sherr, 1983. Subtidal food webs in a Georgia estuary: δ13C analysis. - J. Exp. Mar. Biol. Ecol. 67: 227-242.

Hunt, J.M., 1966. The sigfnificance of carbon isotope variations in marine sediments. - In G.D. Hobson & G.C. Speers (eds.): Advances in organic geochemistry, 1966, pp. 27-35. Pergamon Press, Oxford.

Incze, L.S., L.M. Mayer, E.B. Sherr & S.A. Macko, 1982. Carbon inputs to bivalve mollusks: a comparison of two estuaries. - Can. J. Fish. Aquat. Sci. 39: 1348-1352.

Johnson, R.W. & J.A. Calder, 1973. Early diagenesis of fatty acids and, hydrocarbons in a salt marsh environment. - Geochim. Cosmochim. Acta 37: 1943-1955.

Kaplan, I.R., 1975. Stable isotopes as a guide to biogeochemical processes. - Proc. R. Soc. Lond. B 189: 183-211.

Kaplan, I.R., 1983. Stable isotopes of sulfur, nitrogen and deuterium in recent marine environments. - In Stable isotopes in sedimentary geology, pp. 2.1-2.108. Soc. Econ. Paleontologists Mineralogists, Tulsa, Okla.

Kennicutt, M.C., C. Barker, J.M. Brookes, D.A. DeFreitas, & G.H. Khu, 1987. Selected organic matter source indicators in the Orinoco, Nile and Changjiang deltas. - Org. Geochem. 11: 41-51.

Kerr, R.A. & J.G. Quinn, 1980. Partial chemical characterization of estuarine dissolved organic matter. - Org. Geochem. 2: 129-138.

Kitting, C.L., B. Fry & M.D. Morgan, 1984. Detection of inconspicuous epiphytic algae supporting food webs in seagrass meadows. - Oecologia (Berlin) 62: 145-149.

Krouse, H.R., 1980. Sulphur isotopes in our environment. - In P. fritz & J.Ch. Fontes (eds.): Handbook of environmental geochemistry Vol. 1, pp. 435-471. Elsevier Sci. Publ., New York.

Land, L.S., J.C. Lang & B.N. Smith, 1975. Preliminary observations on the carbon isotopic composition on some coral reef tissues and symbiotic zooxanthellae. - Limnol. Oceanogr. 20: 283-287.

LaZerte, B.D., 1983. Stable carbon isotope ratios: implications for the source of sediment carbon and for phytoplankton carbon assimilation in Lake Memphremagog, Quebec. - Can. J. Fish. Aquat. Sci. 40: 1658-1666.

LeBlanc, C.G., & M.J. Risk, 1985. Peatland contribution to a Nova Scotia estuary measured by stable isotopes and organic geochemistry. - Abstract, ASLO 48th Annual Meeting, June 18-21, 1985.

Létolle, R., 1980. Nitrogen-15 in the natural environment. - In P. Fritz & J. Fontes (eds.): Handbook of environmental geochemistry Vol. 1, pp. 407-433. Elsevier Sci. Publ., New York.

Létolle, R., A. Lorre, A. Mariotti, J.C. Marty, A. Saliot, P. Scribe & J. Tronczynski, 1987. The application of isotope and biogeochemical markers to the study of the biogeochemistry of organic matter in a macrotidal estuary, the Loire, France. Manuscript in preparation.

McConnaughey, T., 1978. Ecosystems naturally labeled with carbon-13: applications to the study of consumer food-webs. - M.S. Thesis, Univ. Alaska, Fairbanks.

McConnaughey, T. & C.P. McRoy, 1979a. ^{13}C label identifies eelgrass (Zostera marina) carbon in an Alaskan estuarine food web. - Mar. Biol. 53: 263-269.

McConnaughey, T. & C.P. McRoy, 1979b. Food web structure and the fractionation of carbon isotopes in the Bering Sea. - Mar. Biol. 53: 257-262.

McMillan, C., P.L. Parker & B. Fry, 1980. $^{13}C/^{12}C$ ratios in seagrasses. - Aquat. Bot. 9: 237-249.

Macko, S.A., 1981. Stable nitrogen isotope ratios as tracers of organic geochemical processes. Ph.D. Diss., Univ. Texas, Austin.

Macko, S.A., 1983. Source of organic nitrogen in mid-Atlantic coastal bays and continental shelf sediments of the United States: isotopic evidence. - Carnegie Inst. Wash. Year Book 82: 390-394.

Macko, S.A., W.Y. Lee & P.L. Parker, 1982. Nitrogen and carbon isotope fractionation by two species of marine amphipods: laboratory and field studies. - J. Exp. Mar. Biol. Ecol. 63: 145-149.

Macko, S.A., M.L.F. Estep & W.Y Lee, 1983. Stable hydrogen isotope analysis of food webs on laboratory and field populations of marine amphipods. - J. Exp. Mar. Biol. Ecol. 72: 243-249.

Macko, S.A., L. Entzeroth & P.L. Parker, 1984. Regional differences in nitrogen and carbon isotopes on the continental shelf of the Gulf of Mexico. - Naturwissenschaften 71: 374-375.

Mayer, L.M.,S.A. Macko, W.H. Mook & S. Murray, 1981. The distribution of bromine in coastal sediments and its use as a source indicator for organic matter. - Organic Geochem. 3: 37-42.

Minson, D.J., M.M. Ludlow & J.H. Troughton, 1975. Differences in natural carbon isotope ratios of milk and hair from cattle grazing tropical and temperate pastures. - Nature 256: 602.

Myers, E.P., 1974. The concentration and isotopic composition of carbon in marine sediments affected by a sewage discharge. - Ph.D. Diss., California Inst. Techno., Pasadena.

Newman, J.W., P.L. Parker & E.W. Behrens, 1973. Organic carbon isotope ratios in Quaternary cores from the Gulf of Mexico. - Geochim. Cosmochim. Acta 37: 225-238.

Nishimura, M. & E.W. Baker, 1986. Possible origin of n-alkanes with a remarkable even-to-odd predominance in recent marine sediments. - Geochim. Cosmochim. Acta 50: 299-305.

Nissenbaum, A., 1974. Deuterium content of humic acids from marine and non-marine environments. - Mar. Chem. 2: 59-63.

Nissenbaum, A. & I.R. Kaplan, 1972. Chemical and isotopic evidence for the in situ origin of marine humic substances. - Limnol. Oceanogr. 17: 570-582.

Northam, M.A., D.J. Curry, R.S. Scalan & P.L. Parker, 1981. Stable carbon isotope ratio variations of organic matter in Orca Basin sediments. - Geochim. Cosmochim. Acta 45: 257-260.

O'Leary, M.H., 1981. Carbon isotope fractionation in plants. - Phytochem. 20: 553-567.

Parker, P.L., 1964. The biogeochemistry of the stable carbon isotopes in a marine bay. - Geochim. Cosmochim. Acta 28: 1155-1164.

Parker, P.L., 1971. Petroleum - stable isotope ratio variations. - In D.W. Hood (ed.): Impingement of man on the oceans, pp. 431-444. Wiley and Sons, New York.

Parker, P.L. & J.A. Calder, 1970. Stable carbon isotope ratio variations in biological systems. - In D.W. Hood (ed.): Organic matter in natural waters, pp. 107-122. Inst. Mar. Sci. Univ. Alaska Publ. 1, College, Alaska.

Petelle, M., B. Haines & E. Haines, 1979. Insect food preferences analysed using $^{13}C/^{12}C$ ratios. - Oecologia (Berlin) 38: 159-166.

Peters, K.E., R.E. Sweeney & I.R. Kaplan, 1978. Correlation of carbon and nitrogen stable isotope ratios in sedimentary organic matter. - Limnol. Oceanogr. 23:598-604.

Peterson, B.J. & R.W. Howarth, 1983. Sulfur and carbon isotopes as tracers of organic matter flow in salt marshes. - Estuaries 6: 305. (Abstr.).

Peterson, B.J., R.W. Howarth & R.H. Garritt, 1985. Multiple stable isotopes used to trace the flow of organic matter in estuarine food webs. - Science 227: 1361-1363.

Pocklington, R. & J.D. Leonard, 1979. Terrigenous organic matter in sediments of the St. Lawrence estuary and the Saguenay Fjord. - J. Fish. Res. Bd. Can. 36: 1250-1255.

Pocklington, R. & F. Tan, 1983. Organic carbon transport in the St. Lawrence River. - In E.T. Degens, S. Kempe & H. Soliman (eds.): Transport of Carbon and Minerals in Major World Rivers, Part 2, pp. 243-251. Mitteilungen aus dem Geologisch-Paläontologischen Institut der Universität Hamburg, Heft 55, Hamburg.

Rashid, M.A. & G.E. Reinson, 1979. Organic matter in surficial sediments of the Miramichi Estuary, New Brunswick, Canada. - Estuar. Coast. Mar. Sci. 8: 23-36.

Rau, G., 1982. The relationship between trophic level and stable isotopes of carbon and nitrogen. - In W. Bascom (ed.): Coastal water research project biennial report, pp. 143-148. Southern Calif. Coast. Wat. Res. Proj., Long Beach, Calif.

Rau, G.H., R.E. Sweeney, I.R. Kaplan, A.J. Mearns & D.R. Young, 1981. Differences in animal ^{13}C, ^{15}N, and D abundance between a polluted and an unpolluted coastal site: likely indicators of sewage uptake by a marine food web. - Estuar. Coast. Shelf Sci. 13: 701-707.

Rau, G.H., A.J. Mearns, D.R. Yound, R.J. Olson, H.A. Schafer & I.R. Kaplan, 1983. Animal $^{13}C/^{12}C$ correlates with trophic level in pelagic food webs. - Ecology 64: 1314-1318.

Reimers, R.S., 1968. A stable carbon isotopic study of a marine bay and domestic waste treatment plant. - M.S. Thesis, Univ. Texas, Austin.

Rodelli, M.R., J.N. Gearing, P.J. Gearing, N. Marshall & A. Sasekumar, 1984. Carbon sources used by organisms in Malaysian mangrove swamps and nearshore waters as determined by ^{13}C values. - Oecologia (Berlin) 61: 326-333.

Rounick, J.S. & M.J. Winterbourn, 1986. Stable carbon isotopes and carbon flow in ecosystems. - BioScience 36: 171-177.

Sackett, W.M., 1964. The depositional history and isotopic organic carbon composition of marine sediments. - Mar. Geol. 2: 173-185.

Sackett, W.M., W.R. Eckelmann, M.L. Bender & A.W.H. Bé, 1965. Temperature dependence of carbon isotope composition in marine plankton and sediments. - Science 148: 235-237.

Sackett, W.M., B.J. Eadie & M.E. Exner, 1973. Stable isotope composition of organic carbon in recent Antarctic sediments. - In Advances in organic geochemistry, 1973, pp. 661-671. Technip.

Sackett, W.M. & R.R. Thompson, 1963. Isotopic organic carbon composition of recent continental derived clastic sediments of eastern gulf coast, Gulf of Mexico. - Bull. Amer. Assoc. Petrol. Geol. 47: 525-531.

Saino, T. & A. Hattori, 1985. Variation of ^{15}N natural abundance of suspended organic matter in shallow oceanic waters. In A.C. Sigleo & A. Hattori (eds.): Marine and estuarine geochemistry, pp. 1-13. Lewis Pub., Chelsea, Michigan.

Salomons, W. & W.G. Mook, 1981. Field observations of the isotopic composition of particulate organic carbon in the southern North Sea and adjacent estuaries. - Mar. Geol. 41: M11-M20.

Schell, D.M., 1983. Carbon-13 and carbon-14 abundances in Alaskan aquatic organisms: delayed production from peat in Arctic food webs. - Science 219: 1068-1071.

Schiegl, W.E. & J.C. Vogel, 1970. Deuterium content of organic matter. - Earth Planet. Sci. Lett. 7: 307-313.

Schoeninger, M.J. M.J. DeNiro & H. Tauber, 1983. Stable nitrogen isotope ratios of bone collagen reflect marine and terrestrial components of prehistoric human diet. - Science 220: 1381-1383.

Schwarcz, H.P., 1969. The stable isotopes of carbon. - In K.H. Wedepohl (ed.): Handbook of geochemistry, pp. 6B1-6B15. Springer-Verlag, Berlin.

Schwinghamer, P., F.C. Tan & D.C. Gordon, Jr., 1983. Stable carbon isotope studies in the Pecks Cove mudflat ecosystem in the Cumberland Basin, Bay of Fundy. - Can. J. Fish. Aquat. Sci. 40: 262- 272.

Shadskiy, I.P., Y.A. Romankevich & Y.I. Grinchenko, -1982. Isotopic composition of carbon in lipids from suspended matter and bottom sediments east of the Juril Islands. - Oceanology 22: 301-305.

Sherr, E.B., 1982. Carbon isotope composition of organic seston and sediments in a Georgia salt marsh estuary. - Geochim. Cosmochim. Acta 46: 1227-1232.

Schultz, D.J. & J.A. Calder, 1976. Organic carbon $^{13}C/^{12}C$ variations in estuarine sediments. - Geochim. Cosmochim. Acta 40: 381-385.

Sigleo, A.C. & S.A. Macko, 1985. Stable isotope and amino acid composition of estuarine dissolved colloidal material. In A.C. Sigleo & A. Hattori (eds.): Marine and estuarine geochemistry pp. 29- 46. Lewis Pub., Chelsea, Michigan.

Simenstad, C.A. & R.C. Wissmar, 1985. ^{13}C evidence of the origins and fates of organic carbon in estuarine and nearshore food webs. - Mar. Ecol. Prog. Ser. 22: 141-152.

Smith, B.N., 1972. Natural abundance of the stable isotopes of carbon in biological systems. - BioScience 22: 226-230.

Smith, B.N. & S. Epstein, 1970. Biogeochemistry of the stable isotopes of hydrogen and carbon in salt marsh biota. - Plant Physiol. 46: 738- 742.

Smith, B. & S. Epstein, 1971. Two catagories of $^{13}C/^{12}C$ for higher plants. - Plant. Physiol. 47: 380-384.

Sofer, Zvi, 1984. Stable carbon isotope composition of crude oils: application to source depositional environments and petroleum alteration. - Amer. Assoc. Petrol. Geol. Bull. 68: 31-49.

Spies, R.B. & D.J. DesMarais, 1983. Natural isotope study of trophic enrichment of marine benthic communities by petroleum seepage. - Mar. Biol. 73: 67-71.

Spiker, E.C. & P.G. Hatcher, 1984. Carbon isotope fractionation of sapropelic organic matter during early diagenesis. - Org. Geochem. 5: 283-290.

Spiker, A.C. & C. Kendall, 1983. ^{13}C and ^{15}N as source indicators of sedimentary organic matter in the Chesapeake Bay system. - Estuaries 6:305. (Abstr.).

Spiker, E.C. & L.E. Schemel, 1979. Distribution and stable isotope composition of carbon in San Francisco Bay. - In T.J. Conomos (ed): San Francisco Bay: the urbanized estuary, pp. 195-212. Pacific Div. AAAS, San Francisco.

Stephenson, R.L. & G.L. Lyon, 1982. Carbon-13 depletion in an estuarine bivalve: detection of marine and terrestrial food sources. - Oecologia (Berlin) 55: 110-113.

Stephenson, R.L., F.C. Tan & K.H. Mann, 1984. Stable carbon isotope variability in marine macrophytes and its implications for food web studies. - Mar. Biol. 81: 223-230.

Sweeney, R.E. & I.R. Kaplan, 1980. Tracing flocculent industrial and domestic sewage transport on San Pedro shelf, southern California, by nitrogen and sulfur isotope ratios. - Mar. Environ. Res. 3: 215-224.

Sweeney, R.E., E.K. Kalil & I.R. Kaplan, 1980. Characterization of domestic and industrial sewage in southern California coastal sediments using nitrogen, carbon, sulphur, and uranium tracers. - Mar. Environ. Res. 3: 225-243.

Tan, F.C. & P.M. Strain, 1979a. Organic carbon isotope ratios in recent sediments in the St. Lawrence estuary and the Gulf of St. Lawrence. - Estuar. Coast. Mar. Sci. 8: 213-225.

Tan, F.C. & P.M. Strain, 1979b. Carbon isotope ratios of particulate organic matter in the Gulf of St. Lawrence. - J. Fish. Res. Bd. Can. 36: 678-682.

Tan, F.C. & P.M. Strain, 1983. Sources, sinks, and distribution of organic carbon in the St. Lawrence estuary, Canada. - Geochim. Cosmochim. Acta 47: 125-132.

Thayer, G.W., P.L. Parker, M.W. LaCroix & B. Fry, 1978. The stable carbon isotope ratio of some components of an eelgrass, Zostera marina, bed. - Oecologia (Berlin) 35: 1-12.

Thode, H.G. & C.E. Reese, 1970. Sulphur isotope geochemistry and Middle East oil studies. - Endeavour 29: 24-38.

Torgersen, T., A.R. Chivas & A. Chapman, 1983. Chemical and isotopic characterisation and sedimentation rates in Princess Charlotte Bay, Queensland. - Bur. Miner. Resour. J. Aust. Geol. Geophys. 8: 181-200.

Torgersen, T. & A.R. Chivas, 1985. Terrestrial organic carbon in marine sediment: a preliminary balance for a mangrove environment derived from ^{13}C. - Chem. Geol. (Isot. Geosci. Sec.) 52: 379-390.

van der Marwe, N.J., 1982. Carbon isotopes, photosynthesis, and archaeology. - Amer. Sci. 70: 596-606.

VanVleet, E.S., W.M. Sackett, F.F. Weber, Jr. & S.B. Reinhardt, 1983. Input of pelagic tar into the northwest Atlantic from the Gulf Loop Current: chemical characterization and its relationship to weathered IXTOC-I oil. - Can. J. Fish. Aquat. Sci. 40: 12-22.

Virginia, R.A. & C.C. Delwiche, 1982. Natural ^{15}N abundance of presumed N_2-fixing and non-N_2-fixing plants from selected ecosystems. - Oecologia (Berlin) 54: 317-325.

Wada, E., 1980. Nitrogen isotope fractionation and its significance in biogeochemical processes occurring in marine environments. - In E.D. Goldberg, Y. Horibe, J.K. Saruhashi (eds.): Isotope marine chemistry, pp. 375-398. Uchida Rokakuho, Tokyo.

Williams, P.M., 1968. Organic and inorganic constituents of the Amazon River. - Nature 218: 937-938.

Williams, P.M. & L.I. Gordon, 1970. Carbon-13:carbon-12 ratios in dissolved and particulate organic matter in the sea. - Deep-Sea Res. 17: 19-27.

Wong, W.W. & W.M. Sackett, 1978. Fractionation of stable carbon isotopes by marine phytoplankton. - Geochim. Cosmochim. Acta 42: 1809-1815.

Zieman, J.C., S.A. Macko & A.L. Mills, 1984. Role of seagrasses and mangroves in estuarine food webs: temporal and spacial changes in stable isotope composition and amino acid content during decomposition. - Bull. Mar. Sci. 35: 380-392.

TIDAL FLAT AREAS

H. Postma
Netherlands Institute for Sea Research
P.O. Box 59, 1790 AB Den Burg, Texel
The Netherlands

1. INTRODUCTION

Tidal flat areas are parts of estuaries without macrophyte growth which emerge at low tide, excluding beaches which differ from tidal flats by the dominance of surf. Macrophytes mostly form the high tide and tidal creeks the low tide boundary.

Because of their barren aspect tidal flats look much the same all over the world. This uniformity also holds for the inhabitants, although different species of mostly worms and shells occur in different climate zones. Wherever they live, such species must be able to withstand strong environmental stress, from emergence to large variations in salinity and temperature, storms, sediment load, shifting sands and anaerobic conditions.

This paper describes a number of specific aspects of tidal flats, mainly with the purpose of discovering possible new avenues of investigation. Because of its limited length it does not give full credit to the very extensive literature on the subject.

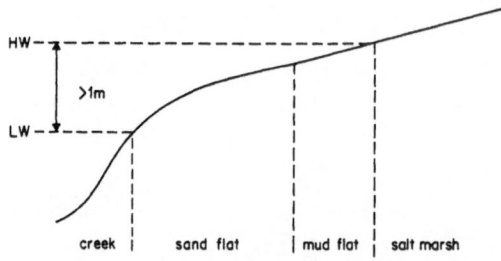

Fig. 1. Cross-section of a tidal flat system.

Lecture Notes on Coastal and Estuarine Studies, Vol. 22
B.-O. Jansson (Ed.), Coastal-Offshore Ecosystem Interactions.
© Springer-Verlag Berlin Heidelberg 1988

2. TIDES

A schematic cross section of a tidal flat system is shown in Figure 1. For the presence of tidal flats the vertical tide must be at least approximately one meter, but there is no limit to its height. Perhaps the most extensive tidal flats occur in the region around "Mont St. Michel" in northern France where the average tidal range is about 12 meters. The tide of the Channel enters here over the sands with high speed as a "bore". This is an extreme demonstration of a tidal wave, almost sinusoidal offshore, becoming asymmetrical in shallow water (Dronkers, 1986; Collins et al., 1981). This is due to the fact that the velocity, c, of the wave depends on water depth, h, according to the formula $c = \sqrt{gh}$, g being the gravity constant of about 10 m sec^{-2}. Consequently the crest of the tidal wave tends to overtake the trough.

The asymmetry of the vertical tide causes an asymmetry in the horizontal tide which is characterized by a fast start of the flood current, a prolonged period of high water, slack and a fast finish of the ebb current (Fig. 2). This current pattern is best developed in small tidal creeks adjacent to the flats. Since the sides of these creeks are steep, the fastest rise of water level and the strongest currents have passed before the flats are submerged. The same situation occurs at the end of the ebb tide. As a result current velocities on tidal flats are much lower, mostly below 50 cm s^{-1}, than in the creeks although in the last water masses draining the flats the current velocities may again increase since these flow down the slopes.

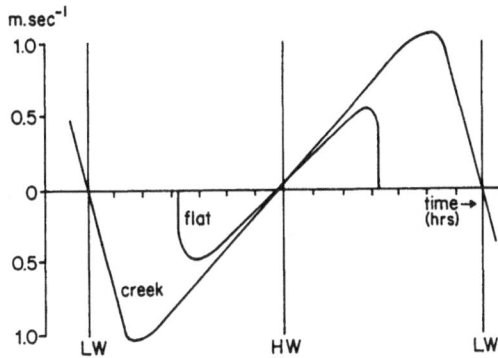

Fig. 2. Asymmetry of tidal currents in a tidal flata system assuming an approximately sinusoidal shape of the incoming tide.

Directions of flood and ebb currents over tidal flats are not simply away from and back to the tidal creeks, but change with the increase and decrease of water depth since such directions are more sensitive to topography when the depth is shallower. Moreover, whereas tidal currents on the flats stop almost immediately at high water they often continue to proceed in the flood direction in the creeks where they may turn up to half an hour after high water. It follows that when the ebb is already running on the flats the flood may still be continuing in the creeks. This effect generates so-called "tide rips" or "tidal fronts" on the edges of the flats which, for reasons to be discussed in paragraph 6, are often visible as foam lines.

Obviously flood and ebb currents do not simply move water back and forth between creeks and tidal flats. On a somewhat larger scale, creeks are not filled or emptied simultaneously by the main tidal channels to which they are connected, since timing depends on the progression of the tidal wave from the inlet of an estuary through these main channels. Moreover, in these channels, horizontal tide residuals are sometimes quite important (Zimmerman, 1986).

As a result of the above effects a residual current pattern mostly exists over tidal flats in addition to the back and forth water movement. Taken together, "new" water is introduced with every tide and the "residence times" of water masses which are at high tide on a tidal flat are relatively short, 50% water renewal taking place in only one or a few tidal periods (Zimmerman, 1976). How much of this new water is ocean and how much estuarine water depends on the distance of a tidal flat from the entrance of the estuary.

3. WAVES

Since tidal currents on tidal flats are generally slow, bottom friction caused by waves is more important than tidal friction. In fact, the morphology of tidal flats is determined by the interplay of these two forces. Assuming, for the sake of the argument, that all height differences in a coastal plain estuary were obliterated; such differences would rapidly be restored by the tidal streams, carving channels into the plain and building up the shallows. With sufficient sand available these shallows would grow above sea level until further growth is stopped by wave erosion. We return to this problem in the next paragraph.

Wave activity over tidal flats is of course quite variable, since it
depends directly on wind stress and direction. Waves break only on the
steep edges of tidal flats directly exposed to large wave motion from
outside the estuary. These edges then have beach characteristics. On
the tidal flats waves have wave lengths up to only a few meters. Sand
is chiefly moved close to the bottom. Small sand ripples are formed of
which the patterns change even in one single tidal period. Considering
the combined effects of waves and tidal currents, the bottom stress by
waves is generally larger than that exerted by currents (Fig. 3).

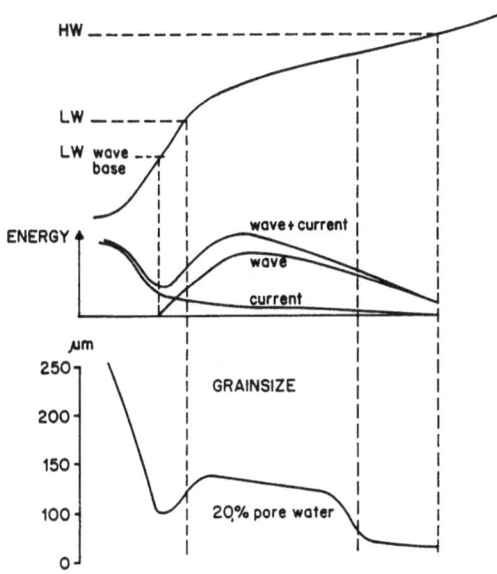

Fig. 3. Schematic representation of the combined effects of wave and
current energies on grainsize distribution in a cross-section of a
tidal flat system.

4. SALINITY AND TEMPERATURE

These properties often show fast and extreme changes on tidal flats,
especially in cold climates. As regards salinity, in cases of river
input and salinity stratification, tidal flats tend to be less saline
than the permanently covered parts of an estuary. Since the flood
current continues to flow through a creek after high tide (paragraph
2), bringing salt water further inward, sharp salinity gradients may
develop on the transition between the creek and the tidal flat at high

water. This is one of the causes of tidal fronts (Postma and Dijkema, 1982). At these fronts the first ebb water from the flats tends to overrun the last flood water in the creek, pushing the foam formed on the tidal flat with it. A parallel situation occurs for temperature in times of the year that the estuary is warmer than the adjacent sea.

Where no river water is supplied and temperatures are high, sea water will evaporate, especially on the shallow flats, and salinities will increase far above that of normal sea water. Under conditions of little or no rainfall, as in the subtropics, supersaturation develops and salt flats instead of marshes form the inner limit of a tidal flat system. In the submerged part of the estuary a so-called negative estuarine circulation will be generated that carries salt water along the bottom to the sea. Such a circulation hampers trapping of silt, since this requires a bottom flow in the opposite direction.

Salinities over tidal flats thus range between near zero to saturation, depending on geographical setting. Also the variation in temperature is very large. High extremes will certainly be found in tropical tidal pools, but in summer also in tidal pools in cold climates. The rising flood tide may then cause a sudden drop of temperature of several tens of degrees. Extremely low temperatures have been observed under ice covers in winter. Freezing may cause formation of brine in which the temperature can drop far below the freezing point.

The back and forth movement of ice floes over tidal flats appears to be able to erode the top of the sediment down into the anaerobic layers causing an increase in estuarine turbidity and mobilization of the nutrients in the pore water. After thawing the surface of a tidal flat is often a hilly landscape because of pockets left behind by melted ice floes.

5. GRAIN SIZE DISTRIBUTION AND SEDIMENT TRANSPORT

A distinction must be made between sand flats and mud flats (Fig. 1). The difference is chiefly caused by difference in wind, that is wave, exposure, mud flats occurring in relatively sheltered areas. Grain size is mostly expressed as medium. On sand flats this size varies between the narrow limits of about 100 and 300 um. On a specific tidal flat these limits are even narrower, since sand movement over the flat tends to reduce differences.

Nevertheless, grain size distribution over a tidal flat is not completely homogeneous and there appears to be two main classes of flats (Postma, 1957). In the first class, median grain size decreases over the flat away from the tidal channel. This is mainly caused by loss of energy of the waves generated in the channel and travelling over the flats. This situation is found on extensive flats larger than about 2 km which are mainly those bordering the land. At some distance from the coast the sand flat becomes a mud flat. Even in the mud flat there are often still some coarse sand grains present which have travelled there during extreme high tides and storms. On the other hand, there is always in a sand flat, an admixture of mud which has been buried there by bioturbation (see paragraph 5). In the second class, the grain size increases away from the tidal channel. This situation occurs when in that direction the wave stress on the bottom increases because of a rapid decrease in water depth. This happens mostly within a distance of one km or less from the tidal channel. On small tidal flats between two creeks the coarsest median grain size is under these conditions found on top of the tidal flats. Many tidal flats consist of a mixture of both classes in the way shown in Figure 3. The combined effect of waves and currents, moreover, frequently has an energy minimum in the transition zone between a tidal creek and the adjacent flat, also shown in Figure 3, located at the low water wave base. If the energy is low enough mud is also accumulated in this zone.

Like beach sands, tidal flat sands are extremely well sorted (Postma, 1957; Collins et al., 1981). This is explained by the fact that all grains are moved back and forth by oscillatory water movements and not in one direction. In a logarithmic grain size scale, grain size distributions approach perfect Gaussian curves with the same degree of sorting.

Little sand movement occurs on flats in periods without wind. During storms, however, much sand is churned up and some of it is transported to deeper water with the ebb tide. In temperate regions tidal flat profiles are therefore probably somewhat lower in winter, as also happens on beaches. Since tidal channels are deep and often have steep sides only strong currents can lift the sand sufficiently high to be carried back on the flats. This process seems to be restricted to spring tide periods which, of course, occur during the whole year.

Inward sand movement in tidal channels and subsequent lifting of sand grains on adjacent tidal flats may be caused by tidal asymmetry, but this asymmetry is certainly much more important for the transport of fine-grained material.This material is much more evenly distributed in the water column than sand and, therefore, easily transported from a tidal creek onto the sand flats. It is subsequently carried over these flats until it reaches a much flat where it settles at high tide.

Mud accumulation is enhanced by the fact that small particles, after settling, stick together. This cohesion is assisted by rapid loss of water from a new mud layer. Thus, the lowest current velocity which can carry fine-grained material towards a mud flat, generally of the order of 10 cm s^{-1}, is much smaller than the (ebb) current velocity required for resuspension, generally more than 30 cm s^{-1}.

Additional factors which promote mud accumulation are a prolonged period of high water, slack (see paragraph 2) and organic matter which binds particles together. Biological activity of small animals moving on and burrowing into the mud may, on the other hand, loosen the mud structure.

Depending on factors such as the rate of sand and mud supply and long term rise of fall of sea level, mud flats may ultimately grow above high tide and be transformed into marsh land that is only flooded occasionally. Sand flats seem to reach above high tide only where the ocean surf brings sand to a beach and wind accumulates this sand in dunes.

6. INTERSTITIAL WATER

Sand tidal flats, owing to their very large degree of sorting, have a homogeneous pore water content of about 20% by volume. Compared to any other type of deposit, except beach sediment of which the pore water may be drained once a tide, the rate of water renewal in interstitial water of sand flats must be very fast. However, since this renewal is caused by a number of processes, it is not easily measured. Part of the water percolates out of the sediment at low tide. A rough estimate of this part can be made by measuring nutrient and (negative) sulphate contents of water collecting in small creeks at the end of the ebb tide and in pools of water staying behind on the flats at low water. Another way is to measure bioturbation in the benthic layer by burrowing animals, for example the lugworm <u>Arenicola</u> (Cadee, 1976; Baumfalk, 1979), and the amount of water pumped by these animals. Still another method is to measure how fast salinity of interstitial water follows changes of the overlying water.

A change between a wet and dry period of the year in the Wadden Sea, with an interval of about three months, is shown in Figure 4. Parts of the changes, however, are not caused by changes in salinity of the estuary. In the upper figure, low salinities are partly caused by precipitations on the flat. In the lower figure, high salinities are due to evaporation. To arrive at useful exchange values, series of very frequent measurements are essential.

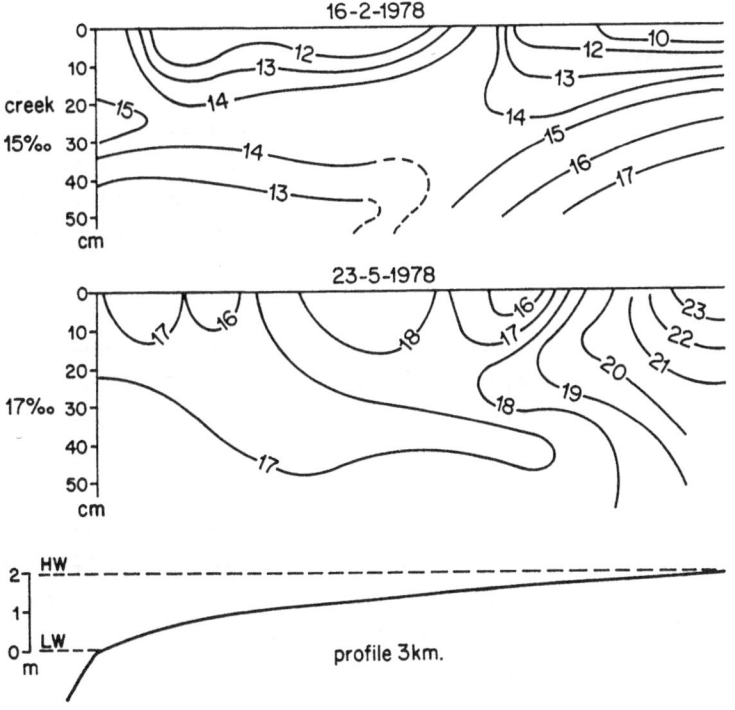

Fig. 4. Changes of salinity, expressed as chlorinity, in the interstitial water along a profile over a tidal flat in the Dutch Wadden Sea. The upper figure shows conditions in a period with excess precipitation, the lower with excess evaporation (Postma, 1982).

All possible methods have been at some point tried out and although no convincing general answer can be given yet, water renewal in the upper 20 cm of sandy flats seems to be closer to a week than to a month (Fig. 4). We are even less sure about seasonal change. In temperate areas in summer, animal pumping activities are obviously more important than in winter; inversely in winter scouring during storms or by large ice floes are influential.

On mud flats interstitial water exchange is much slower, since fine-grained deposits retain pore water more efficiently than sand. Moreover, most animals are living on, and not in, the mud. Vertical profiles of solutes released in the interstitial water often show the form characteristic for simple diffusion processes. Mud flats are mostly closer to the high water line.

7. CYCLE OF ORGANIC MATTER

In principle a closed carbon cycle could exist on a tidal flat, consisting of the microphytobenthos living on the flat as the producer and the benthic fauna and micro-organisms living on and in the sediment as the consumer of organic matter and the suppliers of nutrients to the phytobenthos.

This simple scheme is alluring because it seems often quite possible to cover the food requirements of the main benthic population by local phytobenthos production. It has also been shown that the microphytobenthos population can export organic matter by net transport from tidal flats to adjacent creeks (BOEDE, 1985). On the other hand, microorganisms living in the sediment often mineralize large amounts of organic matter. Moreover, measurements in tidal creeks often show a net export of nutrients with the ebb and an input of oxygen and sometimes sulphates with the flood, indicating that the system cannot be self-sufficient (Figs. 5 and 6).

In the following, conditions on tidal flats in the Wadden Sea are taken as a main example. Much of the literature is found in the Netherlands Journal of Sea Research and the BOEDE report (1985). There are three sources of organic matter for the tidal flats. The first, already mentioned, is the microphytobenthos, the second phytoplankton and the third allochthonous organic particles. The relative importance of these three sources can differ greatly.

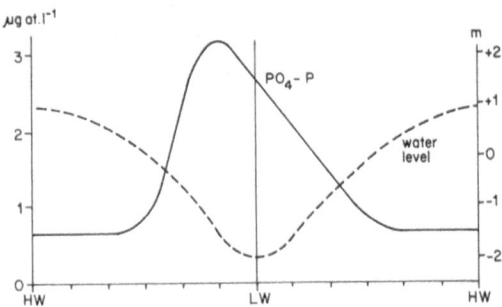

Fig. 5. Changes in phosphate concentration with the tide in a creek in the Wadden Sea showing export from an adjacent tidal flat in the second half of the ebb. Measurements in summer after Lillelund et al., 1985.

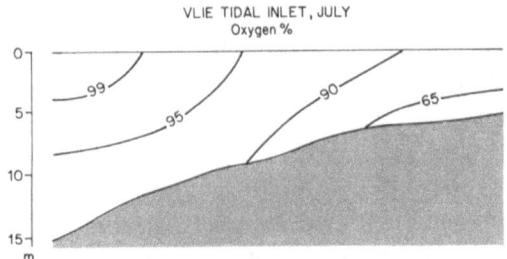

Fig. 6. Oxygen distribution, expressed in percentage of saturation along a section following a channel from the open sea into a tidal flat system. Data of the Wadden Sea in summer (de Groot and Postma, 1968).

For benthic microalgae the best living conditions are found on the highest parts of a tidal flat which, at high tide, is covered with clear sea water. The presence of nutrients in the overlying water is of secondary importance as long as sufficient amounts are provided by the sediment underneath. Production on one specific flat may vary an order of magnitude, depending on elevation.

Phytoplankton also flourishes best in transparent water masses, but needs sufficient nutrient supply. Since in estuaries this supply is mainly from the land and also because the share of the euphotic part of the water column, with respect to the dark part, becomes more important in shallow water, phytoplankton production per m^2 increases away from the inlet towards the land until the euphotic zone touches the bottom on the tidal flats. From then on production decreases because the water column shortens, and phytobenthos production takes over.

However, in addition to local production, in most estuaries import occurs from outside. This input may be organic matter from the open sea, from rivers or from adjacent marshes. Depending on their location in an estuary, tidal flats will receive smaller or larger amounts of this import. Tidal flats near river turbidity maxima will benefit from the organic particulate matter in these maxima. Tidal flats near the shore may profit from organic matter carried inward by tidal asymmetry trapping of mud as described in paragraph 5 and from organic plant debris brought down from adjacent marshes. On the other hand, tidal flats opposite to inlets and offshore tidal deltas will receive much less organic matter. It must be pointed out that estimates of total organic carbon in tidal flat deposits are of little use, as long as no distinction is made between newly produced and inert carbon. The latter can be several hundred years old and frequently accounts for most of the carbon present.

The input of allochthonous organic matter occurs together with the input of fine-grained inorganic particulate matter. These substances together can cause considerable decrease in water transparency and of the depth of the euphotic zone. The latter may decrease from many meters near the entrance to a few decimeters nearshore or in turbidity maxima. This causes a decrease in primary production in the same direction (Fig. 7). Phytoplankton is more sensitive to this change than phytobenthos since the latter emerges at low tide above the turbid water.

Estuaries could have a higher, an equal or a lower total supply of organic matter per square meter than the adjacent sea. The impression is that a higher supply prevails. In turbid estuaries the concentrations of organic matter per unit volume of water is mostly an order of magnitude larger than in the adjacent coastal water; roughly, oceans contain 20 mgC m^{-3}, shelf seas 200 mgC m^{-3} and estuaries 2,000 mgC m^{-3}; of the latter amount however, only a small part is living phyto- and zooplankton. In the Wadden Sea the living percentage is around 20% in summer, but almost zero in winter. The annual average is only 5% (Manuels and Postma, 1974).

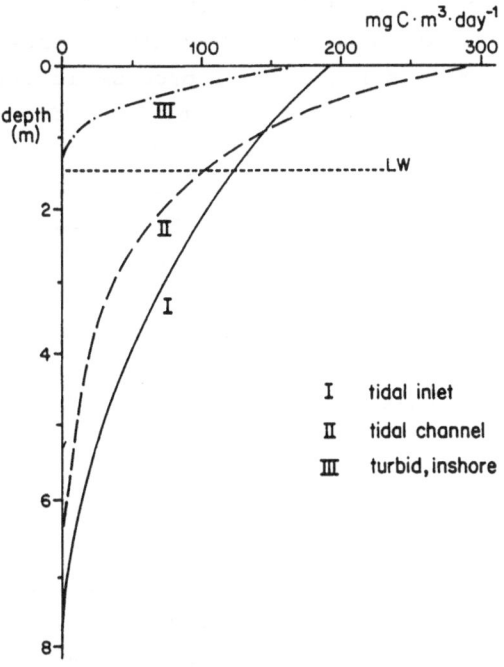

Fig. 7. Changes in primary production in the water column from the inlet of a tidal system (I) to a creek (II) and a nearshore tidal flat (III) in the Wadden Sea. Annual averages (Postma and Rommets, 1970).

8. PROCESSES IN THE SEDIMENTS

Most organic material accumulating on tidal flats is processed in the sediments and not in the water column. At first sight this seems amazing, especially for sand flats where fine-grained material does not permanently settle. Although some organic matter may be buried directly beneath the upper sand cover by wave motion, the main reason must be that it is caught by benthic organisms which filter the overlying water and bury part of it. An estuarine water body may be filtered several times a year. On a typical tidal flat there may be a macrofauna assemblage representing 10-20 grams of organic carbon (dry weight) per m^2 which needs about five times this amount of food (de Wilde and Beukema, 1984). They collect, however, much more than they need and select only food with specific properties out of the total quantity available. The remainder is buried by animals which live at various levels in the sand, down to about 3 decimeters.

For the extraction of organic matter, deeply buried deposit feeders pass large amounts of sand through their digestive tracts and flush these to the sediment surface. This process is somewhat selective in particle size, the relatively fine material being preferred, but this is of little consequence on a well sorted tidal flat. Moreover, the same sand is completely turned around once or twice a year to the depth mentioned above. The maximum depth of bioturbation is often characterized by the presence of a pavement of coarse shell debris.

In addition to this bioturbation of sand,the deposit feeders,in order to move the sand and to obtain oxygenated water, pump much larger quantities of water than they move sand. The latter process is, there fore, of much greater importance for the ventilation of the tidal flat; this process together with other processes, as already stated in paragraph 6, causes a rapid renewal of pore water.

Most of the organic matter that is not used by the macrofauna is mineralized by bacteria. In the upper layer of one or two centimeters this is done by aerobic bacteria, but below this layer oxygen is depleted and anaerobic bacteria, reducing nitrate, nitrite, iron, sulphate or carbon dioxide, in this sequence, take over (Fig. 8).

The reduction of iron and sulphate and the formation of iron sulphide causes the black color of intertidal flats and the escape of free H_2S causes their peculiar smell at low tide. In deeper layers, FeS is transformed into FeS_2. Nitrate reduction yields nitrogen gas as an end product, causing losses of the nutrient from the system.

ZONATION OF ELECTRON ACCEPTOR USE

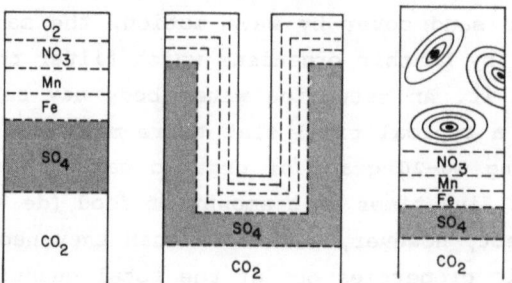

Fig. 8. Zonation of electron acceptor use in a layered sediment, a sediment irrigated by burrowed animals pumping water and in a bioturbated sediment (Aller, 1982).

In general on tidal flats, anaerobic decomposition does not go beyond sulfate reduction but in cases with a very high supply of organic matter methane gas may be formed by reduction of CO_2. Escape of methane accelerates the escape of nitrogen gas.

Interesting methods to measure directly bacterial biomass and activity are those by means of ATP and ETS techniques (Vosjan and Olanczuk-Neyman, 1977). Such measurements indicate that indeed the amount of organic matter required for the carbon budget is mineralized and also that much more material is mineralized in the sediment than in the water column. It appears, moreover, that roughly half of the activity is aerobic and the other half anaerobic (Howarth, 1984), the anaerobic share of course being divided over a much larger part of the sediment column. This is demonstrated by Figure 9 which, however, probably exaggerates the anaerobic part.

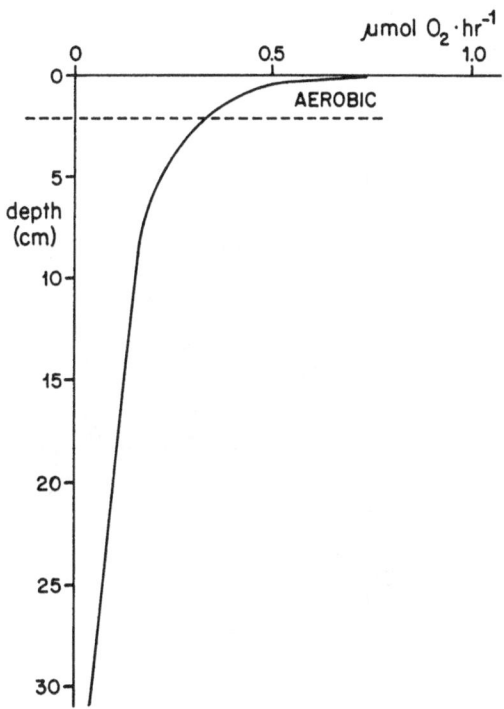

Fig. 9. Oxygen consumption in a bioturbated tidal flat sediment measured by means of ETS activity. This method yields maximum values (Vosjan, 1979; Vosjan and Olanczuk-Neyman, 1977).

It should be stressed that the anaerobic share in the total carbon cycle is much higher on tidal flats than in probably any other cycle because of the rapid burial of new organic matter and the rapid removal of end products. In this sense the intertidal flat system is unique. Another important effect is denitrification, i.e. the conversion of nitrate, and ammonia to nitrogen gas. This means that in cases where the estuarine water and organic matter have the "normal" ratio between P and N of 16, nitrate will become a limiting factor earlier than phosphate. However, this effect is mostly lost by other effects changing atomic ratios, such as eutrophication.

Figure 10 gives an example of fluxes of phosphate and nitrogen compounds through the sediment water interface of a sandy Wadden Sea tidal flat, showing light-dark and seasonal differences but as overall export of phosphate, amounting to about 3 gP m^{-2} y^{-1} and as overall import of nitrogen of about 8 gN m^{-2} y^{-1}. (Hendriksen et al., 1984). The export of P demonstrates the importance of excess mineralization (to which phytobenthos consumption of P should be added, inside the flat. The opposite transport of nitrogen is partly due to denitrification in the sediment.

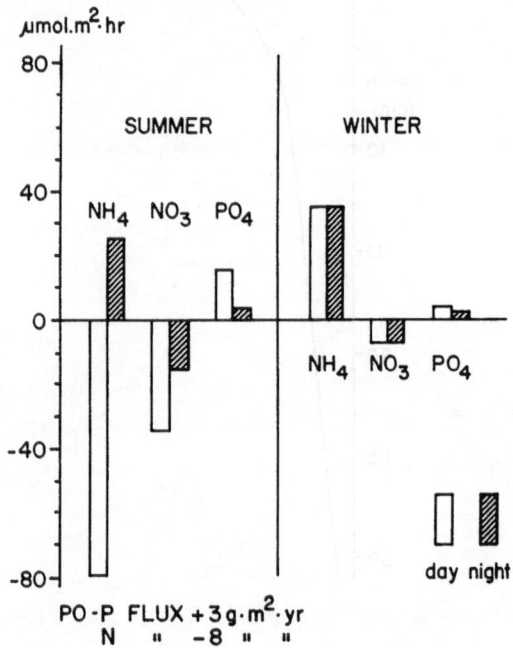

Fig. 10. Example of fluxes of phosphate and nitrogen compounds through the sediment-water interface of a sandy tidal flat in the Danish Wadden Sea (Hendriksen et al., 1984).

9. EFFECTS OF EUTROPHICATION

Eutrophication is defined here as an increased load of nutrients and organic matter, the first causing the latter, or vice versa. Such an increased load may be due to direct input into an estuary, or via eutrophication of coastal water outside the estuary. In the case of a direct input the consequences can be more severe as far as the organic load is concerned, since it will consist of material strange to the marine environment. In the second case the extra organic matter may chiefly consist of additionally formed marine phytoplankton and marine detritus, in eutrophied coastal water.

In the first case, often only the bacteria benefit since the material is unsuitable for the macrofauna. The organic load in the sediment will increase. In the second case, the surplus organic matter carried into an estuary may not only be beneficial for micro-organisms but also for the macrofauna. The higher bacterial activity will cause a thinner aerobic layer and an easier supply of end products to the water column. Since the rate of activity greatly depends on temperature such end products and nutrients will especially be liberated in summer. This causes a double peak in nutrient concentration, one in winter and one in summer, which may be typical for eutrophied estuaries and probably occurred less frequently in the past. The summer peak will enhance _in situ_ phytoplankton and, to a lesser degree, phytobenthos productivity. In the Wadden Sea the summer peak is absent in an early survey in 1950/51 but had developed in 1970/71 (Fig. 11).

The increase in elements in the microfauna and macrofauna has been demonstrated in at least one case, the western Wadden Sea for the period 1970-1980 (Fig. 12). It indicates that the secondary productivity of an estuary is limited by the amount of suitable food available. Since total particulate organic carbon is always very abundant, but the living fraction only a few percent of the total, this might imply that only "fresh" food i.e. phytoplankton and phytobenthos, is suitable as food source for most of the microbenthos. Organic detritus must then, as in the case of river detritus, chiefly be mineralized by micro-organisms such as bacteria (which themselves, in turn, may be a source of fresh food).

The benthic system as a whole would function at an accelerated pace; more macrobenthos would cause more bioturbation which would bury more organic matter, which would in turn increase both aerobic and anaerobic mineralization. Such an accelerated rate would thus not essentially change the state of equilibrium of the system. Oxygen and sulphate supply could not easily be limiting factors because of the rapid water exchange on tidal flats.

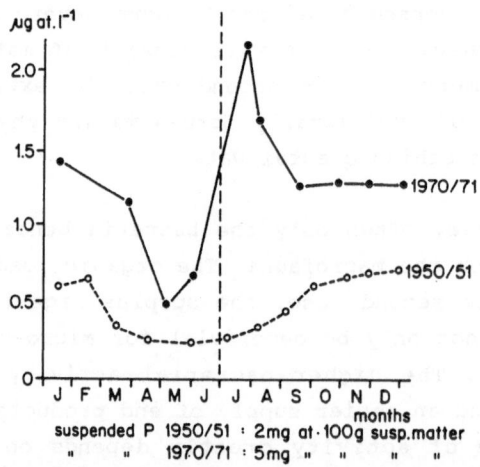

Fig. 11. Seasonal changes of suspended phosphorus per 100 g of suspended matter in the Dutch Wadden Sea in 1950/51 and 1970/71 showing the effect of increased eutrophication in the western Wadden Sea (de Jonge and Postma, 1974).

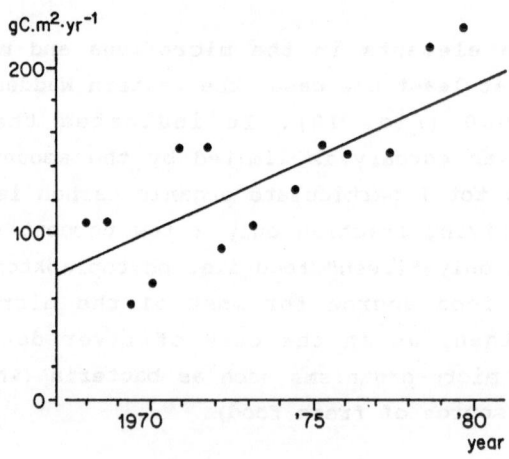

Fig. 12. Increase in microfauna production on tidal flats in the western Wadden Sea over the period 1968-1980 (Cadee, 1984).

POSSIBLE WADDEN SEA TIDAL FLAT BUDGET

		1970	1980
GAINS g C·m^{-2}·yr^{-1}	BENTHIC ALGAE PHYTOPLANKTON	100[1] 70	200[1] 115[2]
	TOTAL PR. PROD IMPORT	170 240	315 300[3]
	TOTAL GAINS	410	615
LOSSES	DISS. ORG. C ZOOPL. + BACT. MEIOFAUNA MACROBENTHOS SEDIMENT AEROB. SEDIMENT ANAEROB.	60 40 14 60 118? 118?	90 60? 14 100 225? 225?
	TOTAL LOSSES	410	615

1) ASSUMING NO LOSSES TO CREEKS
2) VIA CHLOROPHYLL INCREASE
3) LAKE IJSSEL ADDED

Table 1. Possible carbon budget for a Wadden Sea tidal flat budget; modified after de Wilde and Beukema, 1984 and various other sources.

10. CONCLUSIONS

Tidal flats are described in this paper as systems which, compared to permanently submerged environments, are characterized by a relatively fast rate of water renewal both of the overlying and interstitial water. Notwithstanding this fast ventilation, sediments are in an anaerobic condition below the upper few centimeters. It is assumed that, in order to maintain these conditions, local production of organic matter is mostly insufficient and considerable input of organic matter from outside, either from the land or from the ocean, necessary. A high percentage, possibly higher than anywhere else, of this organic matter, is metabolized by anaerobic bacteria which receive organic matter by bioturbation. However, in cases of, for example, high phyto-benthos production, and relatively low macrobenthos density on the flat, no organic matter import may be needed and export to the channels cannot be excluded.

This general picture needs more observational and experimental support in all its aspects. Key issues are more estimates of sediment ventilation and reworking on different tidal flat environments of metabolic rates of the main components, aerobic and anaerobic bacteria, other microbenthos, macrobenthos, microphytobenthos and phytoplankton. Since tidal flat organic matter budgets are rare, there is a great demand for comparative studies both in temperate regions and the tropics.

REFERENCES

Aller, R.C., 1982. The effects of macrobenthos on chemical properties of marine sediment and overlying water. - In Animal-sediment relations, pp. 53-101. Plenum Publ. Corp.

Baumfalk, Y.A., 1979. On the pumping activity of Arenicola marina - Neth. J. Sea Res. 10: 422-427.

Biological Research Ems-Dollard Estuary (BOEDE). 1985. Communications Rijkswaterstaat, The Hague. No. 40, 182 pp.

Cadee, G.C., 1976. Sediment reworking by Arenicola marina on tidal flats in the Dutch Wadden Sea. - Neth. J. Sea Res. 10: 440-460.

Cadee, G.C., 1984. Has input of organic matter into the western part of the Dutch Wadden Sea increased during the last decades? - Neth. Inst. Sea Res. Publ. Ser. 10: 71-82.

Cadee, G.C. 1986. Increased phytoplankton primary production in the Marsdiep area. - Neth. J. Sea Res. 20: 285-290.

Collins, M.B., C.L. Amos & G. Evans, 1981. Observations of some sediment-transport processes over intertidal flats, the Wash, U.K. - Spec. Publs. int. Ass. Sediment 5: 81-98.

Dronkers, J., 1986. Tidal assymmetry and estuarine morphology. - Neth. J. Sea Res. 20: 117-131.

Groot de, S.J. & H. Postma, 1968. The oxygen content of the Wadden Sea. - Neth. J. Sea Res. 4: 1-10.

Henriksen, K., A. Jensen & M.B. Rasmussen, 1984. Aspects of nitrogen and phosphorus mineralisation and recycling in the northern part of the Danish Wadden Sea. - Neth. Inst. Sea Res. Publ. Ser. 10: 51-69.

Howarth, R.W., 1984. The ecological significance of sulfur in the energy dynamics of salt marsh and coastal marine sediments. - Biochemistry ?: 5-27.

Jonge de, V.N. & H. Postma, 1974. Phosphorus compounds in the Dutch Wadden Sea. - Neth. J. Sea Res. 8: 139-153.

Lillelund, K., R. Berghahn & R. Dierrking, 1985. Ver{nderungen im Phosphatgehalt in einem Prielsystem im Wattgebiet nahe der Nord-Strander Bucht (\stliche Nordsee) im Verlauf einer Tide. - Int. Rev. ges. Hydrobiol. 70: 101-112.

Manuels, H.W. & H. Postma, 1974. Size frequency distribution of sands in the Dutch Wadden Sea. - Arch. neerl. de Zool. 12: 319-349.

Postma, H., 1985. Eutrophication of Dutch coastal waters. - Neth. J. Zoöl. 35: 348-359.

Postma, H. & K.S. Dijkema, 1982. Hydrography of the Wadden Sea: movements and properties of water and particulate matter. - Rep. Wadden Sea Working Group, 2. 75 pp.

Postma, H. & J.W. Rommets, 1970. Primary production in the Wadden Sea. - Neth. J. Sea Res. 4: 470-493.

Vosjan, J.H. & K.M. Olanczuk-Meijman, 1977. Vertical distribution of mineralization processes in a tidal sediment. - Neth. J. Sea Res. 11: 14-23.

Vosjan, J.H., 1979. Microbiologische afbraak in de Wadbodem. - Natuur in Techniek 47: 232-247.

Wilde de, P.A.W.J. & J.J. Beukema, 1984. The role of zoobenthos in the consumption of organic matter in the Dutch Wadden Sea. - Neth. Inst. Sea Res. Publ. Ser. 10: 145-148.

Zimmerman, J.T.F., 1986. The tidal whirlpool. - Neth. J. Sea Res. 20: 133-154.

Zimmerman, J.T.F., 1976. Mixing and flushing of tidal embayments in the western Dutch Wadden Sea. - Neth. J. Sea Res. 10: 149-191.

PATTERNS OF ORGANIC CARBON EXCHANGE BETWEEN COASTAL ECOSYSTEMS
The Mass Balance Approach in Salt Marsh Ecosystems

Charles S. Hopkinson, Jr.
University of Georgia Marine Institute
Sapelo Island, GA 31327, USA

1. INTRODUCTION

The role played by individual ecosystems as components of the earth's biosphere is a fundamental ecological question. With respect to organic material, in addition to knowing the rate of carbon fixation and respiration within the ecosystem, it is equally important to know how the ecosystem interacts with its adjacent ecosystems. Is it autotrophic and subsidizing adjacent regions or is it heterotrophic and dependent on allochthonous organic inputs from surrounding areas? Questions such as these have constituted a central theme for much research conducted over the past thirty years in tidally coupled coastal systems.

In the lush marshes at Sapelo Island, Georgia during the mid 1950's, a group of scientists, including among others E.P. Odum, L. Pomeroy, R. Ragotzkie and J. Teal, initiated the first measurements of metabolic processes in salt-marsh-dominated estuaries. They observed that the productivity of the marsh grass was comparable to the most heavily subsidized agricultural crops, that little plant material appeared to accumulate in the sediments, and that relatively little was degraded or consumed by higher trophic levels. It was concluded that considerable amounts of plant material were probably exported or flushed to adjacent creeks and bays by tides which bathe the marshes twice daily (Teal, 1962).

Based on these early conclusions, E.P. Odum developed an hypothesis of outwelling (Odum, 1968) which presumes that net primary production of marsh-macrophyte-dominated estuaries greatly exceeds local degradation and storage of carbon, and that the excess organic material is exported to the adjacent ocean where it is finally degraded and incorporated into the offshore food web.

Although consistent with the evidence, the outwelling hypothesis was based on limited information. Haines (1977) found little direct evidence of marsh-derived plant material in tidal bays adjacent to salt

Lecture Notes on Coastal and Estuarine Studies, Vol. 22
B.-O. Jansson (Ed.), Coastal-Offshore Ecosystem Interactions.
© Springer-Verlag Berlin Heidelberg 1988

marshes, adding doubt to the validity of the outwelling hypothesis. Nixon (1980) summarized the first twenty years of research of salt marshes and concluded that although there often is an export of organic matter from marshes, and the export may contribute substantially to open water production in adjacent creeks and bays, there is not any greater production of fish than in coastal areas without marsh supplements.

In this paper various approaches employed in evaluating the importance of salt marshes as sources of organic carbon for coastal regions are summarized. The mass balance approach is critically analyzed in terms of its advantages and disadvantages. Results of this approach are presented for a number of estuarine regions along the east coast of the U.S. in which salt marshes are a major feature. An attempt is made to determine the factors controlling the degree to which salt-marsh-dominated estuaries are coupled to adjacent coastal regions.

2. EVALUATING ESTUARINE-COASTAL COUPLING - THE APPROACHES

Two major approaches have been taken to examine the extent to which organic carbon fixed in marshes and estuaries is exported to adjacent systems (see Fig. 1). The DIRECT FLUX approach attempts a direct measure of fluxes of various components between adjacent systems. With the Eulerian direct flux technique, the advective flux leaving a system fixed in space is the integral of the product of the velocity of water, the channel cross-sectional area and the concentration of the component of interest over a cross section. Water movement through a cross section has been measured a number of ways including the use of tide gauges, hypsographic curves, and array(s) of current meters. Samples have been collected at intervals ranging from continuous to a tidal cycle. The direct flux approach is difficult for three reasons. 1) The velocity distribution in a cross section is usually quite complex. 2) Tidal cycles vary considerably in amplitude. Small differences from cycle to cycle (the diurnal inequality) lead to large differences in the net fluxes over a cycle. 3) Material concentrations vary considerably over the typical cross section. In a turbulent system it is difficult to achieve a transport estimate with a precision of greater than 20%. Kjerfve (1975) and Kjerfve and Proehl (1979) have fully described this direct flux technique.

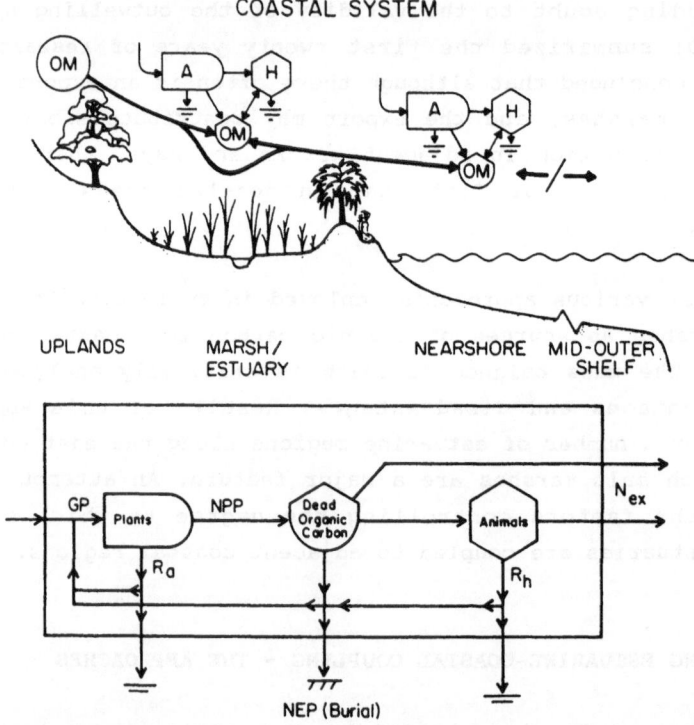

Fig. 1. Top: Conceptualization of a coastal ecosystem illustrating the coupling of marsh/estuary and nearshore subsystems with their corresponding autotrophic and heterotrophic communities with the terrestrial and oceanic environments. Modified from Hopkinson and Hoffman (1984). Bottom: Conceptualization of the production equations model. These forms are the basis for the diagrams in the following case studies (Figs 2-7).

The Lagrangian direct flux technique described by Imberger et al. (1983) avoids some problems of the advective flux technique. With this approach a site is chosen where the tidal excursion is much shorter than the estuarine length so that the average advective loss from the site becomes zero. The instantaneous advective fluxes add to yield a longitudinal diffusive flux. With the proper spatial and temporal sampling resolution, bounds can be derived for the internal turnover times, boundary exchange rates and export fluxes of particular substances.

MASS BALANCE is the second major approach taken to estimate the movement of material between adjacent coastal systems. The difference
between gross primary production, total system respiration and
accumulation within a system is a measure of the extent to which the
system is dependent on, or subsidizing adjacent systems. It is the
approach that was utilized by the first investigators studying the
function of a salt marsh at Sapelo Island. As the calculation of mass
transfer is the balance between production and consumption, it is only
an indirect estimate of movement. In the process of constructing a
mass balance, additional information of a useful nature is provided
concerning flows and cycles of carbon internal to the system. The mass
balance budget then becomes a standard against which the significance
of specific metabolic processes can be compared.

3. THE MASS BALANCE APPROACH

Appraisal of cross-system fluxes of organic matter begins with an
analysis of fluxes internal to the system. It is not necessary to know
a priori whether the system of interest imports or exports organic
carbon. The production equations, first used by Woodwell and Whittaker
(1968) in analysis of a forest, are a useful model for describing the
major internal and external exchanges (Fig. 1).

$$NPP = GP - R_a \qquad (1)$$

$$N_{ex} = GP - (R_a + R_h) - NEP \qquad (2)$$

where NPP= net primary production, GP= gross primary production, R_a=
respiration of autotrophs, R_h= respiration of heterotrophs, NEP= net
ecosystem production (= burial) and N_{ex}= net exchange across system
boundaries. Equation 1 deals only with autotrophs and states that net
primary production is the balance between gross primary production and
respiration of the autotrophs. Equation 2 deals with the entire
ecosystem and states that net ecosystem exchange is equivalent to gross
primary production minus the respiration of autotrophs and heterotrophs
minus the amount of material that is stored or accumulated within the
ecosystem. In a closed system, NEP is the annual increment in total
biomass plus the annual buildup of organic carbon in the soil. In salt
marshes, it is generally assumed that the system is in steady state
with respect to biomass, hence NEP represents burial of organic matter
in the flooded sediments. N_{ex} will be either positive or negative

depending on the balance between allochthonous inputs and outputs. A positive balance indicates a net export from the system of interest, while a negative balance indicates import. Thus, export from a salt marsh to the coast would be represented by a positive value of N_{ex} if the equations are solved for the salt marsh (or donor system), or as a negative value if the equations are solved for the adjacent ocean (or recipient system).

While several of the terms in the production equations can be assessed directly, others must be determined indirectly by difference. Techniques are available for assessing NPP, total respiration ($R_a + R_h$), NEP and occasionally N_{ex} (if the DIRECT FLUX approach is utilized). Frequently NPP and R_h are determined for individual populations within a community and then summed to determine the rate for the entire system. It must then be assumed that all parts have been identified and that the sum of the parts equals the whole. Techniques are presently unavailable for measuring R_a, R_h and GP (the ^{14}C technique for measuring aquatic primary production may give an estimate of GP under certain conditions, see for example Peterson, 1980). The goal of the MASS BALANCE approach is to estimate N_{ex} without actually measuring it; unfortunately there are frequently too many unknowns to completely specify the two production equations. In such case a modified operational equation is used to evaluate N_{ex}

$$N_{ex} = NPP - R_h - NEP$$

Although it is technically impossible to totally separate respiration of autotrophs from heterotrophs in natural systems, it is often possible to minimize the autotrophic contribution so that perhaps a reasonable approximation of R_h can be made. R_a can then be determined as $(R_a + R_h) - R_h$.

3.1. Potential bias and weakness of the mass balance approach

There are a number of problems inherent in the mass balance approach, the effects of which must be considered during the interpretation of results. One problem is that the accuracy and precision of the estimated value of N_{ex} for the entire system is only as good as the accuracy of measurements of specific system processes. As the accuracy of measurements for any specific process is seldom better than $\pm 20\%$, the addition and multiplication of several terms, each with large variances, yields an overall estimate of N_{ex} with tremendously wide confidence intervals.

A second problem is that by definition, any material unaccounted for is considered to be exchanged with the adjacent system. Unmeasured or poorly measured metabolic fluxes or rates of burial are interpreted as cross system exchange. In a salt marsh, where the complexity of the heterotrophic community greatly exceeds that of the autotrophic community, it is common to have more complete accounts of NPP than of R_h. As a result, the estimate of N_{ex} may be higher than it should be.

An additional problem associated with the mass balance approach is that R_h is frequently measured for individual populations when isolated from the system. In a salt marsh system, Montague (1980) showed that respiration measurements of specific populations can be grossly underestimated when those populations are uncoupled from macroorganism feedback effects.

Many problems encountered when employing the mass balance approach can be minimized by taking direct measurements of whole system fluxes rather than by summing measurements of many autotrophic and heterotrophic populations/communities within the system.

3.2. The mass balance approach - case studies

Five marsh/estuarine systems were chosen to illustrate the mass balance approach for estimating estuarine-coastal ecosystem couplings: Sippewissett Marsh, Massachusetts; Narragansett Bay, Rhode Island; Flax Pond, New York; Duplin River, Georgia; and Barataria Basin, Louisiana. Direct flux estimates of coastal exchange from North Inlet, South Carolina are included for comparison with mass balance exchange estimates. These systems were chosen because sufficient information is available to allow at least crude application of the production equations.

Sippewissett Marsh and Flax Pond are pocket marshes, with restricted passageways to the ocean that are typical of New England marshes. Both are small systems that drain almost entirely at low tide. Narragansett Bay is a drowned valley estuarine system proximate to Sippewissett and Flax Pond. It is much larger and deeper than its neighbors and is

Table 1. Factors potentially influencing the movement of materials between estuarine and coastal waters.

FACTOR	ESTUARINE SYSTEM					
	Sippewissett	Narragansett	Flax Pond	North Inlet	Duplin River	Barataria Bay
Morphology	restricted, pocket marsh	open, drowned valley	restricted, pocket marsh	semi-restricted lagoon marsh	semi-restricted lagoon marsh	open, lagoon marsh/bay
Circulation	1A	NA	1A	1A	1A	1A
Total area	$4.8 \cdot 10^5$ m^2	$3.3 \cdot 10^8$ m^2 uplands: $4.7 \cdot 10^7$ m^2	$5.7 \cdot 10^5$ m^2 uplands: $1.1 \cdot 10^6$ m^2	$3.2 \cdot 10^7$ m^2	$1.3 \cdot 10^7$ m^2	$5.1 \cdot 10^9$ m^2
Land area	$3.1 \cdot 10^5$ m^2	$\approx 3.2 \cdot 10^6$ m^2	$3.0 \cdot 10^5$ m^2	$2.5 \cdot 10^7$ m^2	$1.1 \cdot 10^6$ m^2	$3.0 \cdot 10^9$ m^2
Water area	$1.7 \cdot 10^5$ m^2	$3.3 \cdot 10^8$ m^2	$2.5 \cdot 10^5$ m^2	$7 \cdot 10^6$ m^2	$1.7 \cdot 10^6$ m^2	$2.1 \cdot 10^9$ m^2
Land:water	66:34	<1:99	53:47	78:22	87:13	59:41
Creekbank: high marsh	8:92	7:93 within marsh	75:25	20:80	8:92	NA
Percent ponded	<2%	NA	NA	<5%	<5%	50-95%
Total length	1 km	$\approx$45 km	$\approx$1.2 km	6-12 km	12.4 km	$\approx$142 km
Average distance to creek	NA	NA	NA	NA	Max. 50 m	NA
Drainage density	NA	NA	NA	NA	18-75 km·km^{-2}	1-2.5 km·km^{-2} (major creeks only)
Tidal excursion	1 km	NA	1.2 km	$\approx$4 km	4.6 km	1-10 km?
Tidal excursion: total length	1:1	NA	1:1	0.4:1	0.33:1	<.05:1
Mean depth	0.26 m	8.3 m bay	1.1 m	1.7 m MLW	5.1 m	2.0 m
Tidal range	1.1 m	1.25 m	1.8 m (2.3 m in L.I.Sound)	1.6 m	1.4-3.2 m	0-0.3 m (diurnal)
Tidal prism	$1.8 \cdot 10^5$ m^3	$4.1 \cdot 10^8$ m^3	$8.5 \cdot 10^5$ m^3	$2.1 \cdot 10^7$ m^3	$4-18 \cdot 10^6$ m^2	$2.4 \cdot 10^8$ m^3

Low water volume	$4.5 \cdot 10^4$ m^3	$2.5 \cdot 10^9$ m^3	$2.8 \cdot 10^4$ m^3	$2.0 \cdot 10^7$ m^3	$8.4 \cdot 10^6$ m^3	$4.2 \cdot 10^9$ m^3
Rainfall	1.2 m	1.0 m	1.0 m	1.2m	1.5 m	1.5 m
Upland runoff and groundwater	1.3-$6.7 \cdot 10^3$ m^3 per tide	$4.1 \cdot 10^6$ m^3 per tide	insignificant	$4.3 \cdot 10^4$ m^3 per tide	insignificant?	locally great
Net water surplus	$7.5 \cdot 10^3$ m^3/tide	$4.5 \cdot 10^6$ m^3/tide	780 m^3/tide	$9.5 \cdot 10^4$ m^3/tide	$2.6 \cdot 10^4$ m^3/tide	$9.6 \cdot 10^6$ m^3/tide
Water surplus: tidal prism (20/16)	0.004/tide	0.011/tide	0.0009	0.005/tide	0.002/tide	0.04/day
Water surplus:low water volume (20/16)	0.06/tide	0.002/tide	0.00004/tide	0.004/tide	0.0018/tide	0.002/day
Tidal prism:low water volume (16/17)	4	0.16	30.4	1	1.33	0.06
Effective horizontal diffusion	NA	NA	NA	NA	38 m$^2 \cdot$sec^{-1}	NA
Net advection	≈0	NA	≈0	≈0	≈0	NA
Net advection/ estuarine length	≈0	NA	≈0	≈0	0	NA
Sediment accretion	0.15 cm/yr 90 gC/yr	6 gC/yr	0.5 cm/yr 200 gC/yr	0.25 cm/yr	0.2-0.5 cm/yr 29 gC/yr	0.5-2.0 cm/yr 100-231 gC/yr
Concentration gradient	1-4 g DOC-C/m^3	NA	NA	NA	5 g DOC-C/m^3	6 g DOC-C/m^3
Winds - orientation	NA	NA	west to mouth	NA	mouth to head mostly	mouth to head
Other	-	embayment sill 26 d residence time	inlet sill	-	-	land loss/ erosion:cold fronts drop water level

Notes:Data collected from a variety of sources including Nixon and Oviatt (1973) Kremer and Nixon (1978) Nixon and Pilson (1984), Pilson (1985), Kremer (pers. comm.), Valiela et al. (1978) Kaplan et al. (1979), Valiela and Teal (1979), Peterson (pers. comm.), Woodwell and Pecan (1973), Woodwell et al. (1979), Houghton and Woodwell (1980), Dame, McKellar and Kjerfve (pers. comm.), Kjerfve and Proehl (1979), Chalmers et al. (1985), Imberger et al. (1983), Wadsworth, J. (1980), Ragotsky and Bryson (1955), Pomeroy and Wiegert (1981), Hopkinson and Hoffman (1984), Hopkinson (1979), Gael and Hopkinson (1979), Happ et al. (1977, Day et al. (1982), and Day (pers. comm.). NA = not applicable or not available.

largely a phytoplankton-dominated system, rather than a marsh-dominated system. The Duplin River marsh at Sapelo Island, Georgia is a lagoonal system much larger than the New England pocket marshes with a semi-restricted connection to the ocean. Due to its size it never empties at low tide. North Inlet marsh in South Carolina is similar to the system at Sapelo, but is more restricted in its connection with the ocean. Barataria Basin is one of the largest estuarine systems in America. Although it is a lagoon, connection to the ocean is essentially unrestricted.

The mass balance approach has been applied at a number of different hierarchical levels within these overall coastal systems. The coastal system portrayed in Figure 1 can be hierarchically decomposed into several levels: the entire system can be separated into offshore and inshore components; the inshore into wetland and aquatic subsystems; the aquatic into bay, tidal river and tidal creek; the wetlands into swamp and marsh; marsh into saline, brackish and fresh marsh; saline marsh into creekbank (primarily tall _Spartina alterniflora_),low marsh (primarily short _S_. _alterniflora_) and high marsh (primarily _S_. _patens_ and _Distichlis spicata_ in New England marshes), etc. For a number of reasons (probably primarily a function of the relative area, relative presumed importance of individual subsystems within the whole system and the availability of research funding) research efforts at the various research sites have been focused at one or more of the individual hierarchical levels. Estimates of N_{ex} for the whole system are calculated as the areal-weighted sum of N_{ex} for each subsystem within each hierarchical level.

3.2.1. Sippewissett Marsh

Sippewissett Marsh is one of the more intensely studied marshes in the U.S. It is a relatively small pocket marsh system having a single, restricted connection with the sea (Table 1). Approximately 66% of the entire system is marsh, the remainder being water. Of the marsh system, approximately 92% is low and high marsh. Most research has been focused on the low marsh region. Fluxes of nitrogen, fine particulate organic carbon and groundwater were measured with the direct flux approach. Due to limitations in the availability of data, a mass balance for carbon was applied only to the low marsh; extrapolation to transfer with the sea is therefore approximate.

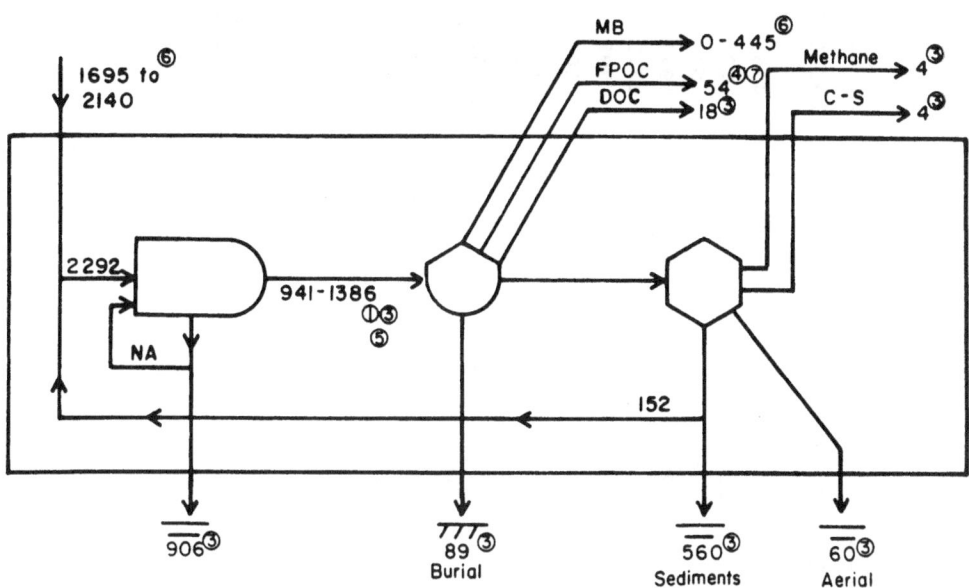

SIPPEWISSETT MARSH

Fig. 2. Conceptual model of major carbon flows and storages within Sippewissett Marsh, Massachusetts. Model primarily summarizes carbon fluxes of the low marsh. Units: g C·m^{-2} low marsh·yr^{-1}.
Notes (numbers in circles): 1. Valiela et al., 1976; 2. Howes et al., 1984; 3. Howes et al., 1985; 4. Valiela and Teal, 1979 (mass flux per 302,000 m^2 marsh); 5. Howes (personal communication that below-/aboveground production ratio for <u>Spartina alterniflora</u> is 3.5:1; 6. Mass balance; 7. Excludes DOC, wrack and fisheries export.

Carbon flux in the Sippewissett system is abstracted in a 3-compartment model (Fig. 2). Directly measured cross boundary fluxes include burial, autotrophic and heterotrophic respiration, methane and volatile C-S emissions, dissolved organic carbon loss (DOC), fine particulate organic carbon (FPOC) flux estimated from Eulerian measurements at the marsh inlet, and net primary production. The greatest uncertainty in budget construction is in the estimation of net belowground production of <u>S. alterniflora</u>. The higher estimate presented for belowground NPP is based on tedious measurements of seasonal changes in the live and dead mass of roots and rhizomes. The lower estimate was determined by mass balance (Howes et al., 1985) assuming all possible cross boundary fluxes had been accurately measured (i.e. solve for NPP rather than N_{ex}. Actually only the flux of FPOC has been rigorously examined at the estuarine inlet).

Table 2. Rates of carbon flow in several coastal estuarine systems with salt marshes Units: $gC \cdot m^{-2} \cdot yr^{-1}$

PROCESS	SYSTEM						
	Sippewissett[1]	Narragansett[2]	Flax Pond[3]	North Inlet[4]	Duplin River[5]	Georgia coast[6]	Barataria[7]
GP	1695-2140	310	1275	—	3941	539	4261
Ra	906	41	696-1190	—	2149	75	2287
Ra/GP	42-53%	13%	40-68%	—	55%	14%	54%
NPP	941-1386	269	535-1029	—	1791	464	1974
Rh	772	189	235-729	—	710	674	1773
Rh/NPP	56-82%	70%	44-68%	—	40%	145%	89%
NPP - Rh	169-614	80	280	—	1081	-210	201
NEP (burial)	89	6	200	—	29	0	174
BURIAL/NP	6-9%	2%	19-37%	—	2%	0	9%
Exp (meas)	80	—	100	339	379	—	9-30*
Exp/Npp	6-9%	—	10-19%	—	21%	—	1%
Nex	80-525	75	100	339	1052	-210	27
Nex/Npp	6-38%	27%	10-19%	—	59%	-45%	1%

Notes: See text for explanation and sources of information
1. Per unit area low marsh
2. Per unit area Narragansett Bay.
3, 4, 5, 7. Per unit area estuary (marsh and open water)
6. Per unit area nearshore region to 3.2 km offshore; based only on the heterotrophic deficit of the nearshore zone; does not include estuarine carbon flowing through the nearshore zone unmetabolized.
*. Lower estimate based on surplus freshwater runoff and TOC concentration while higher estimate is a mean from Happ et al. (1977)
GP. Gross production.
Ra. Autotrophic respiration.
NPP. Net primary production
Rh. Heterotrophic respiration.
Exp (meas). Directly measured material exhange across system boundaries.
Nex. Export calculated in this paper from mass balance considerations (includes directly measured estimates of export).
NA. Not applicable or not available.

In Sippewissett Marsh approximately 50% of the 2000 g $C \cdot m^{-2}$ yr^{-1} of plant GP is respired by the plants themselves (Table 2). The remaining 941-1386 g C is respired by heterotrophs (56-82%), buried (6-9%) or transferred to the sea. By mass balance, total estimated exchange of carbon with the sea ranges from 54 g $C \cdot m^{-2} \cdot yr^{-1}$ (direct Eulerian measurement of FPOC flux) to 445 g C (based on the higher estimate of NPP).

Direct measurements of FPOC export were measured over two years with very different amounts of groundwater input to Sippewissett. During a year when groundwater input was about 16% of that input occurring during a wet year, FPOC export was -14 g $C \cdot m^{-2} \cdot yr^{-1}$ as opposed to 54 g C for the wet year (Peterson, unpubl. pers. comm.).

3.2.2. Narragansett Bay

Narragansett Bay is not a salt marsh dominated system, although it has along its shoreline an area of marsh an order of magnitude greater than that found in either Sippewissett marsh or Flax Pond (Table 1). Water comprises over 99% of the total area of the system and it is rather deep (8.3 m) in comparison to the salt-marsh-dominated systems.

Although considerable research has been conducted in a number of Narragansett Bay subsystems (emergent marsh, tidal marsh embayments and Narragansett Bay itself), little of it is complete enough at the system level to allow solution of the production equations. Consequently it has been necessary to use values from the literature to estimate many rates. In Figure 3 the Narragansett system is decomposed at two levels (top and bottom): 1) the marsh and its adjacent water bodies/tidal creeks (excluding Narragansett Bay proper), and 2) the entire bay and all the adjacent marshes and embayments. On the marsh, estimates of macrophytic NPP (Nixon & Oviatt, 1973) were raised by a factor of 3 to account for an assumed 2:1 belowground-aboveground plant production split (Schubauer & Hopkinson, 1984). It was assumed that all below ground production was respired in place (R_h) or was buried (amount unknown). Export to the adjacent water bodies was calculated from the heterotrophic deficit of the embayment (Nixon & Oviatt, 1973) and mass balance for the marsh. As total system P and R were known for the embayment, export from the embayment to Narragansett Bay was calculated by mass balance to be as high as 628 g C m^{-2} embayment$\cdot yr^{-1}$ (dependent on the mass balance estimate of export from the marsh itself). At the

RHODE ISLAND- MARSHES AND NARRAGANSETT BAY

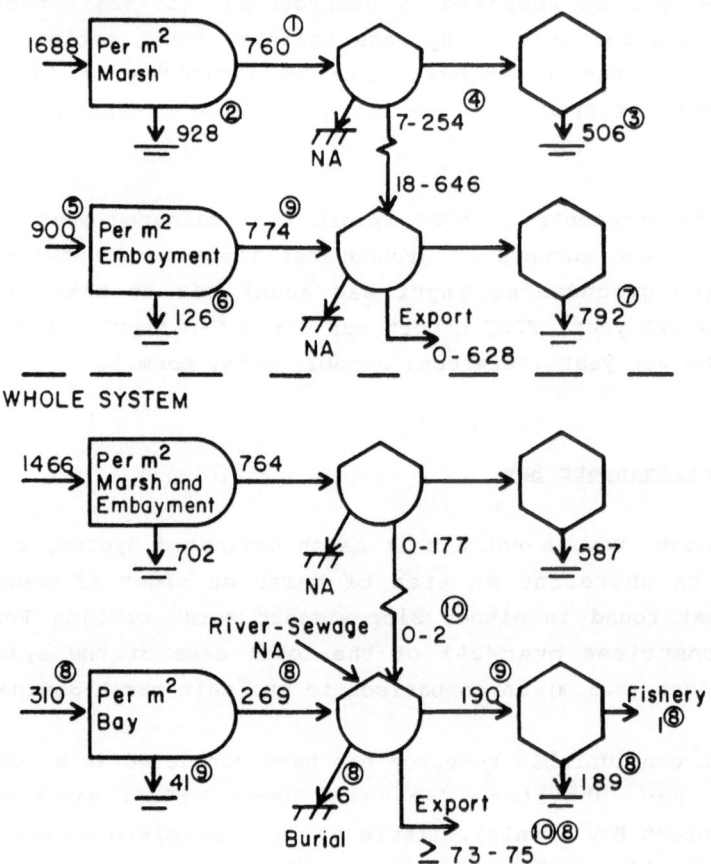

Fig. 3. Two hierarchically different conceptual models of carbon flow and storage for the Narragansett Bay ecosystem and its subsystems, the salt marsh and salt marsh embayments. Fluxes are scaled to a unit area of each system.
Notes: 1. Nixon and Oviatt, 1973 and assuming belowground production of marsh macrophytes to twice aboveground production; 2. Assuming marsh macrophyte respiration is 55% of gross production; 3. Assuming all belowground production is respired by heterotrophs; 4. Nixon and Oviatt, 1973 and mass balance; 5. Nixon and Oviatt, 1973; 6. Assuming respiration is 14% of gross production; 7. Nixon and Oviatt, 1973 after subtracting autotrophic respiration; 8. Nixon and Pilson, 1984. C-14 and O^2 techniques for measuring production; 9. Mass balance; 10. Nixon and Oviatt, 1973 and this paper.

level of the entire bay system, the conceptual models of carbon flow illustrate the relative unimportance of the fringing marshes on the carbon budget of the bay itself. An export of 0-77 g C from each square meter of marsh/embayment represents an input of only 0-2 g C to each square meter of bay.

A carbon budget for the entire bay system (bottom model of Fig. 3) was developed using values generated from a unique stoichiometric model of C, N and P in the Bay (Nixon & Pilson, 1984), and incorporating inputs from the adjacent marshes which were calculated from mass balance. At this scale of integration, the contribution of carbon from phytoplankton takes on an increased importance relative to the contribution from fringing marshes. For Narragansett Bay alone, phytoplankton contribute 269 g $C \cdot m^{-2} \cdot yr^{-1}$ (GP = 310 $gC \cdot m^{-2} \cdot yr^{-1}$). In contrast to macrophyte-dominated systems autotrophic respiration is minor. Of the 269 g C NPP, 70% is consumed by heterotrophs, 2% is buried in the sediments and approximately 25% is exported. The extent to which carbon inputs from rivers and sewage drives this flux has not been determined.

3.2.3. Flax Pond

Flax Pond is a small pocket marsh system very similar to Sippewissett (Table 1). In contrast to Sippewissett however, there is a sill across the tidal inlet which tends to poise water levels and dampen tidal amplitude. There is no significant input of freshwater to the system but there is an extremely high tidal prism volume in comparison to the pond volume at low tide.

The Flax Pond studies are one of the best examples of ecosystems level research in estuaries. The goal of the Flax Pond research program was designed at the outset to determine the role of the marsh in the carbon budget of adjacent coastal waters. A hybrid of mass balance and direct flux approaches was utilized. As such the results of each can be cross-compared. At the level of the entire Pond ecosystem (marsh and water), two conceptual models based on two different estimates of NPP summarize carbon flow (Fig. 4). Model A' incorporates an estimate of NPP higher than that presented by Woodwell et al. (1979) because I felt that their production estimate of belowground roots and rhizomes was too low (Model A' uses a 2:1 belowground:above-ground production ratio while Model B' uses a 1:2.7 ratio). The effects of the two different estimates of NPP are minor, reflected mainly in the relative importance of autotrophic versus heterotrophic respiration.

FLAX POND

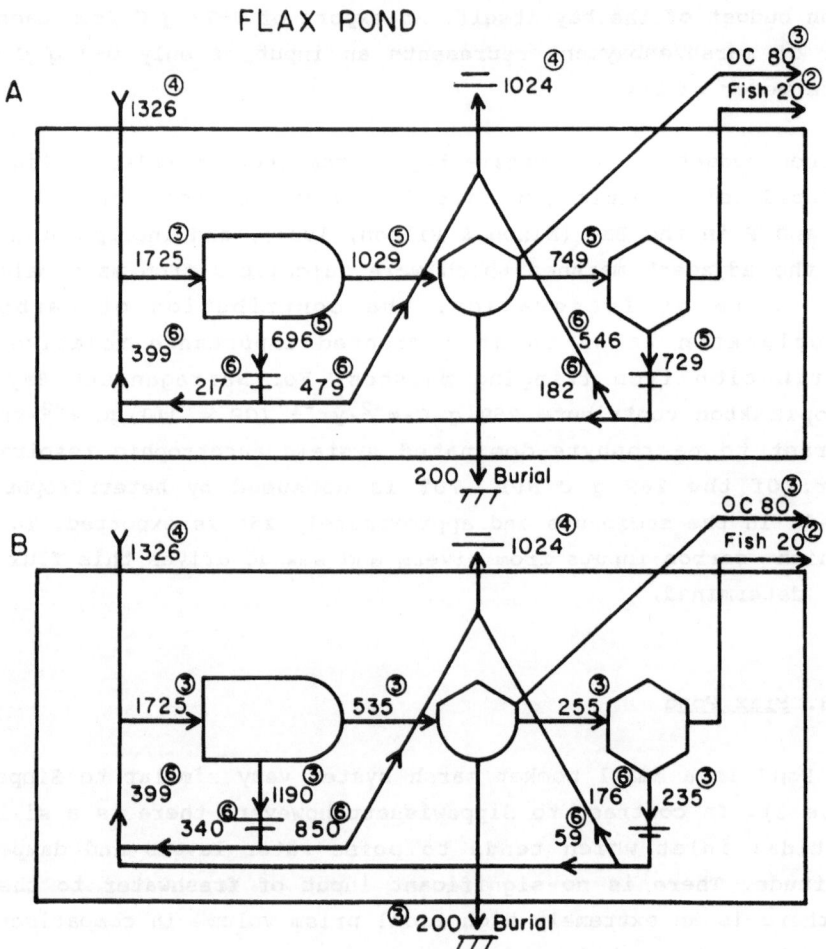

Fig. 4. Conceptual models of major carbon flows and storages for the Flax Pond ecosystem. In the top model, NPP is estimated to be 1029 g C $\cdot m^{-2} \cdot yr^{-1}$ while in the bottom model NPP is estimated at 535 $gC \cdot m^{-2} \cdot yr^{-1}$ See text for explanation of differences.
Notes: 1. Units $gC \cdot m^{-2}$ total ecosystem $\cdot yr^{-1}$; 2. Woodwell et al., 1977; 3. Woodwell et al., 1979; 4. Houghton and Woodwell, 1980; 5. Recalculated assuming NPP of 1029 to account for a 2:1 belowground/aboveground production ratio for S. alterniflora; 6. Mass balance.

Measurements of total CO_2 flux across system boundaries indicated that the sum of burial and N_{ex} was 302 g $C \cdot m^{-2} \cdot yr^{-1}$. Direct measurements of burial (200 g C) and N_{ex} (100 g C including fish migration) were in remarkably close agreement. From the production equations GP was calculated at 1725 g $C \cdot m^{-2} \cdot yr^{-1}$ (Table 2). Depending on the estimated value of NPP, 40-68% of GP was respired by autotrophs and 44-68% of NPP was remineralized by heterotrophs. Burial made up a rather large percentage of NPP (19-37%). The completeness of metabolic flux measurements made it possible to determine the approximate importance of plant refixation of respired CO_2, which amounted to 23% of GP. The remainder was captured from the atmosphere.

3.2.4. <u>Duplin River marshes and inner continental shelf</u>

Research into the structure and function of the salt marsh in Georgia has been continuous since the pioneering work in the 1950's. In recent years, research effort has extended into the nearshore region in an attempt to further evaluate the importance of carbon transfer to the ocean. In contrast to the pocket marshes of the north, the areal expanse of marshes in the south prohibits whole marsh studies. Research has consequently focused on "discrete watersheds" for which tidal drainage patterns can be distinguished. The Duplin River is a major pristine tidal creek with a salt marsh drainage basin of about 1100 ha (Table 1). The river is a 12.4 km long tidal slough in a marsh-filled lagoon adjacent to Sapelo Island. Of the systems analyzed in this paper, the Duplin River system has the largest tidal amplitude. The tidal prism is small, however, relative to the low water volume of the system. The nearshore zone is a shallow-water (<15 m depth) region of the continental shelf in the Georgia Bight bordered by coastal barrier islands to the west and by density fronts 20 to 30 km offshore. Inlets along the coast connect the nearshore zone to the tidal marshlands. The inner 6 km of the nearshore zone is highly turbid from water ejected by the tide from the estuaries.

Considerably more research effort has focused on the marshes of the Duplin system than on the adjacent tidal creeks and sounds. Sufficient information exists however to conceptualize carbon flow in both aquatic and marsh regions (Table 2 and Fig. 5). Net primary production on the marsh amounts to 2025 g $C \cdot m^{-2} \cdot yr^{-1}$. Of this approximately 36% is respired by heterotrophs and about 1% is buried. Attempts to determine the fate of the unaccounted for excess production have been equivocal. Flume studies on the marsh (conducted in an area that has since been identified as probably not being representative of an "average" piece of marsh in the Duplin River drainage basin) indicate that during rain storms approximately 190 g C are eroded off each square meter of marsh

138

GEORGIA - DUPLIN RIVER ESTUARY AND NEARSHORE

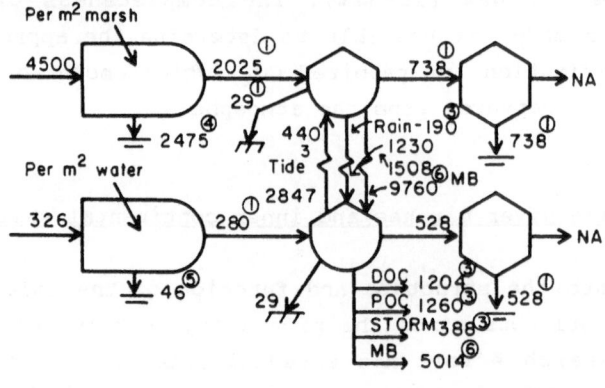

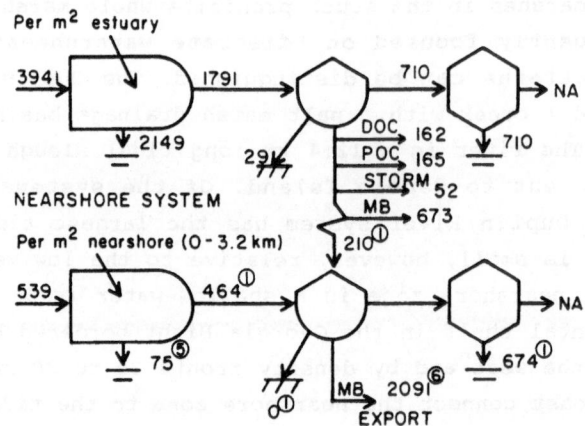

Fig. 5. Conceptual models of Georgia coastal systems, including the
marsh proper (top), adjacent tidal creeks and rivers (second from top),
the entire marsh/estuary complex (next to bottom) and the nearshore
region out to 3.2 km offshore (bottom). Units: g C·m^{-2} of each
respective subsystem·yr^{-1}.
Notes: 1. Hopkinson and Hoffman, 1984; 2. Pomeroy and Wiegert, 1981; 3.
Chalmers et al., 1985; 4. Assuming 55% of gross production; 5. Assuming
14% of gross production; 6. Mass balance; 7. Discontinuous lines
portray scalar differences for donor and recipient systems areas.

surface. During average conditions however,there appears to be a net
transfer onto the marsh of 440 g C·m^{-2}·yr^{-1} (Chalmers et al., 1985).
Thus in addition to the unaccounted for excess macrophytic production
on the marsh, there may be allochthonous input of carbon to some

portions of the marsh during high tide. Mass balance indicates that 1230 g C or 60% of NPP must be exported from the marsh.

The adjacent aquatic system is markedly heterotrophic. Community respiration exceeds community production by a factor of 1.8. Lagrangian studies within the Duplin River indicate that considerably more carbon is being exported to Doboy Sound and the ocean than is actually produced within the aquatic system. Mass balance considerations would suggest that the marsh is the source of this carbon.

The third conceptual model of carbon flow in Fig. 5 integrates marsh and aquatic systems (scaled to an average square meter of estuary which is comprised of roughly 80% marsh and 20% open water). The budget indicates that of the 1791 g C Npp $m^{-2} \cdot yr^{-1}$, 40% is heterotrophically respired, 2% is buried in the sediments and 21% (162 + 165 + 52 g C) is exported during storms and calm weather. A large fraction (673 g C) is unaccounted for, and assumed by mass balance to be exported. Export directly measured (based on calm summer conditions only) and that estimated from mass balance amounts to 1052 g C m^{-2} yr^{-1} or 59% of NPP. This material is presumably transferred to the nearshore region of the continental shelf.

In the nearshore region out to 3.2 km, the ratio of primary production to total community respiration (P/R) averages 0.7:1 over the year. A deficit of 210 g $C \cdot m^{-2} \cdot yr^{-1}$ is calculated from the production equations. For each meter of coastline in Georgia, the total respiratory carbon deficit in the nearshore region out to 3.2 km is 672 kg C (210 $g \cdot m^{-2}$ * 3200 m). Also for each meter of coastline, the autotrophic surplus in the estuary (the amount of carbon calculated by mass balance in the previous paragraph to be exported to the ocean) is 7364 kg C(1052 g $C \cdot m^{-2}$ * 7000 m wide band of marsh/estuary between barrier island and mainland). Hopkinson and Hoffman (1984) calculated that carbon entering the coastal zone from rivers is 1852 kg C per m of coastline. As there is clearly more than enough carbon potentially available from either riverine or estuarine sources (1852 and 7364 kg C, respectively) to satisfy the nearshore demand (672 kg C) it is difficult to evaluate the nearshore deficit solely in terms of the estuarine carbon budget. However, using the ratio of carbon potentially available in the estuary to that in rivers (4:1), I calculate that the offshore deficit can be met (along with riverine inputs) by an output of 88 g C (77 g C for the entire estuary) from

each square meter of marsh. That is equivalent to only 5% of marsh
NPP. Mass balance calculations indicate that there must be a
considerable amount of carbon entering the nearshore region from the
marsh/estuary and rivers that passes through the region unmetabolized
(7364 + 1852 - 672 = 8544 kg C per meter coastline out to 3.2 km
offshore).

These results show the utility of the mass balance approach when
applied to the presumed recipient system (the ocean) to calculate donor
system (the marsh) outputs. The interpretation of results is different
for donor and recipient sites however. Applied to the recipient system,
the mass balance approach indicates that 77 g C must pass from each
square meter of adjacent marsh/estuary to the nearshore region. Applied
to the donor system however, mass balance suggests that over 1000 g C
is exported from each square meter of marsh/estuary to the ocean. Can
we presume that of the 1000 g C exported (mass balance of the estuary),
77 g C (mass balance of the nearshore out to 3.2 km and scaled to the
area of the marsh and estuary) is respired within the nearshore region
and the remainder is carried away by coastal currents?

The intensity of measurements conducted in the Duplin River estuarine
system give some idea as to the importance of scale in designing
experiments to evaluate ecological processes. For logistic reasons,
detailed, controlled experimentation with many system components is
usually only possible in small portions of the larger system.
Unfortunately, due to spatial variability, extrapolation of results
from a small area to the much larger, overall system can give
misleading indications of total system function. Witness the apparent
paradox found from flume measurements in only a small portion of the
Duplin River system. Results from the Duplin system suggest that
overall patterns of exchange with adjacent systems are best determined
from replicated measurements of total community metabolism in a number
of the major subsystems of both receiving and donor ecosystems. Rates
evaluated at this level can form a framework upon which more detailed,
supplemental measurements of specific processes, mechanisms and
controlling factors can be evaluated.

The hierarchical level at which a research question about an ecosystem
is addressed should influence the degree to which the system is
subsequently decomposed and spatially scaled for investigation.
Although the precision of measurements of specific low hierarchical
level processes is usually greater than that observed for high

hierarchical level process, the cumulative precision of many low level process measurements when they are balanced to evaluate/calculate high hierarchical level phenomena can be poor. To minimize variance at the ecosystem level it is probably best to conduct ecological measurements at a level as close to the level of the question being addressed as possible. For example, the outwelling question is probably better evaluated by directly measuring total community metabolism (1 measurement with medium precision) than by measuring and balancing P and R of many different populations (many measurements with high precision).

3.2.5. Barataria Basin

In general, Barataria Basin and the marshes in Louisiana are the extreme opposite of the New England pocket marshes. In area, Barataria Basin is 4 orders of magnitude larger than Sippewissett or Flax Pond and includes not only saline marsh but also brackish and fresh marsh and freshwater swamp forest. The only freshwater entering the region is rainfall. However, due to the size of the wetland drainage basin there is considerable surplus runoff from the basin to the Gulf of Mexico. Tidal range decreases from about 0.3 m near the mouth of the basin to nothing about the 10 ppt isohaline. Water level is influenced more by wind than by tide.

Due to the large number of habitats within Barataria Basin it was necessary to construct conceptual models of carbon flow for each of the subsystems including wetland and aquatic subsystems in the swamp, and fresh, brackish and saline regions (Fig. 6). The shape of each compartment in Fig. 6 indicates the net autotrophic or heterotrophic nature of each subsystem. All wetland systems are autotrophic, while only the brackish and saline aquatic regions are. Material moves with the increasing flow of surplus water from wetland to adjacent water and from the head of the estuary to the mouth (Fig. 6). Export from wetlands was measured in the swamp and estimated from a simulation model of the salt marsh. It was hypothesized that export as a percentage of NPP for fresh and brackish marshes was between that of the swamp and saline systems (Day et al., 1982). In the downstream direction, organic carbon inputs from upstream systems become increasingly important relative to autochthonous production. Mass balance calculations indicate that total export to the Gulf of Mexico is 13.6 10^{10} g C·yr^{-1}. A separate estimate of export is calculated as

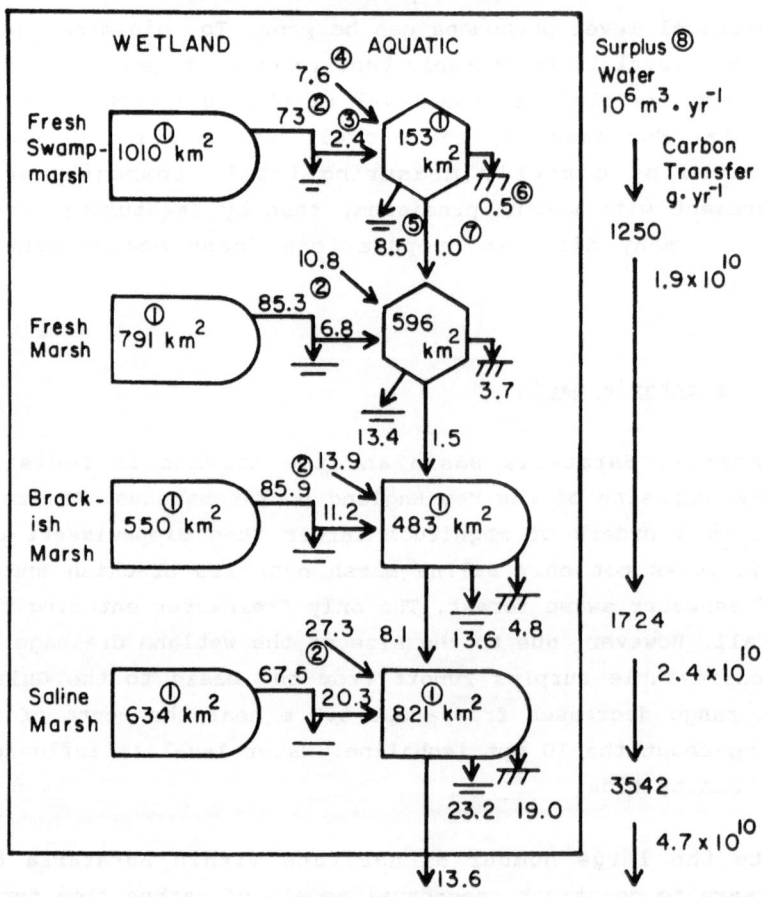

Fig. 6. Conceptual model of carbon flow in Barataria Basin, Louisiana, illustrating the upstream-downstream coupling of wetlands and their adjacent water bodies from the head of the estuary to its mouth. Also portrayed is the amount and flow of surplus rainwater through Barataria Basin. A transfer of organic carbon is associated with the movement of the surplus water through the basin. Bullet shaped objects indicate a net autotrophic system while hexagons indicate a net heterotrophic system. Units: $gC \cdot 10^{10} \cdot yr^{-1}$ or $10^6 \; m^3$ water $\cdot yr^{-1}$.

Notes: 1. Areas from Day et al., 1982 (units km^2); 2. Net aboveground production of emergent macrophytes from Day et al., 1982; 3. Export of organic carbon from wetland to adjacent aquatic system from Day et al., 1982; 4. Net production of aquatic system from Day et al., 1982; 5. Total respiration of aquatic system from Day et al., 1982; 6. Burial from Day et al., 1982; Surplus rain water budget showing annual runoff through system from Baumann, personal communication; 9. Carbon movement associated with surplus water runoff - Calculated as the product of total organic carbon concentration in the water column and mass flux of water - from Day, personal communication.

the product of the net surplus water runoff through the aquatic systems of Barataria Basin (Fig. 6) and the concentration of dissolved and particulate organic carbon in the saline water column. Export estimated with this technique is 4.7 10^{10} g C·yr^{-1}.

Conceptual models are also presented for unit areas of total wetlands, of total water bodies and for the entire basin (Fig. 7). With respect to the wetland subsystem, although net primary production of the wetland macrophytes is high, about 90% of it is metabolized in place by heterotrophs. About 6% of NPP is buried and only 4% is exported to adjacent water bodies. Overall, the aquatic subsystem within Barataria Basin is heterotrophic; less organic matter is exported to the Gulf of Mexico than is imported from adjacent marshes. At the level of the entire basin (lower diagram in Fig. 7), a single model integrating wetland and aquatic subsystems indicates that NPP is strongly influenced by the high levels of production of wetland macrophytes. Mass balance calculations indicate that almost 90% of NPP is heterotrophically respired within the system, 9% is buried and about 1% is exported to the Gulf of Mexico.

4. PATTERNS IN COUPLING BETWEEN ESTUARINE AND COASTAL WATERS

The mass balance approach consistently indicates that there is a net transfer of organic carbon from estuarine to coastal waters. Estimated levels of export range from about 30 g C from Barataria Basin to over 1000 g C·m^{-2} total estuary yr^{-1} from the Duplin River estuary.

As there are great differences in the size of systems compared in this study, mass balance budgets scaled to a unit area of estuary do not demonstrate the magnitude of total estuarine export to the adjacent ocean. The relative importance to the nearshore region of estuarine inputs estimated by mass balance is best exemplified at the scale of the oceanic receiving system. System size increases several orders of magnitude from the northern pocket marshes to the southern lagoonal systems; the absolute transfer of material to the sea increases concomitantly. Estimated annual export ranges from $5.7 \cdot 10^7$ to $1.4 \cdot 10^{11}$ g C per estuary for Flax Pond and Barataria Basin, respectively. Another perspective from which to demonstrate this export is to consider the net exchange per unit length of shoreline. From each meter of coastline in Georgia or across Barataria Basin, approximately $7 \cdot 10^6$ and $1 \cdot 10^6$ g C are exported annually, respectively.

LOUISIANA BARATARIA BASIN

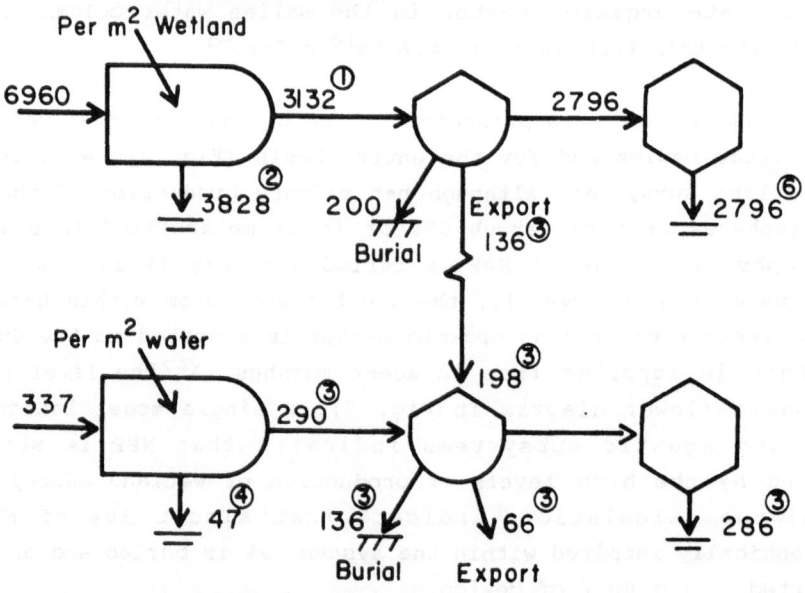

ESTUARINE SUBSYSTEMS
Per m² Wetland

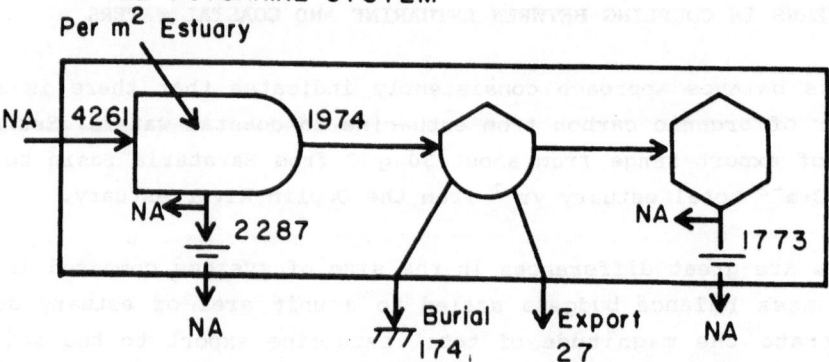

WHOLE ESTUARINE SYSTEM
Per m² Estuary

Fig. 7. Conceptual models of Barataria Basin carbon flows and storages integrated for all wetlands (Top), all aquatic systems (Middle) and for wetland and aquatic systems combined (Bottom). Units: g C·m⁻² of each respective subsystem·yr⁻¹.

Notes: 1. Net above- and belowground production of marsh macrophytes assuming a 1:2 production ratio - Aboveground production from Day et al., 1982; 2. Assuming marsh macrophytic respiration is 55% of gross production; 3. Day et al., 1982; 4. Assuming autotrophic respiration is 14% of gross production; 5. Day, personal communication; 6. Mass balance.

Export of this magnitude can be tremendously important locally. One way to quantify the potential importance is to determine the offshore distance to which heterotrophic demands in excess of local production could be sustained by allochthonous organic inputs from the estuary. Assuming the nearshore zone off Barataria Basin is as heterotrophic as that found in the Georgia nearshore zone (Hopkinson, 1985), the carbon input from marshes can sustain a heterotrophic deficit to 40 km in Georgia and 10 km in Louisiana. This was calculated using a 0.75 heterotrophic deficit (PR = 0.75) and a P of 539 and 279 g $C \cdot m^{-2} \cdot yr^{-1}$, in Georgia and louisiana, respectively. Because of the limited area of northern pocket marshes relative to shoreline length, estuarine export to northern shelf waters must be orders of magnitude less important than that from the much larger southern marshes, perhaps unimportant to shelf waters.

The mass balance approach consistently indicates wetland subsystems to be net autotrophic, adjacent tidal creeks to be net heterotrophic and larger bays such as Barataria Bay and Narragansett Bay to be net auto- trophic. Whether using the donor-export or the recipient-import view point, the mass balance approach indicates a substantial transfer of estuarine carbon to the nearshore region. There have been few studies other than those cited earlier providing the information necessary to determine the autotrophic-heterotrophic nature of nearshore regions. Teal (1967) noted that coastal waters of Woods Hole, Massachusetts were over saturated with respect to CO_2 and concluded that allochthonous inputs were required to sustain respiration. In contrast, at a larger scale Riley (1956) found no evidence for net heterotrophy in Long Island Sound. The size of Long Island Sound is so large relative to the are of surrounding estuaries however, that estuarine inputs would be greatly overshadowed by local production. In Narragansett Bay, there is little evidence of export from adjacent marshes even though the ratio of marsh to water is considerably larger than in Long Island Sound.

When all forms of organic carbon flux have been measured, the direct flux and mass balance approaches show reasonable agreement with respect to patterns of coupling between estuaries and adjacent coastal systems. Estimates of export measured by direct flux for North Inlet agree within a factor of 3 with mass balance estimates for the geomorphically similar Sapelo Island marshes (Table 2). For the Flax Pond system direct flux vs. mass balance estimates of export are within 1% of each other. Three estimates of export are now available for Barataria Basin and they all agree within 30 g $C \cdot m^{-1} \cdot yr^{-1}$ of each other.

5. FACTORS INFLUENCING MASS BALANCE ESTIMATIONS OF COUPLING

A complex suite of physical and biological factors acts to control each of the major ecosystem processes described in the production equations model. It is the net effect of all these factors working in unison on different parts of the ecosystem which determines the extent to which one landscape is coupled to another. For there to be a net export of organic material from a marsh system, net primary production must first exceed the rate of organic matter burial within the system and secondly the organic material remaining after burial must exceed heterotrophic consumption. A mechanism must exist to remove organic material from the marsh and estuary before it is totally buried and respired. The medium of transport is known; the water couples the adjacent subsystems of the coastal zone. Largely unknown however, are the mechanisms/processes that move dissolved and particulate material from the marsh and through the water and the factors that influence those mechanisms.

Several hypotheses have been proposed to explain the export character-istics of estuaries. Most of these hypotheses are unsatisfactory because they integrate a number of factors influencing export and do not identify or explain the actual operative mechanisms. Odum et al. (1979) proposed that the geomorphic shape of the estuary determined import/export characteristics. The idea was that the more restricted the connection to the sea, the greater the tendency for the system to import. Tidal assymetries (duration and velocity) have also been pro-posed to explain direction of movement. Postma (1967) proposed that systems import when maximum tidal velocities occur near low-slack while Boon (1975) proposed that systems export when maximum velocities occur near high slack. Boon and Byrne (1981) further hypothesized that geological age/maturity of a system influenced the tidal velocity assymetries. Mature, sediment-full basins were proposed to be ebb dominant and exporting systems, while immature, open systems were proposed to be flood dominant and importing systems.

These hypotheses deal mainly with particulate movement; they do not deal with turbulence, mixing or water residence times. As such they do not explain the mechanisms for dissolved constituent movement within estuaries. A distinct advantage of the Lagrangian flux technique em-

ployed by Imberger et al. (1983) in the Duplin River was that it
described the importance of turbulence (effective horizontal diffusion)
and concentration gradients in the movement of dissolved constituents.

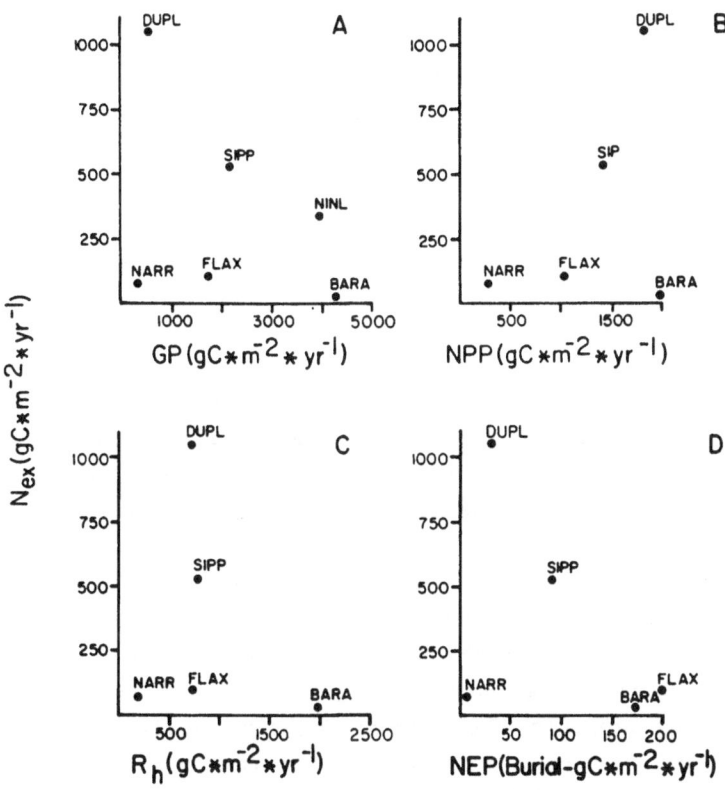

Fig. 8. Relation between rates of ecosystem metabolism and burial and
the amount of organic carbon exported (N_{ex}) for several estuaries. a)
GP vs Ne_x, b) NPP vs N_{ex}, c) R^h vs N_{ex}, and d) NEP vs N_{ex}. System
abbreviations are Sipp - Sippewissett, Narr - Narragansett Bay, Flax-
Flax Pond, Ninl - North Inlet, Dupl - Duplin River, Bara - Barataria
Basin.

A number of biological, geomorphic, hydrologic and hydrodynamic factors
potentially influence the nature of coastal system coupling. Biological
factors which may strongly influence N_{ex} are GP, R_a, NEP, and R_h. It
might be presumed that the more material produced within the estuary,
or in converse, the less material consumed within the estuary, the
greater the potential for export. However, there appears to be little
relationship between rates of organic matter production or consumption

within the estuary and the estimated level of export to the adjacent ocean (Fig. 8). No significant relations were found between N_{ex} and GP, NPP, R_h or NEP, unless Narragansett Bay and Barataria Basin are excluded from the analysis (Fig. 8). If attention is confined to pocket and lagoonal salt marsh-dominated systems (Flax, Sippewisett and Duplin), there is a significant relation between N_{ex} and GP and NPP. A strong negative relation exists between estimated export and NEP or burial for all the salt marsh-dominated systems (Fig. 8). As production increases or storage decreases, the level of export increases. Apparently there is an underlying basic difference in the function of salt marsh-dominated systems and that of phytoplankton-dominated, drowned river valleys and lagoonal, deltaic marsh systems.

Export of organic matter from estuaries does not appear to be related to overall estuarine morphology (i.e. circulation or geological origin - see Tables 1 and 2), although internal estuarine morphology does appear to be related to export. Although the relative proportion of creekbank versus non-creekband marsh (Fig. 9A) shows no relation with export, the relative proportion of intertidal wetland to water area within marsh-dominated estuaries (Fig. 9B) is strongly correlated with the level of export. Export generally increases as the relative area of marsh increases. This relation suggests that much of the material being exported must originate on the marsh, presumably from the marsh macrophytes. Phytoplankton-dominated Narragansett Bay behaves quite differently; it does not fit the export-relative marsh area relation. As suggested from the mass balance considerations, Narragansett Bay exports organic material largely originating from phytoplankton rather than from marsh macrophytes. There also may be a weak inverse relation between the level of export and total system size. Perhaps large systems tend to achieve P/R balance because internal exchanges among subsystems are more likely to be able to meet imbalances, whereas such flows in small systems must cross system boundaries.

Hydrologic and hydrodynamic factors must ultimately determine the ex port characteristics of estuaries. In the absence of tide, diffusion, turbulence and freshwater flow through a system (e.g. surplus rainwater runoff, river flow), no mechanism exists to move material across system boundaries except for the active migration of fish or other macrofauna. Unfortunately little scientific attention has been focused on describing or quantifying these physical characteristics in the estuaries for which the biology and ecology has been intensely studied. Of the estuaries considered in this paper, the effective horizontal

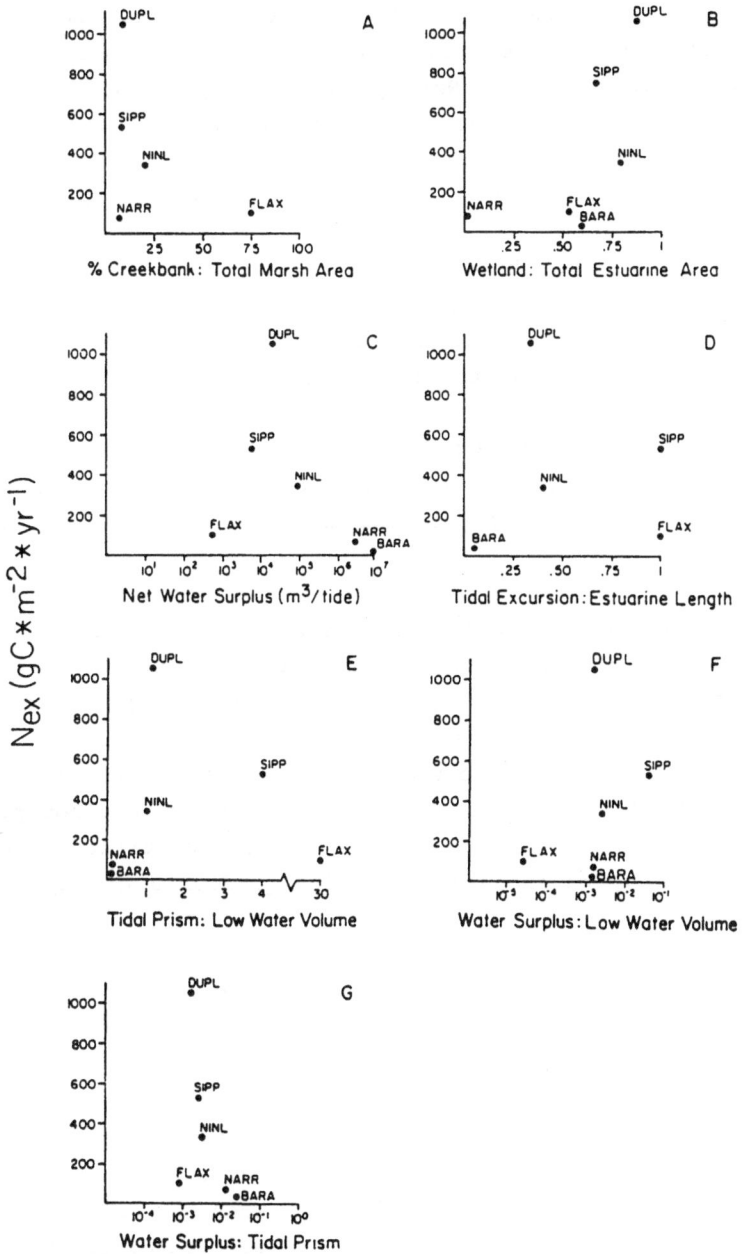

Fig. 9. Relation between estuarine physical characteristics and the level of organic carbon export to the sea (N_{ex}) for several estuaries. a) area of creekbank marsh relative to low and high marsh, b) relative areas of marsh to water within the estuarine system, c) the net freshwater input to a system, d) the length of the tidal excursion relative to the length of the estuary, e) volume of the tidal prism relative to the low water volume of an estuary, f) volume of freshwater inputs relative to estuarine low water volume, and g) volume of freshwater inputs relative to tidal prism volume. System abbreviations as in Figure 8.

diffusion has been measured only once and only in the Duplin River marsh/estuary (Imberger et al., 1983). Water residence time has been measured adequately only in Narragansett Bay (Pilson, 1985). Detailed rainfall-surplus runoff water budgets are available for only Barataria Basin. Most of the physical information available consists of estimates of river inputs, tidal ranges, tidal prisms and morphometric measurements such as system size and depth. As can be seen in Figs. 9C-G, these general physical characteristics show no relation to the export characteristics of the various estuaries compared. There is no tendency for export to increase with increasing freshwater input (9C), with increasing penetration of the tidal excursion into the estuary (9D), with increasing volume of tidal water relative to low water volume (9E) or with increasing volume of freshwater inputs relative to low water volume ( )F). Further, export does not appear to be related to the interaction of freshwater inputs and the size of the tidal prism (Fig. 9G).

None of the general physical characteristics identified in Table 1 or rates of ecosystem metabolism/storage (Table 2) can be used singly or in combination to predict the level of estuarine organic carbon export as determined by mass balance considerations. Does this mean that each estuary is unique and/or controlled by factors which may or may not have been identified in Table 1? Or perhaps export was not predictable because the true controlling factors were not identified or quantified (e.g. turbulence, mixing, concentration gradients of exportable substances, etc.).

After completing this review of organic carbon export from coastal marshes and estuaries, I believe each of the estuaries examined is unique; for each estuary examined, there appears to be a factor which can be used to explain the export characteristic. In Sippewissett marsh, groundwater input seems to be of paramount importance in controlling export. During a year when groundwater import was high, the export of FPOC was 54 g $C \cdot m^{-2} \cdot yr^{-1}$, but when groundwater input was low there appeared to be a net import of FPOC (Peterson, pers. comm.). In Narragansett Bay, it seems as if the combination of freshwater input and turbulence (which leads to a short 26 day residence time for water in Narragansett Bay) determines the export characteristic. In Flax Pond there is a strong tendency for the system to be a net importer of organic carbon because of the presence of a sill and an active mussel community at the system inlet. However, prevailing westerly winds tend to push floating macrophytic material out the inlet at the eastern end of the system, causing the system to be a net exporter of C. Some of the export characteristics of North Inlet appear to be related to a net throughput of freshwater, in from the Winyah Bay side of the estuary and out through the oceanic, North Inlet side (Dame, pers. comm.) In the Duplin River system, in spite of prevailing winds directed primarily into the system, turbulence caused by channel morphology and strong tidal currents is high. A high degree of effective horizontal diffusion and a strong concentration gradient of POC and DOC within the estuary act to move estuarine carbon to the ocean. Barataria Basin is controlled in a manner somewhat similar to Narragansett Bay. High rainfall and a large basin area lead to a large flow of freshwater through the system. Residence time of water and dissolved constituents are consequently fairly low in the basin and organic carbon is exported to the Gulf of Mexico.

6. CONCLUSIONS

On the basis of a mass balance evaluation of patterns of coupling between salt marsh-dominated estuaries and the ocean, several conclusions can be drawn.

1. Salt marshes are net autotrophic: more carbon is produced than consumed and stored within them.

2. Tidal creeks adjacent to marshes appear to be net heterotrophic: more carbon is degraded than is locally produced.

3. Larger bays and sounds (e.g. Barataria and Narragansett) appear to be net autotrophic.

4. Although marsh-dominated estuaries generally appear to export organic carbon to coastal waters, such export is usually only of local importance.

5. Like coral reefs, rates of gross primary production and internal recycling are extremely high. The ultimate source of nutrients to estuaries which enables them to be net autotrophic and hence exporting systems is unknown.

Our understanding of the functioning of salt marsh systems continues to be only rudimentary. Many questions remain to be answered, including:

1. What is the role of marshes in supporting coastal fisheries production? Are marshes important as sources of food for fish and shellfish, as habitat for fish, or both?

2. What is the source of "new", allochthonous nutrients that enables marsh/estuarine systems to be net autotrophic?

3. What controls the high rate of internal carbon cycling within estuarine systems?

4. Do macrophytic systems buffer adjacent downstream systems from inorganic nutrient inputs from urban systems? For example in the Caribbean region, mangroves are frequently adjacent to seagrass systems which are adjacent to coral reefs. Are these systems coupled in a dynamic sense? With increasing urbanization in these regions, will the loss of mangroves or seagrasses deleteriously affect proximate seagrass and coral reef systems?

5. What is the relation between cumulative impacts and marsh/estuarine fish production? What is the relation between fish production and a) the percentage of original marsh remaining or b) the percentage of original marsh primary production? Is the relation linear, or is there a threshold level below which fish production crashes? These questions and others need to be evaluated so that the importance of natural estuarine systems can be balanced against increasing urban- developmental pressures.

ACKNOWLEDGEMENTS

Thanks are extended to Alice Chalmers for her critical and constructive criticism of a preliminary draft of this manuscript, to an anonymous reviewer for additional constructive criticism and to Cindy Holcombe for her attractive preparation of figures. This is contribution No. 574 from the University of Georgia Marine Institute, Sapelo Island, GA 31327, USA. This work is the result of research sponsored by NOAA, Office of Sea Grant, Department of Commerce, under grant No. NA80AA-D-00091. The U.S. government is authorized to produce and distribute reprints for governmental purposes notwithstanding any copyright notation that may appear hereon.

REFERENCES

Boon, J., 1975. Tidal discharge asymmetry in a salt marsh drainage system. - Limnol. Oceangr. 20: 71-80.

Boon, J., R. Byrne, 1981. On basin hypsometry and the morphodynamic response of coastal inlet systems. - Mar. Geol. 40: 27-48.

Day, J.W., C.S. Hopkinson & W.H. Conner, 1982. An analysis of environmental factors regulating community metabolism and fisheries production in a Louisiana estuary. - In V.S. Kennedy (ed.): Estuarine comparisons, pp. 121-138. Academic Press.

Chalmers, A., R. Wiegert & P. Wolf, 1985. Carbon balance in a salt marsh: interactions of diffusive export, tidal deposition and rainfall-caused erosion. - Estuar. Coastal Shelf Sci. 21: 757-771.

Gael, B.T. & C.S. Hopkinson, 1978. Drainage density, land-use and eutrophication in Barataria Basin, Louisiana. - In J. Day, D. Culley, R. Turner & A. Mumphrey (eds.): Proc. Third Coastal Marsh and Estuary Management Symposium, pp. 147-163. Louisiana State University of Continuing Education, Baton Rouge, Louisiana.

Haines, E.B., 1977. The origins of detritus in Georgia salt marsh estuaries. - Oikos 29: 254-260.

Happ, G., J. Gosselink & J. Day, 1977. The seasonal distribution of organic carbon in a Louisiana estuary. - Estuar. Coastal Mar. Sci. 5:695-705.

Hopkinson, C.S., 1979. The relation of man and nature in Barataria Basin, Louisiana. - Ph.D. Diss. Louisiana State Univ., Baton Rouge, Louisiana. 236 pp.

Hopkinson, C.S., 1985. Shallow-water benthic and pelagic metabolism - evidence of heterotrophy in the nearshore Georgia Bight. - Mar. Biol. 87: 19-32.

Hopkinson, C.S. & R. Hoffman, 1984. The estuary extended - A recipient-system study of estuarine outwelling in Georgia. - In V. Kennedy (ed.): The estuary as a filter, pp. 313-330. Academic Press, New York.

Houghton, R & G.M. Woodwell, 1980. The Flax Pond ecosystem study: exchanges of CO_2 between a salt marsh and the atmosphere. - Ecology 61:1434-1445.

Howes, B., J. Dacey & G. King, 1984. Carbon flow through oxygen and sulfate reduction pathways in salt marsh sediments. - Limnol. Oceanogr. 29: 1037-1051.

Howes, B., J. Dacey & J. Teal, 1985. Annual carbon mineralization and belowground production of Spartina alterniflora in a New England salt marsh. - Ecology 66: 595-605.

Imberger, J., T. Berman, R. Christian, E. Sherr, D. Whitney, L. Pomeroy, R. Wiegert & W. Wiebe, 1983. The influence of water motion on the distribution and transport of materials in a salt marsh estuary. - Limnol. Oceanogr. 28: 201-214.

Kaplan, W., I. Valiela & J. Teal, 1979. Denitrification in a salt marsh ecosystem. - Limnol. Oceangr. 24: 726-734.

Kjerfve, B., 1975. Velocity averaging in estuaries characterized by a large tidal range to depth ratio. - Estuar. Coastal Mar. Sci. 3: 311-323.

Kjerfve, B. & J. Proehl, 1979. Velocity variability in a cross-section of a well-mixed estuary. - J. Mar. Res. 37: 409-418.

Kremer, J & S. Nixon, 1978. A coastal marine ecosystem. Springer-Verlag, New York. 217 pp.

Montague, C., 1980. The net influence of the mud fiddler crab, Uca pugnax, on carbon flow through a Georgia salt marsh: the importance of work by macroorganisms to the metabolism of ecosystems. - Ph.D. Diss., Univ. Georgia, Athens. 157 pp.

Nixon, S.W., 1980. Between coastal marshes and coastal waters - A review of twenty years of speculation and research on the role of salt marshes in estuarine productivity and water chemistry. - In P. Hamilton & K. Macdonald (eds.): Estuarine and wetland processes, pp. 261-290. Academic Press, New York.

Nixon, S. & C. Oviatt, 1984. Ecology of a New England salt marsh. - Ecol. Mono. 43:463-498.

Odum, E.P., 1968. A research challenge: evaluating the productivity of coastal and estuarine water. - In Proc. 2nc Sea Grant Conf., Grad. School Oceangr., Univ. Rhode Island, Kingston, pp. 63-64.

Odum, W.E., J. Fisher & J. Pickral, 1979. Factors controlling the flux of particulate organic carbon from wetlands. - In R. Livingston (ed.): Ecological processes in coastal and marine systems, pp. 69-80. Plenum Press, New York.

Peterson, B., 1980. Aquatic primary productivity and the $^{14}CO_2$ method: A history of the productivity problem. - Ann. Rev. Ecol. Systematics 11: 359-385.

Pilson, M., 1985. On the residence time of water in Narragansett Bay. - Estuaries 8: 2-14.

Postma, H., 1967. Sediment transport and sedimentation in the estuarine environment. - In. G. Lauff (ed.): Estuaries, pp. 158-179. AAAS, Washington, D.C.

Pomeroy, L. & R. Wiegert (eds.) 1981. The ecology of a salt marsh. Springer-Verlag, New York. 271 pp.

Ragotskie, R. & R. Bryson, 1955. Hydrography of the Duplin River, Sapelo Island, Georgia. - Bull. Mar. Sci. 5: 297-314.

Riley, G., 1956. Oceanography of Long Island Sound, 1952-1954. Production and utilization of organic matter. - Bull. Bingham Oceanogr. Coll. 15: 324-344.

Schubauer, J.P. & C.S. Hopkinson, 1984. Above- and belowground emergent macrophyte production and turnover in a coastal marsh ecosystem, Georgia. - Limnol. Oceanogr. 29: 1052-1065.

Teal, J.M., 1962. Energy flow in the salt marsh ecosystem of Georgia. - Ecology 43: 614-624.

Teal, J., 1967. Biological production and distribution of pCO_2 in Woods Hole waters. - In G. Lauff (ed.): Estuaries, pp. 336-340. Amer. Assoc. Adv. Sci Publ. No. 83, Washington, D.C.

Valiela, I., J. Teal & N. Persson, 1976. Production and dynamics of experimentally enriched salt marsh vegetation: belowground biomass. - Limnol. Oceanogr. 21: 245-252.

Valiela, I., J. Teal, S. Volkmann, D. Shafer & E. Carpenter, 1978. Nutrient and particulate fluxes in a salt marsh ecosystem: tidal exchanges and inputs by precipitation and groundwater. - Limnol. Oceanogr. 23: 798-812.

Valiela, I. & J. Teal, 1979. Geomorphology and hydrography of the Duplin River estuarine system. - Ph.D. Diss., Univ. Georgia, Athens, GA., USA. 140 pp.

Woodwell, G.M. & E. Pecan, 1973. Flax Pond: an estuarine marsh. - BNL 50397, Brookhaven Nat. Lab., Upton, New York, USA. 7 pp.

Woodwell, G.M., D. Whitney, C. Hall & R. Houghton, 1977. The Flax Pond ecosystem study: exchanges of carbon in water between a salt marsh and Long Island Sound. - Limnol. Oceanogr. 22: 833-838.

Woodwell, G. & R.H. Whittaker, 1978. Primary production in terrestrial ecosystems. - Amer. Zool. 8: 19-30.

Woodwell, G.M., R. Houghton, C. Hall, D. Whitney, R. Moll & D. Juers, 1979. The Flax Pond ecosystem study: the annual metabolism and nutrient budgets of a salt marsh. - In R. Jefferies, & A. Davy (eds.): Ecological processes in coastal environments, pp. 491-511. Blackwell Scientific, Oxford.

COUPLING OF MANGROVES TO THE PRODUCTIVITY OF ESTUARINE AND COASTAL WATERS

Robert R. Twilley

Department of Biology

University of Southwestern Louisiana

Lafayette, LA 70504

1. INTRODUCTION

Most of the tropical coastline between 25° N and 25° S latitude is vegetated by forested wetlands called mangroves (McGill, 1958). These plant communities have received considerable botanical investigation because of their unique taxonomy and ovivipary (Tomlinson, 1986), and the diverse fauna that inhabit these coastal areas (Macnae, 1968;Chapman, 1976). However, the ecology of mangroves is poorly understood, particularly the significance of these ecosystems to the productivity and nutrient cycling of estuarine and adjacent coastal waters. It has been suggested that the high fishery yields of coastal tropical waters are due to the presence of these communities (Macnae, 1974; Turner, 1977; Jothy, 1984), yet there is no evidence of a cause and effect relationship for mangroves and fisheries (Macnae, 1974). Thus the function of these wetlands in supporting secondary productivity continues to be a complex issue. Mangroves may also influence the primary productivity of coastal waters by controlling the fate of dissolved nutrients and suspended sediments. Mangroves are considered a source of organic detritus yet a nutrient sink, contributing to the confusion of their role in coastal processes.

There is a critical need to understand the function of mangroves in tropical ecosystems because of the rate these intertidal areas are being converted to alternate land uses. Mangroves are exploited for forestry products including fuelwood for cooking, fenceposts, charcoal, tannins, pulpwood, chipwood and timber (Polunin, 1983). Areas inhabited by mangroves are also reclaimed for agriculture, aquaculture and residential development. For instance in the Philippines, the construction of ponds for the production of milkfish and penaied prawns has resulted in the loss of 189,000 ha of forest since 1920, which represents nearly 45% of the original total mangrove resource (Jara, 1984). Other major shrimp pond operations have developed in the intertidal zone of Ecuador, Thailand and Panama. In southwest Florida

Lecture Notes on Coastal and Estuarine Studies, Vol. 22
B.-O. Jansson (Ed.), Coastal-Offshore Ecosystem: Interactions.
© Springer-Verlag Berlin Heidelberg 1988

(USA) the reclamation of mangroves for urban development resulted in the loss of 24% of mangrove area in Marco Island from 1952 to 1984 (Patterson, 1986). A projected 200% increase in population in this coastal area by the year 2000 demonstrates the pressure to develop these natural resources. In contrast to foresty, which attempts to maintain some sustainable yield in mangrove ecosystems, reclamation activities such as urban development, agriculture and pond mariculture result in the loss of this resource from the coastal zone.

Conflicting management plans for the intertidal zone in many tropical countries result from desire to preserve mangroves as a natural resource for fisheries, versus their utilization as an economic resource for humans. Mangroves may indirectly support economically important fisheries by providing free services such as habitat, food and good water quality. Thus the loss of these plant communities would negatively impact industries that rely on productive coastal fisheries. However, to settle these conflicts, information is needed that more clearly describes the function of mangroves to better understand the importance of these systems to the fisheries of tropical estuaries. The purpose of this paper is to review our current understanding of the coupling of mangroves to the productivity and nutrient cycling of coastal waters. This analysis should help present a framework for the development of future mangrove research that will address the issue of outwelling from these intertidal ecosystems.

2. APPROACH

A mass balance approach can be used to evaluate our present understanding of the function of mangroves in the productivity and nutrient cycling of coastal ecosystems. The model in Figure 1 establishes the major boundaries of a mangrove-estuarine ecosystem and shows the major processes associated with the exchange of materials within this system and with coastal waters. The boundary between mangroves and estuarine water is distinct, based on the distribution of vegetation. The function of mangroves as either a source or sink of organic matter and nutrients depends on the net flux of materials across this boundary. The estuary-coastal water boundary is somewhat undefined since estuarine water can extend unto continental shelves during periods of high freshwater discharge. The model in Figure 1 will use a physical boundary to define coastal waters as those associated with continental shelves. Thus the coupling of mangroves to estuarine

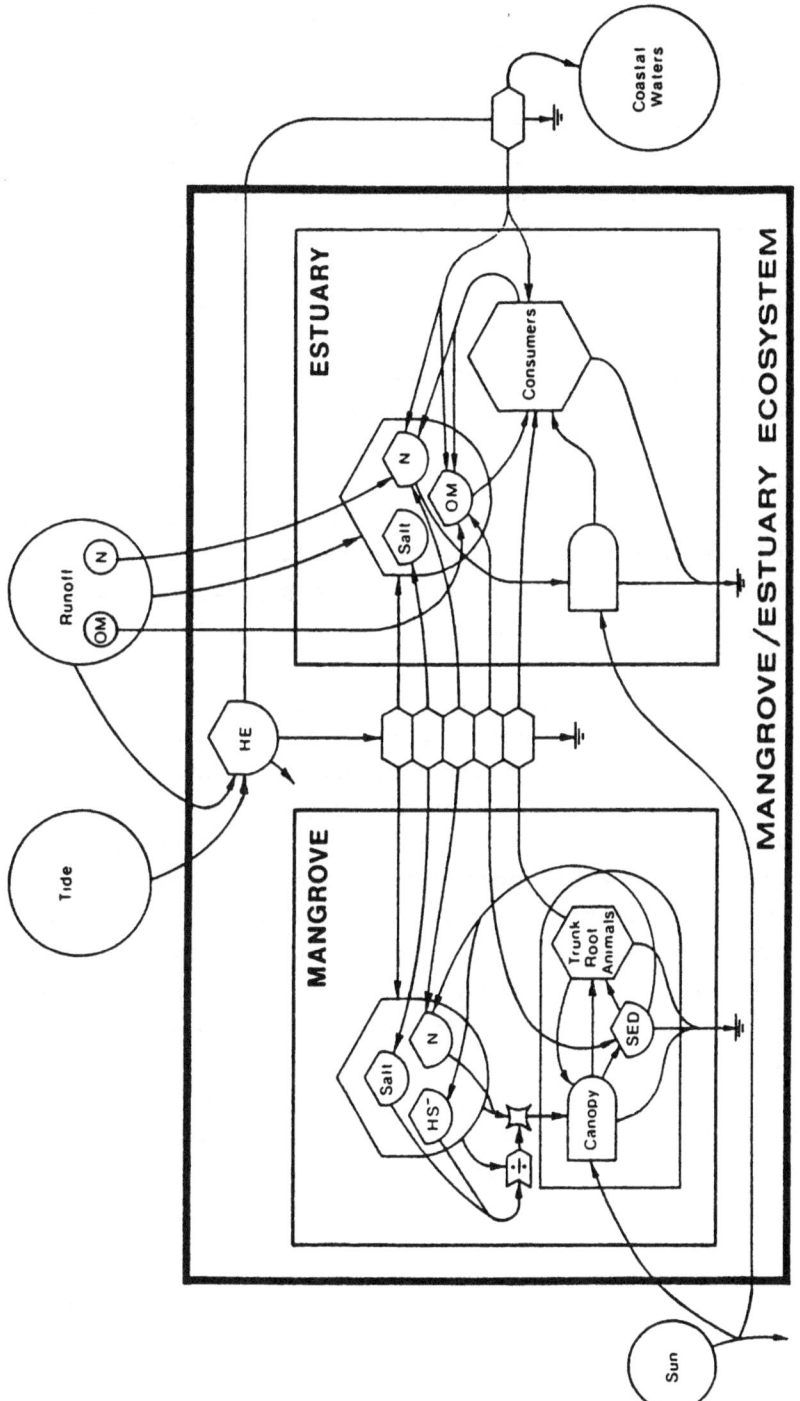

Fig. 1. Model of a mangrove-estuarine ecosystem and its coupling to coastal waters (N = nutrients, OM = organic matter, SED = sediments, HS = hydrogen sulfide, HE = hydrologic energy).

and coastal waters involves exchange across the mangrove-estuary boundary, followed by transport from the estuary to coastal waters.

The conceptual model in Figure 1 indicates that tides and runoff control the exchange of materials across the boundaries of the mangrove-estuarine ecosystem. The input of water from tides, rainfall and runoff relative to losses from evapotranspiration are described by the term hydrologic energy. The amount of water transported among mangrove, estuary and continental shelf is dependent on this potential hydrologic energy and the geomorphology of the region (Odum et al., 1979). These two factors also determine the extent to which the intertidal zone is inhabited by mangrove vegetation (Thom, 1982).

Mangrove forest structure and phytosociology have long been associated with tidal inundation patterns (Watson, 1928; Chapman, 1944; Chapman, 1976). Lugo and Snedaker (1974) classified the structure of mangroves into six community types that related forest physiognomy with their hydrology and geomorphology. These six groups can be combined into three types (riverine, fringe and basin) using the grouping by Brown et al. (1979) for forested wetlands based on frequency and amplitude of water inundation (Fig. 1). The hydrologic energy of riverine mangroves is high since it is dominated by river flow and tidal inundation, while fringe mangroves are influenced mainly by frequent tidal inundations. Basin mangroves have less hydrologic energy since they are located inland of fringe or riverine communities and as a result are less frequently inundated by either tides or river floods. This continuum of hydrologic energy has been associated with distinct types of mangrove forest structure (Lugo and Snedaker, 1974; Pool et al., 1977). It is less clear if the function of these systems is also specific among these types of mangroves along this tidal continuum. The objective of this paper is to review information on the ecology of mangrove-dominated estuaries for evidence that hydrologic energy controls the exchange of materials from mangroves to estuarine and coastal waters.

3.   MASS BALANCES OF THE MANGROVE SUBSYSTEM

3.1.   Organic matter dynamics

Leaf litter on the forest floor represents a major source of organic matter and nutrients for outwelling from mangroves to adjacent estuarine waters. Thus the balance of litter productivity, decomposition and export influence the exchange of these materials at the boundary of the mangrove subsystem. Litter productivity values for mangrove forests worldwide range from 1.20 $t \cdot ha^{-1} \cdot yr^{-1}$ for scrub mangroves in south Florida to 23.4 $t \cdot ha^{-1} \cdot yr^{-1}$ for a 20 yr old managed forest in Malaysia (Twilley et al., 1986). It has been suggested by Pool et al. (1975) that litter production rates in mangroves are a function of water turnover within the forest and the rank of the means of litter production in Figure 2 (riverine > fringe > basin > scrub) supports this hypothesis. In southwest Florida the apparent inverse relation of litter production among five basin mangrove sites and average soil salinity further supports the argument that environments with higher hydrologic energy support higher litter productivity (Twilley et al., 1986). Fewer tides result in higher soil salinity (Cintron et al., 1978) and/or the accumulation of toxic substances (eg. hydrogen sulfide; Carlson et al., 1983; Nickerson and Thibodeau, 1985), which can result in increased stress on these inland mangrove forests (Hicks and Burns, 1975; Lugo et al., 1981). Wharton and Brinson (1979) suggested that the production of forested wetlands was dependent on water movement, not only as a source of silts and clays, but also a supply of nutrient and aeration for optimal growth. Increased hydrologic energy in intertidal areas seems to increase the potential for litter production in the estuary.

Export from mangroves may also be associated with the amount of hydrologic energy within the intertidal zone. Rates of organic carbon export from basin mangroves was dependent on the volume of tidal water inundating the forest each month, and accordingly export rates were seasonal in response to the seasonal rise in mean sea level (Twilley, 1985). Rainfall also increased organic carbon export from mangroves (Twilley, 1985), especially dissolved organic carbon (DOC), similar to results for salt marshes (Harriss et al., 1980). Total organic carbon (TOC) export from infrequently flooded basin mangroves in southwest Florida was 64 $gC \cdot m^{-2} \cdot yr^{-1}$, and nearly 75% of this material was DOC (Twilley, 1985). Particulate detritus export from fringe mangroves in south Florida was estimated at 186 $gC \cdot m^{-2} \cdot yr^{-1}$ (Heald, 1969), compared to 420 $gC \cdot m^{-2} \cdot yr^{-1}$ for a mangrove forest in Australia (Boto and Bunt, 1981). Estimates of tidal amplitude per tide in these three mangrove forests were 0.08 m, 0.05 m and 3 m, respectively. At low tidal amplitude in the basin forests, only 20% of litterfall was exported (Twilley, 1985), compared to 45% in the fringe forest, and at the upper tidal range leaf litter on the forest floor was negligible (Boto and Bunt, 1981). These results indicate that as hydrologic energy increases, both the magnitude of litter produced within mangroves and the proportion of this litter that is exported also increase. Both of these processes enhance the transport of organic matter across the mangrove boundary to estuarine waters.

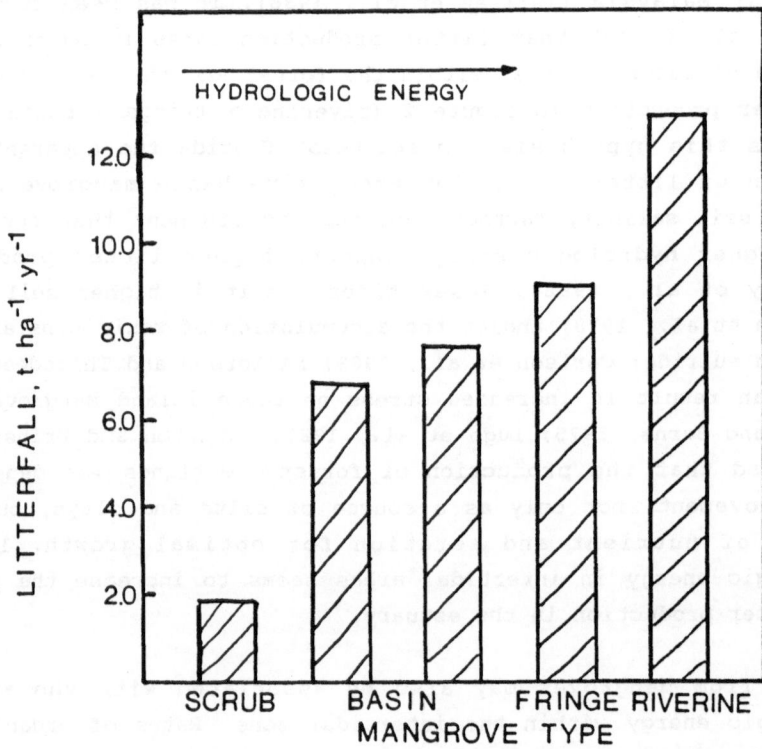

Fig. 2. Rates of litterfall in different types of mangrove forests based on Twilley et al., 1986 (two values for basin mangroves are monospecific and mixed forests).

Organic carbon exchange in mangroves can also be evaluated by looking at the net ecosystem productivity of these forested wetlands. Net ecosystem production (NEP) of mangroves is the balance of inputs from in situ production and allochthonous sources, minus losses from community respiration and export. Mangrove research in southwest Florida and in Puerto Rico have made significant contributions to the development of organic carbon budgets for mangroves in microtidal environments (Table 1). Gross primary productivity(GPP) including canopy and epiphytic algae of fringe and basin mangroves ranged from 2457 to 5074 $gC \cdot m^{-2} \cdot yr^{-1}$ (Table 1). Most of the GPP of these communities was associated with the canopy, however epiphytic algae associated with either prop roots or pneumatophores may also contribute as much as 16% of the total fixation of carbon.

Table 1. Mass balance of carbon flow ($gC \cdot m^{-2} \cdot yr^{-1}$) in mangrove forests in Florida and Puerto Rico.

	Rookery Bay[1,2,3,4]		Puerto[5] Rico	Fahkahatchee Bay[6]		
	Fringe	Fringe	Fringe	Basin	Fringe	Fringe
GPP						
Canopy	2055	3292	3004	3760	4307	5074
Algae	402	26	276			
Total	2457	3318	3280			
Respiration (plants)						
Leaves, stems	671	2022	1967	1172	1416	3084
Roots-AG	22	197	741	146	182	215
Roots-BG	?	?	?	?	?	?
Total	693	2219	2708	1318	1598	3299
NPP	1764	1099	572	2442	2709	1775
Growth		186	153			
Litterfall		318	237			
Respiration (heterotrophs)		197	135			
Respiration (total)		2416	2843			
Export		64	500			
NEP		838	-63			
Burial		?	?			
Growth		186	153			

1 Lugo et al., 1975
2 Twilley, 1985
3 Twilley et al., 1986
4 Twilley, 1982
5 Golley et al., 1962
6 Carter et al., 1973

Respiration losses of gross production from autotrophes ranged from 693 to 3299 $gC \cdot m^{-2} \cdot yr^{-1}$. All of these estimates only considered respiration of the aboveground surface area, and there may be a large sink of carbon transported from the canopy and respired by roots located belowground. Root:shoot ratios for mangrove forests are high, ranging from 0.8 to 1.2 (Golley et al., 1962; Golley et al., 1975), and the turnover rate from this belowground biomass may account for a substantial portion of net primary productivity (NPP). There is a wide variation in total root respiration (prop root) among the six mangrove forest in Table 1 resulting in NPP:GPP ratios ranging from 0.85 to 0.19. Lugo et al. (1975) observed a lower respiration rate per surface area of prop roots compared to pneumatophores in the forests at Rookery Bay, which would account for some difference since <u>Rhizophora</u> dominated the fringe forest and <u>Avicennia</u> dominated the basin forest. However, the greater surface area of prop roots in fringe forests would tend to minimize these differences. The greater surface area of prop roots in Puerto Rico account for the extreme rate in root respiration at this site that resulted in a NPP:GPP of only 0.19. Mangroves in well flushed areas tend to have greater biomass allocated to prop roots that could influence the relative amount of NPP depending on hydrology.

Although autotrophic respiration rates for the fringe forest at Puerto Rico and basin forest in Rookery Bay were similar, heterotrophic respiration (based on sediment respiration) was similar at 135 and 197 $gC \cdot m^{-2} \cdot yr^{-1}$. The ratios of GPP to total respiration (GPP:RT) were 1.4 for the basin and 1.1 for the fringe mangroves. A ratio near one for the fringe mangrove indicates a system near steady state as discussed by Golley et al. (1962), yet a ratio greater than one for the basin forest suggests there is an excess of organic matter that could be exported to adjacent estuarine waters. Direct measurements of organic carbon export from basin mangroves shows that this flux was only about 6% of apparent NPP, while in the fringe forest export was nearly 100% of net production. The amount of carbon exported relative to apparent NPP in the basin mangrove forest is similar to those ratios for temperate salt marsh ecosystems (Hopkinson, this volume).

NEP in the fringe mangrove forest in Puerto Rico was 63 $gC \cdot m^{-2} \cdot yr^{-1}$ indicating a rather balanced ecosystem. However, tree growth in this forest was 153 $gC \cdot m^{-2} \cdot yr^{-1}$ and along with belowground root respiration and peat accumulation (depth of peat was 1 m in the center of this site), there is a large deficit in the carbon budget of this forest. NEP was 838 $gC \cdot m^{-2} \cdot yr^{-1}$ in the basin forest at Rookery Bay, and even with carbon accumulation associated with tree growth of 186 $gC \cdot m^{-2} \cdot yr^{-1}$ there is a large surplus of carbon based on this budget. The depth of peat in this forest is nearly 2 m and accretion in these systems is assumed to maintain steady state with the rise in sea level (Scholl and Stuiver, 1967); yet no direct measurement of organic matter burial in this mangrove forest had been made. Further studies are needed to determine the fate of this excess carbon in basin mangroves that may account for nearly 60% of the apparent NPP. These estimates of NEP between a fringe and basin mangrove suggest that a large proportion of NPP in the more inundated forests is exported, while in basin forests more of the net production is accumulated or utilized within the system. This supports the "open" versus "closed" concept of fringe and basin mangroves, respectively, as proposed by Lugo and Snedaker (1974) in relation to hydrologic energy.

3.2. Nutrient Exchange

The idea that intertidal wetlands may be a nutrient sink is particularly confusing since it contradicts the "outwelling" concept of detritus exchange established above for mangroves. Indications that mangroves may be a nutrient sink come from a study by Walsh (1967) where nutrient concentrations in waters moving through a mangrove in Hawaii decreased in concentration. However, this study did not directly measure nutrient flux across the mangrove boundary, and very few studies have attempted to budget flows of nutrients in mangrove ecosystems. It has been argued for other plant communities that nutrient recycling may be of greater significance than inputs to maintaining productivity. Such processes include resorption or retranslocation of nutrients prior to leaf fall (nutrient use efficiency; Ryan and Bormann, 1982; Vitousek, 1982), the immobilization of nutrients in leaf litter during decomposition (Brinson, 1977; Twilley et al., 1986), and nitrogen fixation. In mangroves these patterns in nutrient recycling processes may influence the exchange of nutrients at the boundary of mangroves, which would determine whether mangroves serve as either a nutrient source or sink to estuarine and coastal waters.

There is some indication that nutrient resorption can vary within a mangrove species in different environments. The mass/nutrient ratio of nitrogen in leaf litter relative to nitrogen return to the forest floor via litterfall is an index of nutrient-use efficiency. High efficiencies of Rhizophora occur in trees located in either riverine or fringe forests (Fig. 3). In riverine and fringe mangrove forests where leaf export is high, higher nutrient recycling efficiency in leaves prior to leaf fall is an advantage to increasing nutrient conservation. Trees with lower efficiencies in nutrient utilization, which are represented by nitrogen loss less than 20 $kg \cdot ha^{-1} \cdot yr^{-1}$ and mass/nutrient ratios less than 200, are located in more inland sites (Fig. 3). Also, nutrient use ratios of Avicennia, which are also located in more inland sites, are lower than for the other mangrove species. Further testing will have to be done to see if in fact nutrient use efficiencies in mangrove species are an adaptive strategy to sites with varying hydrology or fertility.

Increases in nitrogen concentration of leaf litter during decomposition may immobilize nitrogen in basin mangrove forests, effecting within-system nutrient conservation (Brinson, 1977; Vitousek, 1984). Nitrogen enrichment of leaf litter includes absorption and adsorption processes by bacterial and fungal populations (Fell and Master, 1973; Rice and Tenore, 1981; Rice, 1982), as well as nitrogen fixation (Zuberer and Silver, 1978; Gotto et al., 1981). These inputs of nitrogen result in an absolute increase of nitrogen in decomposing litter, and particularly for _Rhizophora_ leaf litter, recycles nitrogen on the forest floor (Twilley et al., 1986). Inland mangrove forests have higher decomposition and mineralization rates of leaf litter and along with infrequent tides, have lower rates of detritus export. Thus high litter turnover may increase the availability of nutrients for reabsorption by roots and prevent the loss of nitrogen from the system.

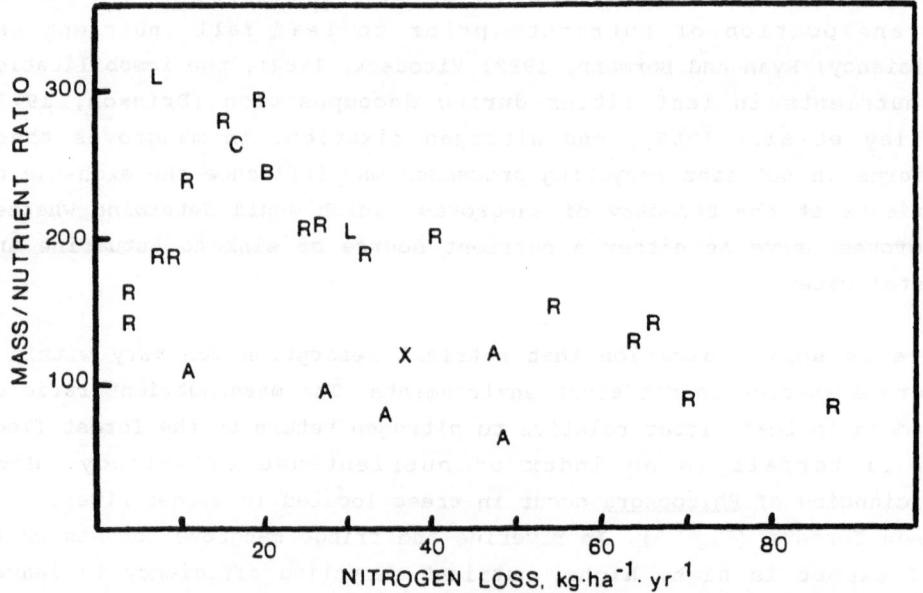

Fig. 3. Dry mass: nutrient ratios for litterfall per unit of nitrogen loss from litterfall in species of mangroves including _Rhizophora_ (R), _Avicennia_ (A), _Laguncularia_ (L), _Ceriops_ (C), _Brugeirra_ (B) and _Xylocarpus_ (X).

These results indicate that mechanisms which conserve nutrients may vary along a tidal continuum. In areas of high tidal frequency, higher

recycling efficiency may occur in the canopy; whereas in lower tidal
activity, nutrient recycling may occur on the forest floor. These
mechanisms are based on limited data for basin mangrove sites in
Rookery Bay and Estero Bay, with little information on riverine or
fringe mangrove forests (Twilley et al., 1986). Also, fluxes associated
with denitrification and nitrogen fixation must be accounted for before
the significance of these recycling processes can be evaluated
(Vitousek, 1984; Twilley et al. 1986). Nedwell (1975) using enclosures
to measure nutrient uptake in mangrove sediments noticed they had a
great capacity to dissimilate nitrate, particularly in areas of
nutrient enrichment from sewage discharge, indicating high
denitrification potentials. Others have noted nitrogenase activity in
decomposing leaves, root surfaces (prop roots and pneumatophores) and
sediment, but none of these studies have interpreted these rates into
areal estimates of nitrogen input to the forest (Kimball and Teas,
1975; Gotto and Taylor, 1976; Zuberer and Silver, 1978; Potts, 1979;
Gotto et al., 1981). There are no estimates of the relative rates of
nitrogen fixation and denitrification on the mass balance of nitrogen
in mangroves, nor any indication that these processes may influence
nitrogen flux at the mangrove-estuary boundary.

4. MASS BALANCES OF THE ESTUARINE SUBSYSTEM
4.1. Organic matter

A mass balance of organic matter in Rookery Bay, a mangrove-estuarine
ecosystem in southwest Florida, indicates that mangrove detritus may be
a significant source of energy for secondary productivity (Fig. 4A).
Mangroves contribute nearly 345 $gC \cdot m^{-2} \cdot yr^{-1}$ to the estuary (per m^2 of
estuary) which is 83% of the allochthonous inputs. Since litter export
from fringe mangroves is nearly double the per area rates for basin
mangroves, they are considered important sources of detritus in south
Florida.   However, the area of basin mangroves is nearly double the
area of fringe forest, thus the contribution of organic carbon to
Rookery Bay is nearly the same for both types of communities. Thus the
much larger land mass occupied by the inland mangroves establishes
these systems as very important sources of detritus to estuarine
ecosystems. Allochthonous inputs of organic carbon together with NPP of
the water column represent the pool of organic matter available for
secondary productivity. Allochthonous inputs accounted for 47% of this
organic pool in Rookery Bay and mangroves alone accounted for 39% of
the total organic matter supply (fringe = 21% and basin = 19%). During

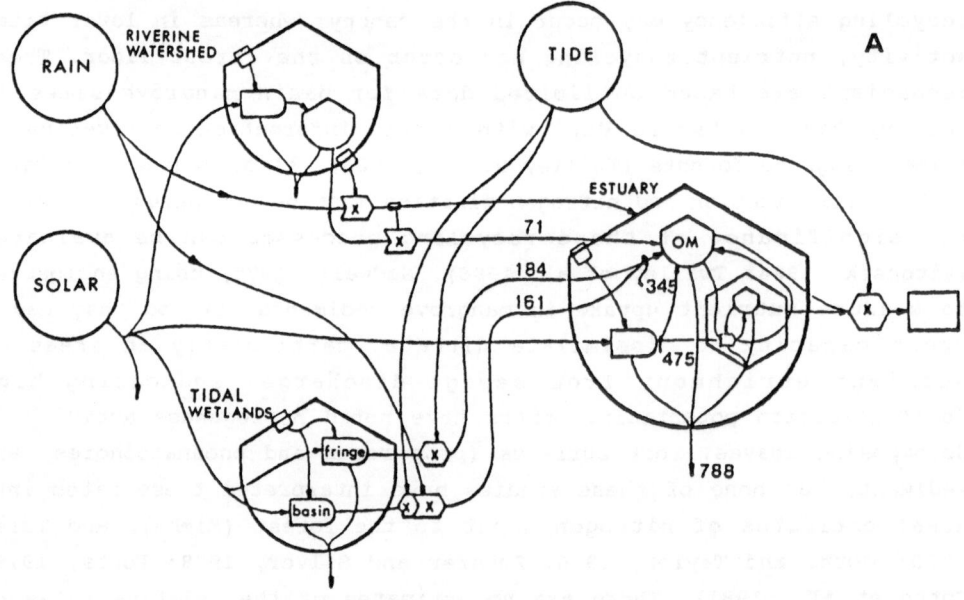

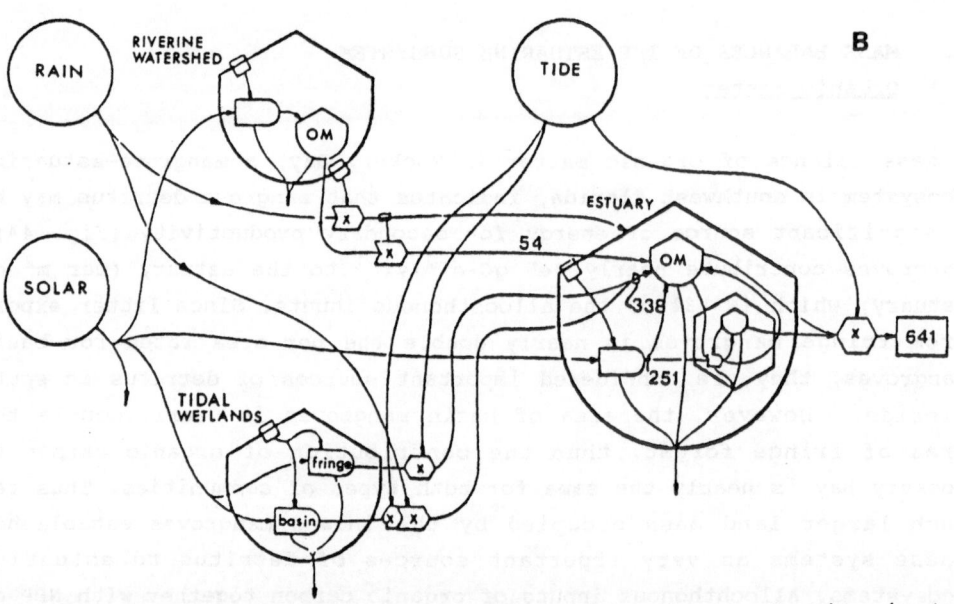

Fig. 4. Models of allochthonous inputs and in situ production in two mangrove estuarine ecosystems in southwest Florida: A) Rookery Bay (Twilley 1982), and B) Fahkahatchee Bay (Carter et al. 1973). Flows are in $gC \cdot m^{-2} \cdot yr^{-1}$ of the estuary area (surface water at mean low tide).

the summer nearly 52% of the organic carbon available for secondary production was allochthonous, indicating the significance of mangroves to this estuary.

A mass balance of organic carbon has also been estimated for a mangrove-estuarine ecosystem 25 km southeast of Rookery Bay (Fig. 4B; Carter et al., 1973). The primary productivity of Fahkahatchee Bay was estimated at about 251 $gC \cdot m^{-2} \cdot yr^{-1}$, and phytoplankton and benthic plants contributed about 73 and 27% of the total, repectively. Input of organic detritus from freshwater discharge to this estuary was not intensively studied, so it was assumed that 40% of the litter produced in wetlands was exported, for an estimate of 338 $gC \cdot m^{-2} \cdot yr^{-1}$ (per m^2 of estuary). Based on this estimate of export, mangroves contribute about 52% of the organic matter supply that is available for secondary productivity in this system. Thus the importance of mangroves to the organic matter budget of Fahkahatchee Bay is similar to results for Rookery Bay (Fig. 4).

Net productivity of phytoplankton in Rookery Bay was 251 $gC \cdot m^{-2} \cdot yr^{-1}$ which is similar to productivity values in Fahkahatchee Bay and the backwaters of Goa, India; yet much higher than the productivity of Biscayne Bay in southeast Florida (Table 2). In Rookery Bay, phytoplankton production was highest during the summer when allochthonous inputs were also greater. This is in contrast to the mangrove backwaters of India (Qasim et al., 1969; Krishnamurthy et al., 1975; Verlencar and Qasim, 1985) and Brazil (Teixeira et al., 1969). In those backwaters phytoplankton production was minimum during freshets when increases in turbidity apparently limited photosynthesis. Several investigators have observed a stimulation of primary productivity in the water column by the addition of DOC from mangroves (Prakash, 1971; Prakash et al., 1973; Cooksey and Cooksey 1978; J.W. Day, pers. comm.). Prakash et al. (1973) demonstrated that there was an upper limit to this stimulation, above which the negative effects of light absorption limited primary productivity. Information in Table 2 suggests that primary productivity in coastal waters offshore from mangrove dominated estuaries is high for continental shelves, due to enhancement from both mangrove detritus and presence of upwelling along the coast. More studies are needed to determine if mangroves may not only contribute to the organic pool of coastal waters via export, but also by enhancing primary productivity.

Table 2. Primary productivity and chlorophyll concentrations of waters in mangrove dominated estuaries and coastal waters.

LOCATION	DISTANCE OFFSHORE (km)	CHLORO-PHYLL (mg/m^{-3})	PRIM. PROD. $mgC \cdot m^{-3} \cdot h^{-1}$	$gC \cdot m^{-2} \cdot yr^{-1}$	REFERENCE
BACKWATERS					
Cochin, India	0	7.3-20.0		124(NPP) 280(GPP)	Quasim et al 1969
Pichavaram, India	0	3.2-30.2	300-600		Krishnamurthy et al 1975; Sundararaj and Krishnamurthy 1973
Kollur, India	0	0.1-1.8	1.9-7.5		Untawale et al 1977
Cananeia, Brazil	0		11.4-91.1		Teixeira et al 1969
Goa, India	0	4.3		223	Verlencar and Quasim 1985
ESTUARINE WATERS					
Vellar Estuary, India	0	0.4-10.5			Santhakumari 1971
Biscayne Bay, Florida	0	0.2-4.2	13-46		Roman et al 1983
Rookery Bay, Florida	0			251	Twilley 1982
Fahkahatchee Bay, Florida	0			183	Carter et al 1973
COASTAL WATERS					
Lacadive Sea	7-15	0.8			Shah 1973
Goa, India	10	1.9		342	Verlancar and Quasim 1985
Phangha Bay, Thailand	1			468	Wium-Anderson 1979

The ratio of gross productivity (P_g) to 24-h community respiration (R24) indicates the ability of autochthonous production to meet the heterotrophic demands of the system. The P_g/R_{24} ratio of Rookery Bay was 1.08 based on six diurnal surveys of changes in dissolved oxygen concentrations (Table 3). During the summer, when net production was higher, P_g/R_{24} ratios were >1.75 on two occasions; afterwards ratios were <1.0 with the lowest value of 0.70 in December. Seasonal differences in P_g/R_{24} ratios have also been observed in other mangrove dominated estuaries, reaching as low as 0.22 in October at Coots Bay in south Florida (Table 3). Untawale et al. (1977) measured a metabolic P/R ratio of 0.77 for a backwater area of Kollur estuary in India. These mangrove dominated estuaries are basically heterotrophic systems except for periods of peak primary production and rely on imported organic matter to support what organic matter is available for net ecosystem production (NEP). In Rookery Bay the sum of gross production and allochthonous inputs minus losses due to respiration was equal to 482 $gC \cdot m^{-2} \cdot yr^{-1}$. It is still unknown how much of this excess material is utilized directly by higher trophic levels, and how much is exported as detritus to adjacent coastal waters.

Table 3. Community metabolism and production:respiration ratios for mangrove dominated estuaries.

STATION	DATE	Production	Respiration	P/R	REFERENCE
		($gO_2 \cdot m^{-2} \cdot d^{-1}$)			
Rookery Bay,	Annual	6.25	5.76	1.08	Twilley 1982
Florida	April	5.12	6.40	0.80	
	May	6.40	7.20	0.89	
	June	7.52	4.40	1.71	
	June	6.68	5.76	1.16	
	July	7.04	4.00	1.76	
	December	5.88	6.79	0.70	
Coots bay,	July	7.50	14.10	0.53	Tabb et al. 1962
Florida	October	1.10	4.97	0.22	
Station 3e	January	1.50	1.30	1.15	
	April	2.44	4.68	0.52	
Station 5e	January	0.81	1.66	0.49	
Whitewater Bay,	April	8.22	7.03	1.11	Tabb et al. 1962
Florida					
Laguna Madre,	Annual	4.30	5.60	0.76	Odum and
Texas					Hoskin 1958
Kollur, India		2.24	2.90	0.77	Untawale et al. 1977

4.2 <u>Nutrients</u>

Preliminary estimates of nutrient dynamics in mangrove-estuaries can be
made using mixing diagrams of nutrient concentration along a salinity
gradient. Nixon et al. (1984) compared mixing diagrams in a mangrove
dominated estuary to an estuary where mangroves had been removed. This
comparison was to determine the linkage of mangroves to estuarine
waters. Curves exhibiting non-conservative behavior of nutrients
suggest that mangroves are a source if the curves are convex, and a
sink if the curves are concave. A conservative nature of nutrients
indicates little net exchange (Liss, 1976). This analysis has several
shortcomings since changes in concentrations of inorganic species other
than by dilution may only represent transformations within the water
column and not a loss or input to the system. Surveys of total nitrogen
and phosphorus, and suspended matter by Nixon et al. (1984) in the
Sangga river estuary, Malaysia, during the dry season generally
exhibited conservative behaviors in the estuary. Concentrations of
these materials were much lower in the natural estuary than in the
estuary where mangroves had been reclaimed. Suspended particulate
materials were especially higher in the estuary that had been diked for
agriculture (Nixon et al., 1984). A nutrient survey of the Gambia River
estuary found that total nitrogen and suspended sediments also
exhibited conservative concentrations with salinity during October and
November. These surveys may approximate the mass balance of nutrient
flux in mangrove-dominated estuaries. However, because of their limited
application, much more work is needed to determine the function of
mangroves in nutrient dynamics in tropical estuarine ecosystems, and
the exchange of nutrients with coastal waters.

5. OUTWELLING TO COASTAL WATERS

Based on a mass balance of Rookery Bay, there was 482 $gC \cdot m^{-2} \cdot y^{-1}$
available for either net ecosystem production or export from the
estuarine ecosystem. There is no direct evidence of organic matter
export from Rookery Bay to adjacent coastal waters except that total
organic carbon concentrations exhibited little seasonal change at the
coastal boundary (Twilley, 1982). Exchange of organic matter between
Fahkahatchee Bay and coastal waters was measured for two tides in
October, 1972. A slough with a drainage basin of 28,961 ha empties into
this system which transports a large volume of material to the coast.
For instance, stations located 7 km offshore have salinities that are

10 ppt lower during periods of peak freshwater runoff in July and August (Carter et al., 1973). An estimate of detritus export was 13,000 kgC/d or about 641 gC·m^{-2}·yr^{-1} (Carter et al., 1973). It was suggested by the authors that the major source of this carbon was from mangrove and freshwater wetlands in the watershed (Fig. 4B).

Table 4. Characteristics of Goa estuaries dominated by mangroves in relation to monsoon season.

CHARACTERISTIC	STATION[3]	PREMONSOON	MONSOON	POSTMONSOON
Time[1]		Feb-May	Jul-Sept	Oct-Jan
Description[1]		Stable	Change	Recovery
Physics/Salt[1]		Well-mixed	Stratified	Stratified/ Mixed
1% Isohaline (Distance upriver)		65 km	11-20 km	
Currents[1]		Flood-dominated	Ebb-dominated	Flood/ dominated
Salinity[1]	S	36.27	31.09	34.45
	M	30.49	5.73	23.64
	Z	30.43	7.97	25.70
"F" (Fresh-water fraction)[1]	M	0.16	0.82	0.31
Dilution factor[1]	M	6.20	1.20	3.2
	Z	6.20	1.30	4.0
Residence time[1] (days)	M Z	50 (longer)	5-6 (longer)	
Phosphates[1] (μmol/l)		LOW	0.42-2.50	
Silicates[1]		LOW	HIGH	
Prim. prod.[2] (mgC·m^{-2}·d^{-1})	S M	938.1 300-1200	– 0-600	938.1 400-1600
Chlorophyll[2] (mg/m^3)	S M	1.94 1.8-6.0	– 0.5-50	1.94 6-11
Detritus[2] (% POC)	M	76	75.9	46
POC:CHL[2]	M	59.40	107.9	74.2

1 Qasim and Gupta 1981
2 Verlencar and Qasim 1985
3 S = Offshore station; M = Mandovi River estuary; Z = Zuari River estuary

The extension of estuaries unto continental shelves in areas of high freshwater discharge are more obvious examples where materials from mangrove-dominated estuaries may be transported to coastal waters. Such is the case of extended estuaries along the west coast of India that are strongly influenced by southwest monsoons that occur from June to September (Qasim and Gupta, 1981). Characteristics of the coastal and estuarine waters of Goa which are surrounded by 1800 ha of mangrove (Untawale et al., 1982), can be identified into time zones of premonsoon, monsoon, and postmonsoon. This temporal sequence has been described as periods of stability, change and recovery, respectively (Table 4). In response to monsoons, salinites in coastal waters 10 km offshore dropped from 36.27 ppt during premonsoon to 31.09 ppt, and the 1 ppt isohaline moved seaward from 65 km upriver to about 15 km upriver. Shah (1973) also found an influence of mangrove backwaters on coastal waters 7-15 km offshore in the seas off Cochin, west India. In these areas coastal water salinities dropped from 35 ppt to values <30 ppt during peak discharge. Associated with this increased discharge was increased stratification, higher nutrient concentrations and turbidity, and lower primary productivity rates (Table 4). Following the monsoon, primary productivity of estuarine and coastal waters were stimulated relative to premonsoon conditions. These sequence of events associated with high freshwater discharge are also characteristic of mangrove backwaters in Cochin (Qasim, 1973).

The significance of detritus from mangrove dominated estuaries on the composition of particulate organic materials (POC) in coastal waters is indicated by seasonal shifts in POC/Chl ratios (Fig. 5). Chlorophyl was 77.3% of the POC pool in estuarine waters based on annual averages of values. Peak POC/Chl ratios were >4400 followed by minimum values in postmonsoon when phytoplankton production was maximum. In the coastal region, the percentage of detritus was greater than in the estuarine region suggesting that this coastal area may act as a sink for materials transported from these mangrove-dominated estuaries (Verlencar and Qasim 1985; Fig. 5). In postmonsoon, most of the POC is from phytoplankton; however, in the monsoon season, allochthonous materials originating from surface runoff are dominate (Verlencar and Qasim, 1985). In the Zaire river estuary in southwest Africa, 50% of the material transported from the river was deposited in the estuary and the other 50% transported to the shelf (Eisma and Kalf, 1984). Other investigators have acknowledged that the coastal waters of the Bering Sea and northern coast of Taiwan (Tanoue and Handa, 1979; Tsu Chang et al., 1981) could be influenced by detritus in areas subjected

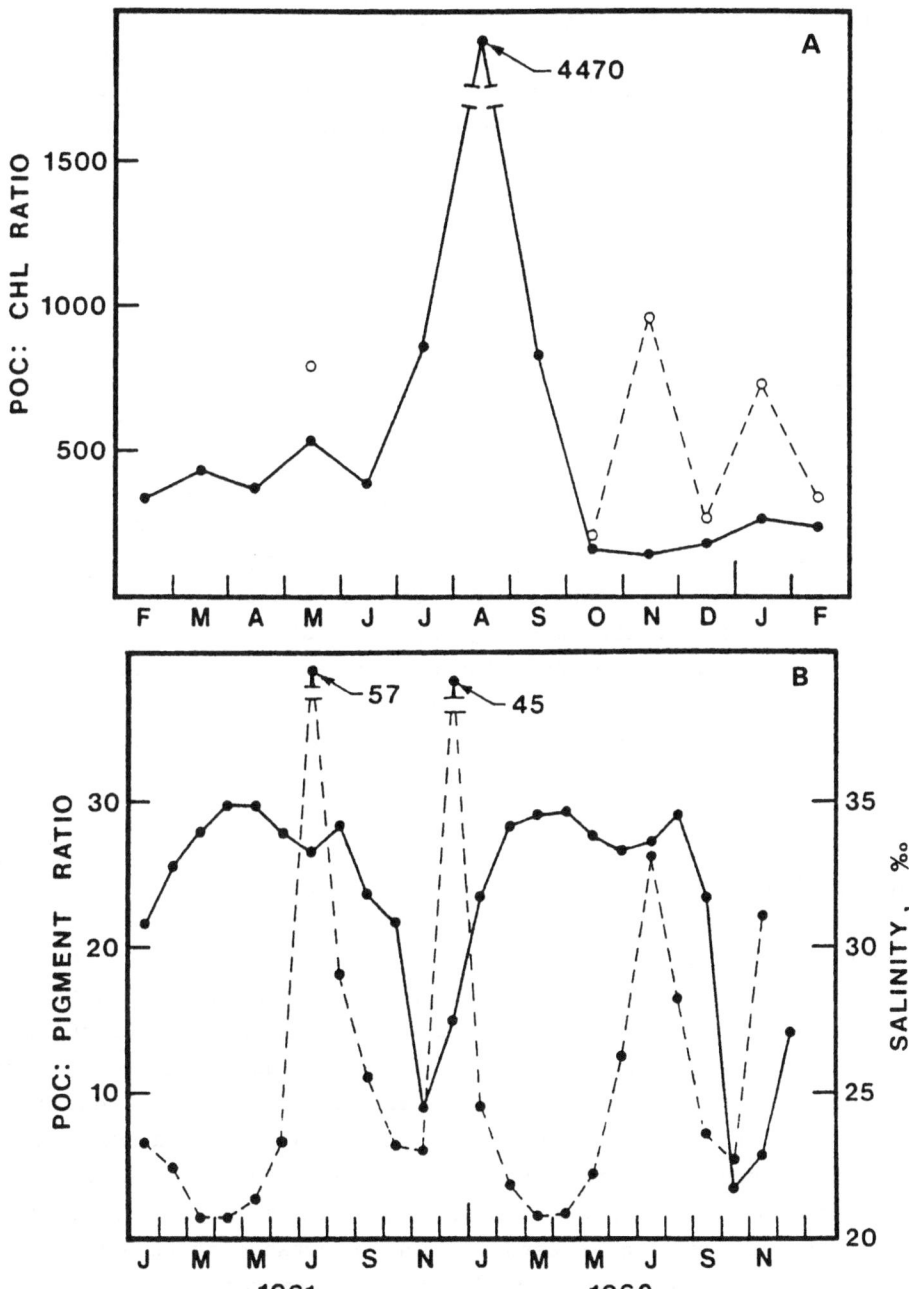

Fig. 5. Ratios of particulate organic carbon concentrations relative to chlorophyll a concentrations for coastal waters offshore the Goa estuaries (A) and Wiltair (B). From Verlencar and Qasim (1985), and Rao and Rao (1975) respectively. In (A) the solid line is for a station at the mouth of the estuary and the dashed line for a station 10 km offshore. In (B) the solid line is salinity and dashed line POC: pigment ratios.

to land drainage (Verlencar and Qasim, 1985). Rao and Rao (1975) also observed maximum POC/Chl ratios (chlorophyll based of Harvey pigment units) during monsoon season at a station 5 km offshore from Waltair, (east coast of India) and minimum values occurred during postmonsoon.

The exact source of this detrital material is unknown, and is not likely from mangroves in areas of extremely high currents such as the Goa estuaries (Untawale et al., 1982). There are no mangroves at the mouth of this estuary apparently due to the rocky substrate that, along with strong currents, prevent the establishment of mangrove seedlings in this area (Untawale et al., 1982). The mangrove:estuary area ratio of this system is only 0.3, and thus freshwater discharge or in situ production may dominate as sources of organic material to coastal seas. Yet in more deltaic estuarine areas where the geomorphology is more suitable for extensive mangrove habitats, such as in northeast India or southeast Asia, exchange of materials from mangroves to coastal waters may be significant. Mangroves in areas of monsoons present a particular case where the offshore coupling may be substantial. Even in microtidal areas with about 1500 mm of rainfall such as in south Florida, detritus from mangroves is significant in coastal waters. More studies are needed to document the magnitude of organic matter exchange in coastal ecosystems under various hydrologic energies.

5. CONCLUSIONS

The objective of this paper was to review information on the flux of organic matter and nutrients between mangroves, and estuarine and coastal waters to focus future research in mangrove ecology. It seems that processes involving the exchange of materials from mangroves varies among tropical estuaries depending on hydrologic energy (tides and/or runnoff) and geomorphology. Hydrologic energy influences the per area productivity and flux of materials in mangroves, and together with geomorphology, these two factors also determine the areal extent of mangroves surrounding an estuary. The importance of this total flux of detritus from mangroves to the organic matter budget of estuarine ecosystems depends on the relative size of both systems. It has been suggested that the ratio of wetland area to the surface water area of estuaries may be indicative of the importance of wetlands to the productivity of aquatic ecosystems (Mann, 1975; Welsh et al., 1982). Macnae (1974) and Turner (1977) have both related shrimp productivity to the areal coverage of mangroves, suggesting the importance of these wetlands to coastal fisheries. Mass balances of organic matter and nutrients are needed from a variety of geographical regions representing high river discharge, high tidal amplitudes, and combinations of these hydrologic energies. Studies on sites other than the low and medium tidal environments reviewed in this paper would further test the hypothesis that hydrology and geomorphology control the function of mangroves in estuarine and coastal ecosystems.

The concept of outwelling from mangroves has been based on less data than for salt marshes, yet it seems more conclusive that there exists a net flux of detritus from these forested wetlands. One reason for this may be related to the greater tidal amplitude and runoff at the sites used to study mangrove export compared to temperate intertidal wetlands. For instance, tidal amplitude for mangroves in Australia is greater than 3 m, and rainfall in south Florida is more than 1500 mm. Also litterfall in mangroves is continuous throughout the year and for Rhizophora, this leaf material is very buoyant. These factors lead to a greater potential for exchange in mangrove ecosystems. However, studies of detritus flux in mangroves are few and are based on relatively simple hypsographic analysis of export. More studies on mangroves are needed that represent sites with different hydrologic energies and geomorphology. Also, experimental designs using flume techniques designed for salt marshes should be investigated to further test the hypothesis that the contribution of detritus from mangroves to estuarine and coastal waters depends on hydrologic energy.

The function of mangroves may be strongly related to forest structure such that processes may be specific to riverine, fringe and basin mangrove forests according to their respective hydrologic characteristics. This conclusion is based mainly on organic matter exchange in mangroves, although there are indications that nutrient recycling may also vary along a continuum in hydrology. There are many processes that need to be studied before generalization of mangrove functions can be made. Studies on the importance of nitrogen fixation, denitrification, fertilization, sedimentation and exchange on nutrient budgets would greatly expand our understanding of nutrient dynamics in mangrove ecosystems. The contradiction that tidal wetlands contribute detritus to coastal ecosystems yet also serve as nutrient sinks has been explained by transformations of nutrients that occur within the wetland. The question of whether mangroves are nutrient sinks has received little attention, and processes associated with these transformations may be important in understanding this function.

Evaluating the importance of mangroves to sustaining secondary productivity in estuarine and coastal ecosystems depends on both the transport and utilization of organic matter from mangroves. Based on mass balances, some mangrove-estuarine ecosystems transport substantial amounts of organic material to estuarine and coastal waters. However, the mass balance approach does not indicate directly the utilization of this material by higher trophic levels. Also, mangroves may benefit several fisheries by providing habitat during certain stages of their life cycle rather than serving only as a source of food. Several excellent analysis of trophic food webs in mangrove ecosystems, such as that by Odum and Heald (1972), have demonstrated that mangroves are utilized for food and habitat. Evidence from stable isotope studies suggest that the utilization of mangrove detritus may vary depending on its quantity relative to other sources of food (Macko and Zieman, 1983), similar to conclusions based on mass balance studies. There are several controversies concerning the interpretation of stable isotope information, yet more research on experimental designs using this technique to determine if mangrove detritus is utilized are warranted. We are still far from establishing a cause and effect relationship between mangroves and fisheries.

ACKNOWLEDGEMENTS

Resources for this review were funded by a Faculty Research Award to the author from the University of Southwestern Louisiana. This project was initiated by discussions with Dr. Ariel Lugo, who has provided much of the framework for several of these ideas in mangrove ecology. G. Steyer, A. Montegut and J. Singleton provided assistance on this project; and M. Richard drafted most of the illustrations.

REFERENCES

Boto, K.G. & J.S. Bunt, 1981. Tidal export of particulate organic matter from a northern Australian mangrove system. - Estuar. Coast. Shelf Sci, 13: 247-255.
Brinson, M.M., 1977. Decomposition and nutrient exchange of litter in an alluvial swamp forest. - Ecology 58: 601-609.
Brown, S., M.M. Brinson & A.E. Lugo, 1979. Structure and function of riparian wetlands. - In R.R. Johnson & J.F. McCormick (eds.): Strategies for protection and management of floodplain wetlands and other riparian ecosystems. Symp. Proc. U.S. Dept. Agric. GTR-WO-12.
Carlson, P.R., L.A. Yarbro, C.F. Zimmermann & J.R. Montgomery, 1983. Pore water chemistry of an overwash mangrove island. - Florida Scientist 46: 239-249.

Carter, M.R., L.A. Burns, T.R. Cavinder, K.R. Dugger, P.L. Fore, D.E. Hicks, H.L. Revells & A.W. Schmidt, 1973. Ecosystem analysis of the Big Cypress Swamp and estuaries. - U.S. Environ. Prot. Agency, Region 4, Atlanta, Ga. EPA 904/9-74-002.

Chapman, V.J., 1944. 1939 Cambridge University expedition to Jamaica. I. A study of the botanical processes concerned in the development of the Jamaican shore-line. - J. Linn. Soc. London Bot. 52: 407-447.

Chapman, V.J., 1976. Mangrove vegetation. - J. Cramer, Germany.

Cintron, G., A.E. Lugo, D.J. Pool & G. Morris, 1978. Mangroves of arid environments in Puerto Rico and adjacent islands. - Biotropica 10: 110-121.

Cooksey, K.E. & Cooksey, B., 1978. Growth-influencing substances in sediment extracts from a subtropical wetland: investigation using a diatom bioassay. - J. Phycology 14: 347-352.

Eisma, D. & J. Kalf, 1984. Dispersal of Zaire River suspended matter in the estuary and the Angola Basin. - Neth. J. Sea Res. 17: 385-411.

Fell, J.W. & I.M. Master, 1973. Fungi associated with the degradation of mangrove (Rhizophora mangle L.) leaves in south Florida. - In L.H. Stevenson & R.R. Colwell (eds.): Estuarine microbial ecology, pp. 455-465. Univ. South Carolina Press, Columbia, South Carolina, USA.

Golley, F.B., J.T. McGinnis, R.G. Clements, G.I. Child & M.J. Duever, 1975. Mineral cycling in a tropical moist forest ecosystem. - Univ. Georgia Press, Athens, USA. 248 pp.

Golley, F.B., H.T. Odum & R.F. Wilson, 1962. The structure and metabolism of a Puerto Rican red mangrove forest in May. - Ecology 43: 9-19.

Gotto, J.W., F.R. Tabita & C.V. Baalen, 1981. Nitrogen fixation in intertidal environments of the Texas gulf coast. - Estuar. Coastal Shelf Sci. 12: 231-235.

Gotto, J.W. & B.F. Taylor, 1976. N_2 fixation associated with decaying leaves of the red mangrove (Rhizophora mangle). Applied and Environm. Microbiol. 31: 781-783.

Harriss, R.C., B.W. Ribelin & C. Dreyer, 1980. Sources and variability of suspended particulates and organic carbon in a salt marsh estuary. - In P. Hamilton and K.B. Macdonald (eds.): Estuarine and wetland processes with emphasis on modeling, pp. 371-384. Marine Science Vol. 11, Plenum Press, N.Y.

Heald, E.J., 1969. The production of organic detritus in a south Florida estuary. Ph.D. Diss., Univ. Miami, Coral Gables.

Hicks, D.B. & L.A. Burns, 1975. Mangrove metabolic response to alteration of freshwater drainage to southwestern Florida estuaries. In G. Walsh, S. Snedaker & H. Teas (eds.): Proceedings of the international symposium on the biology and management of mangroves, pp. 238-255. Inst. Food Agric. Sci., Univ. Florida, Gainsville, Florida, USA.

Jara, R.S., 1984. Aquaculture and mangroves in the Philippines. - In J.E. Ong & W.K. Gong (eds.): Productivity of the mangrove ecosystems: Management implications, pp. 97-107. Unit Pencetakan Pusat, Univ. Sains Malaysia, Penang, Malaysia.

Jothy, A.A., 1984. Capture fisheries and the mangrove ecosystem. - In J.E. Ong & W.K. Gong (eds.): Productivity of the mangrove ecosystem: Management implications. Unit Pencetakan Pusat, Univ. Sains Malayse, Penang, Malaysia.

Kimball, M.C. & H.J. Teas, 1975. Nitrogen fixation in mangrove areas of southern Florida. - In G. Walsh, S. Snedaker & H. Teas (eds.): Proceedings of the international symposium on the biology and management of mangroves, pp. 654-660. Inst. Food Agric. Sci., Univ. Florida, Gainsville, Florida, USA.

Krishnamurthy, K., V. Sundararaj & R. Santhanam, 1975. Aspects of an Indian mangrove forest. - In G. Walsh, S. Snedaker & H. Teas (eds.): Proceedings of the international symposium on the biology and management of mangroves, pp. 88-95. Inst. Food Agri. Sci., Univ. Florida, Gainsville, Florida, USA.

Liss, P.S., 1976. Conservative and non-conservative behavior of dissolved constituents during estuarine mixing. - In J.D. Burton & P.S. Liss (eds.): Estuarine chemistry, pp. 93-130. Academic Press, New York.

Lugo, A.E., G. Evink, M.M. Brinson, A. Broce & S.C. Snedaker, 1975. Diurnal rates of photosynthesis, respiration and transpiration in mangrove forests of south Florida. - In F.B. Golley & E. Medina (eds.): Tropical ecological systems, pp. 335-350. Springer-Verlag, N.Y.

Lugo, A.E. & S.C. Snedaker, 1974. The ecology of mangroves. Ann. Rev. Ecol. Systematics 5: 39-64.

Lugo, A.E. & S.C. Snedaker, 1975. Properties of a mangrove forest in southern Florida. - In G. Walsh, S. Snedaker & H. Teas (eds.): Proceedings of the international symposium on the biology and management of mangroves, pp. 170-212. Inst. Food Agric. Sci., Univ. Florida, Gainsville, USA.

Macko, S.A. & J. Zieman, 1983. Stable isotope composition and amino acid analysis of estuarine plant litter undergoing decomposition. - Estuar. Res. Fed. Biann. Meeting, Virginia Beach.

Macnae, W., 1968. A general account of the fauna and flora of mangrove swamps and forests in the Indo-West-Pacific region. - Advances Mar. Biol. 6: 73-270.

Macnae, W., 1974. Mangrove forests and fisheries. - FAO/UNDP Indian Ocean Programme. IOFC/DEV/7434.

Mann, K.H., 1975. Relationship between morphometry and biological functioning in three coastal inlets of Nova Scotia. - In L.E. Cronin (ed.): Estuarine Research Vol. 1, pp. 634-644. Academic Press, New York.

McGill, J.T., 1958. Coastal land forms of the world. Map supplemented. - In R.J. Russell (ed.): Second coastal geography conference, 1959. Coastal Studies Inst., Louisiana State Univ. 472 pp.

Nedwell, D.B., 1975. Inorganic nitrogen metabolism in a eutrophicated tropical mangrove estuary. - Water Res. 9: 221-231.

Nickerson, N.H. & F.R. Thibodeau, 1985. Association between pore water sulfide concentrations and the distribution of mangroves. - Biogeochem. 1: 183-192.

Nixon, S.W., B.N. Furnas, V. Lee, N. Marshall, O. Jun-Eong, W. Chee-Hoong, G. Wooi-Khoon & A. Sasekumar, 1984. The role of mangroves in the carbon and nutrient dynamics of Malaysia estuaries. Proc. Symp. Mangrove Environments - Res. Managem. pp. 534-544.

Odum, E.P., 1980. The status of three ecosystem-level hypotheses regarding salt marsh estuaries: tidal subsidy, outwelling, and detritus-based food chains. - In V.S. Kennedy(ed.): Estuarine perspectives, pp. 485-495. Academic Press Inc., N.Y.

Odum, H.T. & C.M. Hoskin, 1958. Comparative studies on the metabolism of marine waters. - Univ. Texas Inst. Mar. Sci. Pub. 5: 16-46.

Odum, W.E., J.S. Fisher & J.C. Pickral, 1979. Factors controlling the flux of particulate organic carbon from estuarine wetlands. - In R.J. Livingston (ed.): Ecological processes in coastal and marine systems, pp. 69-80. Mar. Sci. Vol. 10, Plenum Press, N.Y.

Odum, W.E. & E.J. Heald, 1972. Trophic analysis of an estuarine mangrove community. - Bull. Mar. Sci. 22: 671-738.

Patterson, S.G., 1986. Mangrove community boundary interpretation and detection of areal changes on Marco Island, Florida: Application of digital image processing and remote sensing techniques. U.S. Fish Wildl. Ser. Biol. Rep. 86(10) 87 pp.

Polunin, N.V.C., 1983. The marine resources of Indonesia. - Oceanog. Mar. Biol. Ann. Rev. 21: 445-531.

Pool, D.J., A.E. Lugo & S.C. Snedaker, 1975. Litter production in mangrove forests of southern Florida and Puerto Rico. - In G.E. Walsh, S.C. Snedaker & H.J. Teas (eds.): Proceedings of the international symposium on the biology and management of mangroves. Inst. Food Agric. Sci., Univ. Florida, Gainesville.

Pool, D.J., S.C. Snedaker & A.E. Lugo, 1977. Structure of mangrove forests in Florida, Puerto Rico, Mexico, and Costa Rica. - Biotropica 9: 195-212.

Potts, M., 1979. Nitrogen fixation (acetylene reduction) associated with communities of heterocystous and non-heterocystous blue-green algae on mangrove forests of Sinai. - Oecologia (Berlin) 39: 359-373.

Prakash, A., 1971. Terrigenous organic matter and coastal phytoplankton fertility. - In J.D. Costlow (ed.): Fertility of the sea. 2. Proc. Int. Symp. Fertility Sea. Sao Paula, Brazil, pp. 351-368. Gordon and Breach Science Publishers. London.

Prakash, A., M.A. Rashid, A. Jensen & D.V. Subba Rao, 1973. Influence of humic substances on the growth of marine phytoplankton: Diatoms. - Limnol. Oceanogr. 18: 516-524.

Qasim, S.Z., 1973. Productivity of backwaters and estuaries. - In B. Zeitzechel (ed.): The biology of the Indian Ocean. Ecological studies No. 3, pp. 143-154. Springer-Verlag, Berlin.

Qasim, S.Z. & R. Sen Gupta, 1981. Environmental characteristics of the Mandovi-Zuari estuarine system in Goa. - Estuarine, Coastal and Shelf Science 13: 557-578.

Qasim, S.Z., S. Wellerhaus, P.M.A. Bhattathiri & S.A.H. Abidi, 1969. Organic production in tropical estuary. - Proc. Indian Acad. Sci. 69: 51-94.

Qasim, S.Z. & V.N. Sankaranarayanan, 1972. Organic detritus of a tropical estuary. - Mar. Biol. 15: 193-199.

Rao, V.C. & T.S.S. Rao, 1975. Distribution of particulate organic matter in the Bay of Bengal. - J. Mar. Biol. Assoc. India 17: 40-55.

Rice, D.L., 1982. The detritus nitrogen problem: new observations and perspectives from organic geochemistry. Mar. Ecol.-Prog. Ser. 9: 153-162.

Rice, D.L. & K.R. Tenore, 1981. Dynamics of carbon and nitrogen during the decomposition of detritus derived from estuarine macrophytes. - Estuarine, Coastal and Shelf Science 13: 681-690.

Roman, M.R., M.R. Reeve & J.L. Froggatt, 1983. Carbon production and export from Biscayne Bay, Florida. I. Temporal patterns in primary production, seston and zooplankton. - Estuarine, Coastal and Shelf Science 17: 45-59.

Ryan, D.R. & F.H. Bormann, 1982. Nutrient resorption in northern hardwood forests. - BioScience 32: 29-32.

Scholl, D.W. & M. Stuiver, 1967. Recent submergence of southern Florida: A comparison with adjacent coasts and other eustatic data. - Geol. Soc. Am. Bull. 78: 437-454.

Shah, N.M., 1973. Seasonal variation of phytoplankton pigments and some of the associated oceanographic parameters in the Lacadive Sea off Cochin. - In B. Zeitzchel (ed.): The biology of the Indian Ocean, pp. 175-185. Springer-Verlag, New York.

Sundararaj, V. & K. Krishnamurthy, 1973. Photosynthetic pigments and primary production. - Current Science 42: 185-189.

Tabb, D.C., D.L. Subrow & R.B. Manning, 1962. The ecology of northern Florida Bay and adjacent estuaries. - Fla. Bd. Conserv., Tech. Ser. No. 39. 79 pp.

Tanoue, E. & N. Handa, 1979. Distribution of particulate organic carbon and nitrogen in the Bering Sea and northern North Pacific Ocean. - J. Oceanogr. Soc. Japan 35: 47-62.

Teixeira, C., J. Tundsi & J.S. Ycaza, 1969. Plankton studies in a mangrove environment. VI. Primary production, zooplantkon standing stock and some environmental factors. - Internat. Rev. Gesamten Hydrobiol. 54: 289-301.

Thom, B.G., 1982. Mangrove ecology - a geomorphological perspective.- In B.F. Clough (ed.): Mangrove ecosystems in Australia, pp. 3-17. Austr. Nat. Univ. Press, Canberra.

Tomlinson, P.B., 1986. The botany of mangroves. Cambridge Univ. Press, Cambridge.

Tsu-Chang Hung, Shian-Ho Lin & Aileen Chuang, 1980. Relationship among particulate organic carbon, chlorophyll a and primary productivity in the seawater along the northern coast of Thaiwan. - Acta Oceanog. Taiwanica 11: 70-88.

Turner, R.E., 1977. Intertidal vegetation and commercial yields of penaeid shrimp. - Trans. Am. Fish. Soc. 106: 411-416.

Twilley, R.R., 1982. Litter dynamics and organic carbon exchange in black mangrove (Avicennia germinans) basin forests in a southwest Florida estuary. - Ph.D. Diss., Univ. Florida, Gainesville.

Twilley, R.R., 1985. The exchange of organic carbon in basin mangrove forests in a southwest Florida estuary. - Estuarine, Coastal and Shelf Science 20: 543-557.

Twilley, R.R., A.E. Lugo & C. Patterson-Zucca, 1986. Production, standing crop, and decomposition of litter in basin mangrove forests in southwest Florida. - Ecology 67:670-683.

Untawale, A.G., T. Balasubramanian & M.V.M. Wafer, 1977. Structure and production in a detritus rich estuarine mangrove swamp. Mahasagar-Bulletin Nat. Inst. Oceanogr. 10:173-177.

Untawale, A.G., S. Wafer & T.G. Jagtap, 1982. Application of remote sensing techniques to study the distribution of mangroves along the estuaries of Goa. - In B. Gopal, R.E. Turner, R.G. Wetzel & D.F. Whigham (eds.): Wetlands: Ecology and management, pp. 51-67. Nat. Inst. Ecol. Internat. Sci. Publ.

Verlencar, X.N. & S.Z. Qasim, 1985. Particulate organic matter in the coastal and estuarine waters of Goa and its relationship with phytoplankton production. - Estuarine, Coastal and Shelf Science 21: 235-242.

Vitousek, P.M., 1982. Nutrient cycling and nutrient use efficiency. - Am. Nat. 119: 553-572.

Vitousek, P.M., 1984. Litterfall, nutrient cycling, and nutrient limitation in tropical forests. - Ecology 65: 285-298.

Walsh, G.E., 1967. An ecological study of a Hawaiian mangrove swamp. - In G.H. Lauff (ed.): Estuaries, pp. 420-431. Am. Ass. Advance. Sci. 83. Washington, D.C.

Watson, J.G., 1928. Mangrove forests of the Malay peninsula. - Malayan Forest Record 6: 1-275.

Welsh, B.L., R.B. Whitlatch & W.F. Bohlen, 1982. Relationship between physical characteristics and organic carbon sources as a basis for comparing estuaries in southern New England. - In V.S. Kennedy (ed.): Estuarine Comparisons, pp. 53-67. Academic Press, N.Y.

Wharton, C.H. & M.M. Brinson, 1979. Characteristics of southeastern river systems. Strategies for protection and management of floodplain wetlands and other riparian ecosystems. - In R.R. Johnson $ J.F. McCormick (eds.): Symp. Proc. U.S. Dept. Agri., Washington, D.C., pp. 32-40. GTR-WO-12.

Wium-Anderson, S., 1979. Plankton primary production in a tropical mangrove bay at the southwest coast of Thailand. - Ophelia 18: 53-60.

Zuberer, D.A. & W.S. Silver, 1978. Biological nitrogen fixation (acetylene reduction) associated with Florida mangroves. - Appl. Environm. Microbiol. 35: 567-575.

PRODUCTION AND TRANSPORT OF ORGANIC MATTER
IN MANGROVE-DOMINATED ESTUARIES

Proserpina L. Gomez
Biology Department
Mindanao State University
Marawi City, Philippines

1. INTRODUCTION

Past works (e.g. Golley et al., 1962; Odum and de la Cruz, 1967) suggested a net export of detritus from mangrove, swamps and salt marshes. Studies were conducted on mangroves in the Philippines to determine their importance as a source of organic matter for coastal waters.

The mangrove vegetation within the reserves of the Mangrove Research Center of the Philippine Forest Research Institute in Pagbilao, Quezon Province, the study site, represents eight families: Rhizophoraceae, Avicenniaceae, Sonneratiaceae, Combretaceae, Meluceae, Euphorbiaceae, Aegeceraceae and Rubiaceae. Associated plant species belong to seven families: Herculiaceae, Arecaceae, Moraceae, Caesalpinaceae, Acanthaceae, Pteridiaceae and Aizoaceae. The Center for Developmental Studies (Anonymous, 1978) reported the temperature, humidity, wind, cloudiness, mean monthly precipitation, salinity, pH, dissolved oxygen (DO) and primary productivity of the waters.

2. LITTERFALL PRODUCTION

Using the modified litter trap method of Newbould (1970), litter was collected at monthly intervals for a period of one year (Fortes et al. 1982). The litter was sorted to species, as leaves, twigs, fruits and other plant parts. Each fraction was dried to constant weight ($105^{\circ}C$, 3 days) to determine organic matter content.

Table 1 shows the total litter production and Figure 1 the monthly variation for the area sampled (500 m^2) for one year. The table also shows the percent contribution by the different components. The dominant species in the site were Avicennia officinalis, Scyphiphora hydrophyllacea, Ceriops decandra and Osbornia octodonta. A. officinalis

Lecture Notes on Coastal and Estuarine Studies, Vol. 22
B.-O. Jansson (Ed.), Coastal-Offshore Ecosystem Interactions.
© Springer-Verlag Berlin Heidelberg 1988

SPECIES	LEAVES	BUD SCALES	FLOWERS	SEEDS/ FRUITS	BARK WOODY ITEMS	TOTAL
A. officinalis (api-api)	197,037.10	-	17,446.07	31,357.78	15,279.53	261,120.48
	75.46%	-	6.68%	12.01%	5.85%	
	3.94	-	0.34	0.62	0.30	5.22
C. decandra (malatangal)	162,412.79	7,403.60	1,680.14	34,772.18	8,018.59	214,287.30
	75.79%	3.45%	0.78%	16.23%	3.74%	
	3.2	0.15	0.03	0.69	0.16	4.28
S. hydrophyllacea (nilad)	196,293.16	-	4,730.78	73,213.68	3,558.55	277,814.17
	70.66%	-	1.70%	26.36%	1.28%	
	3.9	-	0.09	1.46	0.07	5.55
O. octodonta (tawalis)	189,826.33	-	251.27	2,213.19	2,213.19	211,261.69
	89.97%	-	0.12%	1.05%	1.05%	
	3.8	-	0.005	0.04	0.04	4.22

Table 1. Total litterfall (g/500 m^2/yr) of the species in the study sites at Pagbilao, Quezon. Also showing percent distribution among components. Lowermost figures for each species refer to values in t/ha/yr.

produced 522 g dry weight/m^2, $\underline{S}$. $\underline{hydrophyllacea}$ 556 g/m^2, $\underline{C}$. $\underline{decandra}$ 429 g/m^2 and $\underline{O}$. $\underline{octodonta}$ 423 g/m^2.

For most of the components, except seeds and fruit, $\underline{A}$. $\underline{officinalis}$ had a significantly higher production compared to the other species followed by $\underline{S}$. $\underline{hydrophyllacea}$. $\underline{A}$. $\underline{officinalis}$ trees were the oldest, tallest, and had larger trunks compared to the other species, indicating that mature vegetation does produce more litter than young trees (Christensen 1978). However this is not true for $\underline{S}$.$\underline{hydrophyllace}$.

Leaves comprised a large percentage of total litter for all species- from 71% for $\underline{S}$. $\underline{hydrophyllacea}$ to 90% for $\underline{O}$. $\underline{octodonta}$. These figures are lower than those given Golley et al. (1962) in their study of $\underline{Rhizophora\ mangle}$ perhaps because stems were included in their measurements. Leaf fall was continuous throughout the period of study. Leaf fall from $\underline{C}$. $\underline{decandra}$ and $\underline{S}$. $\underline{hydrophyllacea}$ was fairly constant throughout the year, more from $\underline{A}$. $\underline{officinalis}$ and $\underline{O}$.$\underline{octodonta}$. The first peak for $\underline{A}$. $\underline{officinalis}$ was due to heavy leaf fall in February/March. This could be an initial response of the species to a moderate increase in temperature at the onset of the dry months of April and May in order to prevent excessive evapotranspiration. The second peak (July/August) was due to heavy seed/fruit fall after the flowering months which started in April.

3. ORGANIC MATTER TRANSPORT

Mangroves present a situation where surplus production of plant materials drains into a body of water. Hence, the most important source of energy in the coastal waters is not the phytoplankton nor the benthic algae but detritus originating from the trees and the mangrove-associated species. Materials transported by the tides included various plant components and varied animal parts ranging from wings to feces of large invertebrates. However, leaves, seeds and twigs of the mangroves made up the bulk of the moving materials. Studies are needed to determine whether the detritus remains on the ground floor or is brought to the coastal waters.

A study on this matter was conducted for a period of one year (Gomez, 1983). The disturbed area under study included 4-5 hectares of formerly Avicennia-dominated site which was cleared by the Bureau of Forest Development for study purposes. Most of the cut-down parts in different

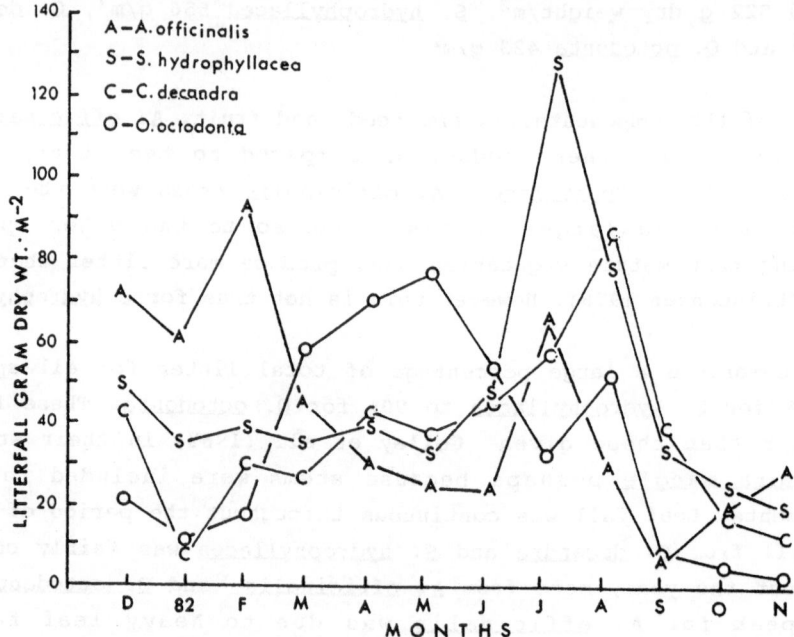

Fig. 1. Litter produced monthly by the four species.

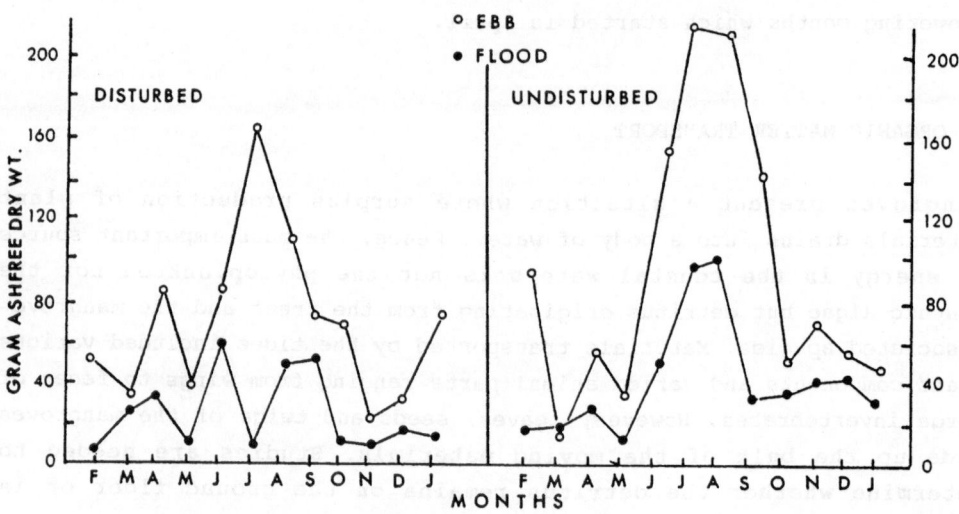

Fig. 2. Graph showing organic matter (AFDW) transported
monthly by tidal flows in disturbed and undisturbed areas.

sizes and degrees of decomposition were still in the area and they made
up the bulk of transported materials, in addition to the freshly-fallen
leaves, twigs, seeds and other plant parts. As to how long the
disturbed area will contribute to the productivity of the marine
environment at a degree that is comparable to the contribution of the
undisturbed area is still open to investigation.

A study of this matter was conducted for a period of one year. To
determine the effects of man-made disturbances, the tidal transport of
organic matter in a relatively disturbed and undisturbed areas was
compared. Organic materials were separated into size classes by
straining successively a volume (totally 264 l) of flood and ebb waters
over a period of one hour at each tidal flow by stretching a 1 mm net
across the water channels draining the established areas of the
mangrove swamp. Course particulate organic matter (CPOM, size greater
than 15 mm), medium particulate organic matter (MPOM, greater than 1 mm
but less than 15 mm), fine particulate organic matter (FPOM, retained
by 0.45 μ millipore filter and dissolved organic materials (DOM).

Data collected for particulate organic detritus carried by the tidal
waters over a one-year period are tabulated in Table 2 and graphically
presented in Figure 2. It shows that peak of transport for particulate
organic matter in the two areas occurred during ebb flows. The ebb
waters had a higher mean value of 84.12 g/tidal cycle compared to the
value of 31.01 g/tidal cycle for flood waters. This indicates a net
export of materials from the estuary to the open seas, a finding
similar to that of Golley et al. (1962) and Odum and de la Cruz (1967)
in a Puerto Rican mangrove and a Georgia salt marsh, respectively.

		EBB	FLOOD
DISTURBED	Dry months	58.43	24.97
	Wet months	85.49	26.58
UNDISTURBED	Dry months	57.75	23.05
	Wet months	113.32	49.44

Table 2. Average organic matter transported by tidal flows in
disturbed and undisturbed areas during the period from February,
1982 to January, 1983 showing seasonal variations (in grams ashfree dry wt.).

More organic materials are transported during the wet months (74.08 g/tidal cycle) than during the dry months (41.05 g/tidal cycle). Monthly values show that greatest transport of detritus in the two areas was during the months of July to August. This higher organic export is related to the peak of litterfall production for some mangrove species in the reserves (see Table 1). Apparently, the materials transported come from the mangroves which continually produce litter throughout the year. Though some of the detritus stay on the ground floor, a great bulk of this is exported to the nearshore areas.

The degree of disturbance did not significantly affect the transport of particulate organic materials. Although the undisturbed site gave a higher mean value of 65.89 g/tidal cycle compared to that of the disturbed area which was 49.24 g/tidal cycle, the difference was not statistically significant.

A comparison of the energy values showed that exported particulate materials from the disturbed area have higher caloric values compared to exported materials from the undisturbed site. Detritus is utilized as energy source by a number of consumers. From the study, export of materials from the two study sites are not statistically significant and energy values are higher for materials from the disturbed site. For the past decade, man has been alarmed by the effects of disturbances on ecosystems. For mangrove communities, there is a need to ascertain and conduct more studies on the impact of forest cutting on secondary productivity as the present study does not suggest a significant loss of energy source from the disturbed site. The obvious problem comes in only when there is complete denudation of the forest.

The dissolved organic matter (DOM) fraction during ebb flows (6.54 g l^{-1}) consistently exceeded the other fractions throughout the period of study. This suggests that litter in the mangroves are decomposed in situ and the less particulate debris are exported to adjacent swamps, a conclusion similar to that of Lugo and Snedaker (1974). The availability of these organic size fractions is important since most organisms are size-selective rather than food selective.

4. CONCLUSIONS

The Philippines Natural Resources Management Center in 1978 estimated the remaining mangroves of the country at 146,140 hectares. Continuous exploitation for firewood, charcoal, forest products extraction and fishpond and harbor development can only diminish mangrove contribution to the biological productivity of adjoining coastal waters.

ACKNOWLEDGEMENTS

The author is grateful to Ms. Rosario Reyes-Bocobo for the permission to use some data; Prof. Miguel D. Fortes of the UP Department of Botany and the Philippine Council for Agriculture and Resources Research Development.

REFERENCES

Anonymous, 1978. Ground inventory and assessment of selected mangrove areas in the Philippines Center for Developmental Studies, Alumni Center, University of the Philippines, Diliman Quezon City.

Christensen, B., 1978. Biomass and primary productivity of R. apiculata Bl. in a mangrove in Southern Thailand. - Aq. Bot. 4: 43-52.

Duke, N.C., S.S. Bunt & W.T. Williams, 1981. Mangrove litter fall in north-eastern Australia. 1. Annual totals by component in selected species. - Aust. J. Bot. 29; 547-553.

Fortes, M.D., P.L. Gomez, R.C. Reyes & G.H. Tiopes, 1982. Litter production, and decomposition, nutrient transport and phenology in a mangrove forest at Pagbilao, Quezon. - A paper read during the mangrove convention/workshop at Covelandia, Kawit, Cavite. Metro Manila.

Golley, F.B., H.T. Odum & R.F. Wilson, 1962. The structure and metabolism of a Puerto Rican mangrove forest in May. - Ecology 43:9-19.

Gomez, P.L., 1983. Transport of organic matter in disturbed and undisturbed mangrove ecosystems at Pagbilao, Quezon. M.S. Thesis., Univ. Philippines in Diliman, Quezon City.

Lugo, E.A. & S.C. Snedaker, 1974. The ecology of mangroves. - Ann. Rev. Ecol. Syst. 5: 39-65.

Newbould, P.J., 1970. Methods for estimating primary production of forests, pp. 32-35. Oxford and Blackwell Scientific Publications.

Odum, H.T. & A.A. de la Cruz, 1967. Particulate organic detritus in a Georgia salt-marsh-estuarine ecosystem. - In G.H. Lauff (ed.): Estuaries, pp. 383-388. AAS Publ. No. 83, Washington, D.C.

ENERGY FLOW THROUGH FJORD SYSTEMS

T.H. Pearson
Dunstaffnage Marine Research Laboratory
P.O. Box No. 3, Oban, Argyll, Scotland

The environmental characteristics of fjordic ecosystems are briefly summarized and the importance of basin morphometry and seasonality are emphasized. General system variability is discussed in terms of four defined fjordic types, and energy flow within subsystems is considered. The information available on mass balance studies in fjordic systems is assessed in terms of nitrogen and carbon budgets and used comparatively to assess the effects of latitude, stagnation and salinity on general energy flow through such systems. Some general conclusions as to the net imports and exports of energy to fjords are drawn.

1. INTRODUCTION

Fjords may be defined as steep sided, silled, overdeepened estuarine basins which, being high latitude features, show strong seasonal variation in their biogeochemical properties (Pearson et al., in press). They therefore share common features with other estuarine systems and with all high latitude marine areas, but their unique topographical characteristics distinguish them as a set of definable systems. Descriptive reviews of fjordic ecology and ecosystems have recently been presented by Pearson et al. (in press) and Pearson (in press), and many aspects of fjord oceanography and biology are discussed in Freeland et al. (1980) and Syvitski et al. (1986). Here a brief discussion of the environmental characteristics responsible for fjordic variability will be presented and the extent of such variation will be examined in the context of a simple model of energy flow through fjordic systems. The information available from mass balance studies of nutrient and carbon flows through fjord systems will be summarized and some suggestions made as to neglected research areas.

2. ENVIRONMENTAL CHARACTERISTICS

The principal forcing functions acting upon ecosystems may be conveniently defined within two groups: 1) those stemming from topographic features, and 2) those related to seasonality. While many, if not most, functions are subject to both topographic and latitudinal influences, initially it is convenient to summarize them separately. The principal physical and chemical characteristics affecting fjord ecology are illustrated in Fig. 1.

2.1. The influence of topography

A "typical" fjord may be said to combine a high aspect ratio (length/breadth) with a glacially overdeepened basin behind a comparatively shallow sill. This combination of characteristics results in a basic system conveniently divided into two sub-systems at the level of the sill; 1) a hydrographically complex upper system, dominated by strong vertical and horizontal temperature and salinity gradients overlying 2) a much more stable basin system below sill level. The surface system is usually dominated by an estuarine circulation with a surface fresh or brackish outflow compensated by a saline inflow at sill level, separated by an intermediate mixed layer. The basin system is subject to deep water renewal which may be driven by aeolian or tidal forces, or be a consequence of seasonal changes in offshore density structures (Gade and Edwards, 1980). The frequency of such renewals and hence the length of stagnation of fjordic basin waters is of fundamental importance to fjordic ecology. The mixing processes consequent on basin water overturn are critical factors controlling the survival and development of all fjordic populations. The distribution of nutrients and carbon between the two major compartments of the fjordic system is largely a result of such turbulent mixing. Diffusion and passive sedimentation across a pycnocline are not normally of sufficient magnitude to sustain highly productive populations. Such populations are more usually maintained by the active vertical and horizontal transfer of such materials. Other influences on the periodicity and strength of the mixing forces are sill depth and the volume of fresh water inflow. While the latter is partially a function of catchment area, it is strongly influenced by seasonal factors and will be discussed in that context. Sill depth is critical in controlling deep-water renewal, and hence basin mixing, in that the shallower the sill the smaller and more infrequent will be the advective inflows of external water into the basin. Deep sills permit basin:offshore water exchange, even along relatively weak density gradients. Shallow sills encourage intermittent stagnation of basin water which, in extreme cases, may result in deep water anoxia (see below).

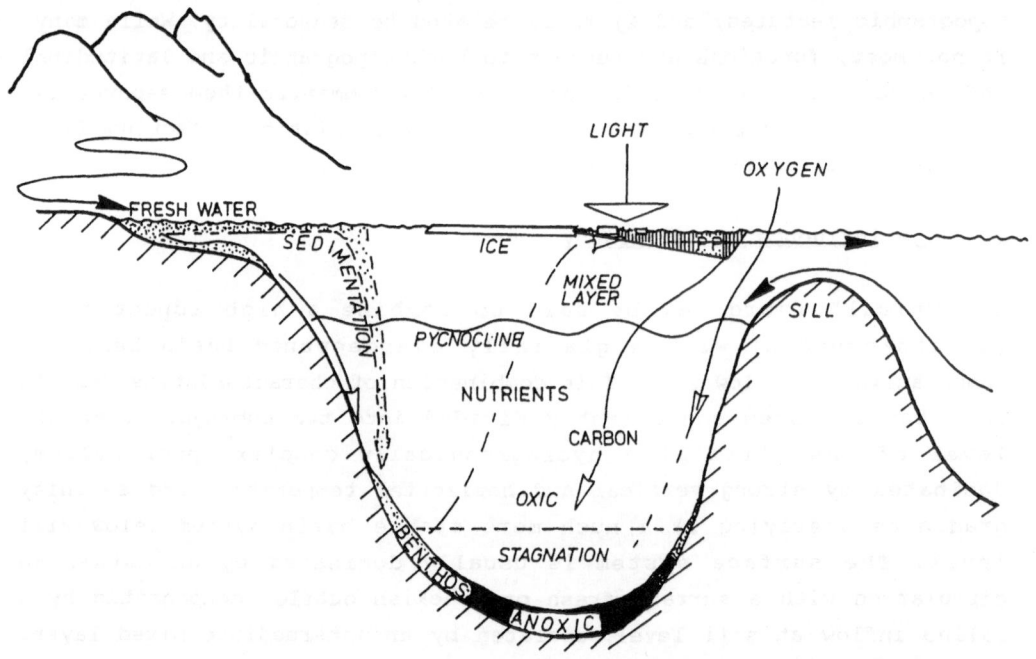

Fig. 1. The principal physical and chemical characteristics affecting fjord ecology. Note the two layer estuarine circulation (black arrows) with freshwater outflow at the surface compensated for by saline deep water upwelling over the sill. Particulate inputs (open arrows) are high in the inner ends of fjords. These may be predominantly mineral from glacial sources in high latitudes, or conversely have a high carbon content from fluvial sources in boreal areas (from Pearson, in press).

2.2 The influence of seasonality

Strong seasonal variability affects nearly all the functions which characterize fjordic ecosystems. Such seasonal signals increase in strength with increasing latitude and have a profound effect on the ecology of polar fjords. Thus productive seasons are short in high latitudes and the effects of icing, and of ice melt and associated turbidity, are particularly marked. Phytoplankton production is controlled by incident light levels in the euphotic zone and this is, in turn, dependent on both ice cover and turbidity. High turbidity originating in glacial melt-water, is a summer feature of many arctic and subarctic fjords and not only reduces light penetration in surface waters, but greatly increases sedimentation rates of mineral particulates, thus radically affecting the benthos. Seasonally episodic freshwater inputs control the near-surface circulation in fjords and impose summer stratification. Melting low-salinity sea ice reinforces such stratification, as do rising summer surface temperatures. Thus, all the seasonal variables contribute to the pycnocline, which inhibits nutrient circulation and thus sharply limits phytoplankton production, although tidal and wind-driven mixing may greatly modify such effects in some areas.

		LOW LATITUDE	HIGH LATITUDE
HYDRODYNAMIC ENERGY INPUT	HIGH	High Biomass High Diversity High Productivity NUTRIENT LIMITED	High Biomass Medium Diversity Low Productivity CARBON LIMITED
	LOW	Low Biomass Low Diversity Intermittently high Productivity CARBON SINKS	Low Biomass Low Diversity Low Productivity NUTRIENT SINKS

Table 1. Generalised characteristics of fjord ecosystems as a function of their overall physical conditions.

3. SYSTEM VARIABILITY

3.1. General

While it is obvious that fjords are as infinitely variable as all other natural systems their special topography, modified by the seasonal effects briefly outlined above, allows some categorization of expected ecosystem types. Pearson (1980) and Pearson et al. (in press) suggested that four basic ecosystem types could be distinguished, dependent on interacting gradients of latitudinal effects and hydrodynamic energy input levels (Table 1). Such types may be regarded as the idealized extremes. Thus, fjords with high hydrodynamic energy inputs tend to be limited by nutrient availability in lower latitudes, where both autochthonous and allochthonous carbon sources are relatively rich. Conversely in high latitudes carbon limitation assumes greater importance. Low hydrodynamic energies, resulting in undisturbed fjordic basins, tend towards anoxic carbon sinks in their bottom waters in lower latitudes, where surface productivities are high. At higher latitudes where ice cover inhibits primary production and terrestrial carbon sources are similarly limited, nutrient pools may exist throughout the system. The general ecological consequences of these variations are summarized in Table 1. In order to examine these in more detail it is convenient to consider a general model of energy flow through fjordic systems (Fig. 2). This emphasizes the division of such systems into upper and lower compartments at sill level and, for the sake of clarity, delineates only the principal pathways and components

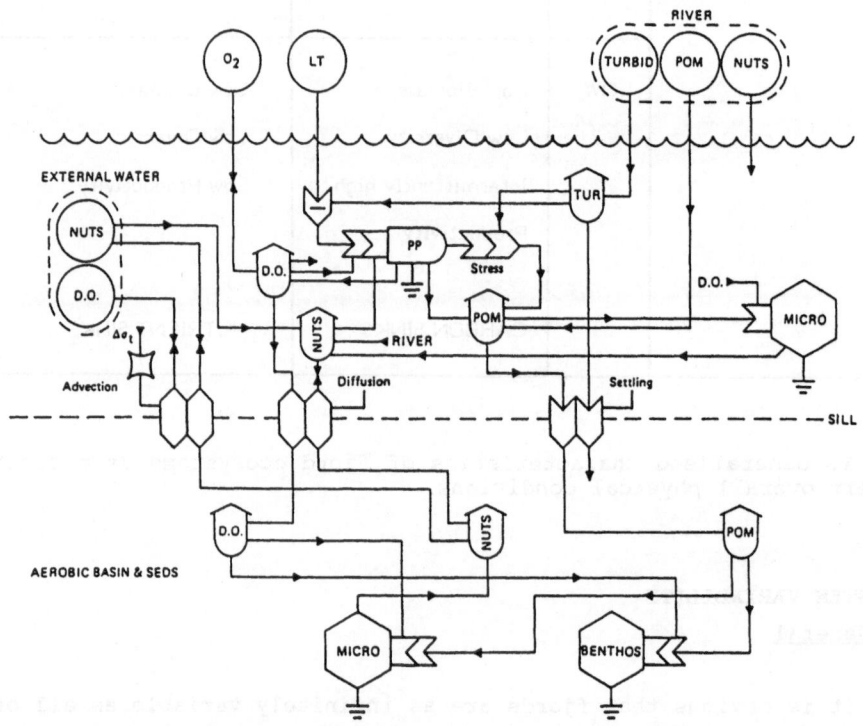

Fig. 2. A general model of energy flow through fjordic systems. Only the principal pathways and components of primary and secondary production are shown (from Pearson et al., in press). LT: light input. TUR and TURBID: turbidity. POM: particulate organic material. NUTS: nutrient inputs. D.O.: dissolved oxygen. PP: primary production. MICRO: microbial populations.

of primary and secondary production. Inputs to the upper system are from fluvial sources or from external water beyond the sill, together with light and oxygen. In boreal fjords, subjected to strong mixing forces, the majority of these inputs are maximized, when assessed over an annual cycle, with the exception of turbidity. Regular mixing of the upper and basin waters ensures a rich supply of carbon, nutrients and oxygen to the basin system. When those functions leading to basin mixing are reduced, that is, low tidal exchange, reduced fluvial input, shallow sills, etc., oxygenation of the basin waters is minimized and advective flows are reduced, but carbon inputs through settlement remain high. These conditions often result in anoxic conditions in deeper basin waters, thus reducing or eliminating all but microbial production. In arctic fjords with high hydrodynamic energy inputs, particulate organic material and nutrients from fluvial sources are minimized and low light levels, for all but a few months, restrict primary production to a single sharp peak. Thus, carbon and nutrient levels to the basin compartment are reduced and production tends to be carbon limited. Reduction of the mixing forces in such areas reduces or minimizes all inputs to the basin water, resulting in a relatively impoverished system.

3.2. Compartmentalisation - division into sub-systems.

The general fjordic ecosystem may be conveniently divided into a set of coupled sub-systems whose internal linkages are generally stronger than their coupled cross linkages. Such sub-systems could be variously defined and refined to accord with the increasingly detailed knowledge now available on energy flow through such systems. For the present purposes three divisions will be considered viz. the pelagic, the phytal and the benthic sub-systems. This accords with the most general divisions described by Jansson et al. (1984) for sub-systems in the Baltic Sea. Such broad divisions necessarily encompass a wide variety of habitat and community types both within a single fjord and, more obviously, between fjords. However, in terms of describing and cate-gorising comparative energy flux such generality is essential. The environmental variables discussed above influence each sub-system some-what differently. Pearson (in press) reviewed the available information on the effect of such variability on the general productivity of fjordic subsystems (Fig. 3). This suggested that the high productivity of boreal mixed basins was reflected through each subsystem but that in stagnant basins benthic productivity was greatly reduced below the

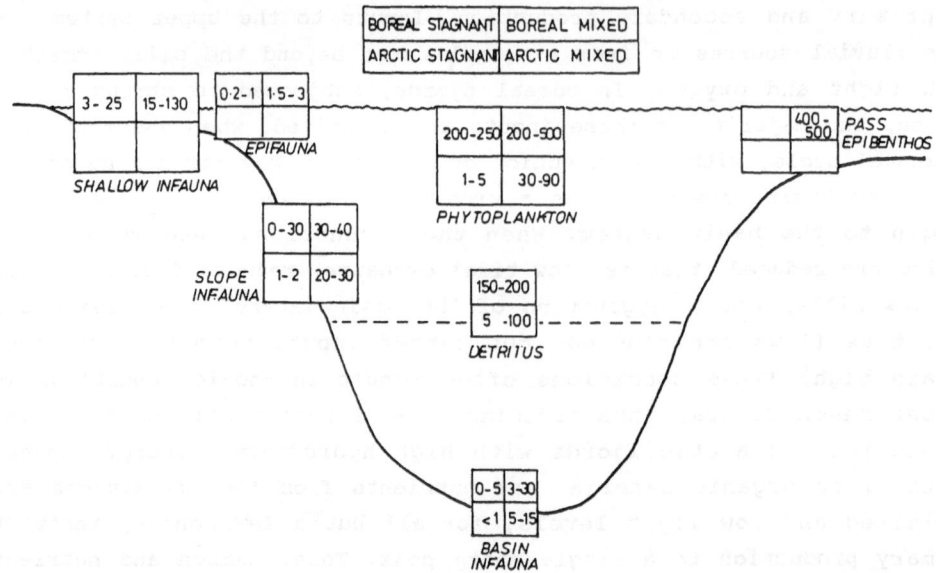

Fig. 3. Summary of available information on the annual productivity of
the principal subsystems within fjordic ecosystems. All values are in g
C m^{-2} yr^{-1}· The four extreme fjordic types are contrasted where
information is available (from Pearson, in press).

pycnocline. The same pattern is common to both high and low latitude
fjords, although information from the former is very limited. In a few
cases the comparative productivity of fjordic subsystems has been
examined. Thus Möller et al. (1985) compared benthic energy flow in
shallow soft bottom areas of fjords in western Sweden. They showed that
infaunal annual production varied between and within habitats with both
physical variables (temperature) and biotic pressures (recruitment
strength, available space and predation). Production of epibenthic
carnivores was stable within habitats, but was highest in vegetated
areas and progressively lower in semi-exposed, exposed and sheltered
habitats. In these shallow inshore habitats much of the production in
the more exposed areas was dependent on phytoplankton produced offshore
and transported by onshore currents as either whole cells or detritus.
Thus infaunal production of up to 354 g ashfree dry weight m^{-2}·yr^{-1} was
considerably greater than the estimated primary production of 180-230
gC·m^{-2}·yr^{-1}. Moreover a significant part of the summer and autumn
production is later transferred to deeper waters through migration and
there provides food for coastal fish populations. It was estimated that
there was an export to deeper waters through the higher carnivores
(fish) of about 90% of the total epibenthic production of the inshore

areas. Thus the subsystem as a whole was strongly coupled to other parts of the system. Energy flows were dominated by different pathways in the different habitats. In vegetated areas detritus formed the principal food source for small epifaunal crustaceans which in turn were taken by the epibenthic predators. In semi-exposed areas, phytoplankton supported filter feeding organisms which, together with deposit feeders, provided the prey for epibenthic carnivores. Nevertheless, epibenthic production levels were similar in both habitats. The effect of variable salinity on subsystem production was also examined. Decreasing salinities were shown to have a progressively depressing effect on the productivity of both infauna and epifauna. Similarly detailed analyses of energy flow through other fjordic subsystems are lacking, but some idea of the structure and variabilty of such flows may be obtained from the studies of nutrient and carbon budgets in fjords discussed below.

4. MASS BALANCE STUDIES

Synoptic studies of fjordic systems are not numerous and those attempting complete or partial budgets for either nutrients or carbon are even fewer. Nevertheless, there is sufficient information from boreal fjords to illustrate the comparative difference in the budgets for stagnant and mixed systems. On the other hand, information from high latitude systems is sparse to non-existant.

4.1 Nutrient budgets

The few studies that have attempted nutrient budgeting in fjords tend to suggest that well-mixed fjords are net exporters of nutrients to coastal waters. Thus Wilmot et al. (1985), in a study of the Himmerfjärd in the central Baltic, showed that there was a net export of nitrogen (N) from the system. A small net input of phosphorous (P) to the fjord was greatly exceeded by loss to the sediments. This area was greatly enriched by nutrient inputs of both N and P from a sewage treatment plant. In studies of a similarly enriched fjord, the Bunnefjord, in Norway, Skei and Melsom (1982) found that there was also a net accumulation of P in the sediments at about 250 mg P $m^{-2} \cdot yr^{-1}$, a figure very similar to the accumulations noted in deep Baltic sediment by Holm (1978). In both areas the bottom waters and sediment were anoxic. In an enriched west coast Scottish fjord Duff (1981) and

Pearson (1982) estimated that a net flux of 8 g $N \cdot m^{-2} \cdot yr^{-1}$ from the bottom sediment was two orders of magnitude greater than that required to maintain the level of primary production in the system, implying that the excess was exported from the fjord. In that tidally flushed system nutrient regeneration from the sediments was thought to maintain primary productivity levels. Davies (1975) in a study of a well-mixed fjordic embayment on the Scottish west coast found that nutrient regeneration from the sediment was sufficient to maintain primary productivity when calculated over the annual productive cycle. The release of nutrients from the sediments tend to be sporadic in that system, and was dependent on wind-driven turbulence. Whether the episodic nature of such nutrient release events leads to subsequent flushing of some nutrients from the system was not estimated. Wassmann and Aadnesen (1984) have assessed the supply, regeneration and distribution of nutrients in a complex poll and fjord system in western Norway. The inner polls of the system are enriched by nutrients from terrestrial sources, much of which were found to be rapidly flushed to outer parts of the system. In the outer fjord areas upwelling and nutrient regeneration were more important nutrient sources than landbased supplies. Nutrient exchange in an unenriched fjordic system has been assessed by Burrell (in press) who found that in Boca de Quadra, a S.E. Alaskan fjord, the exchange of silica (Si) was approximately in balance with the external coastal waters but that significant amounts of N were exported from the system during deep water renewal in spring. During the productive summer months, when deep water exchange was minimal, net N import from the coastal areas took place. The importance of deep water renewal in maintaining nitrogen flux in fjordic basins was emphasized by Takahashi et al. (1977) who ascribed the intermittent summer plankton blooms in Saanich Inlet on the British Columbian coast, to occassional deep water renewal events upwelling nutrient rich water to the surface. Wassmann and Aadnesen (1984) discuss the relative roles of "new" and recycled nutrients (sensu Dugdale and Goering, 1967) in different types of polls and fjords. They suggest that such distinctions are blurred in polls, where the euphotic zone extends below the oxycline and primary production can be partially dependent on "new" production in the nutrient rich bottom water. In general the available literature suggests that in boreal well-mixed fjordic systems the emphasis is on relatively high levels of primary productivity sustained by nutrient renewal from the bottom waters. Such systems may export considerable amounts of nutrients to adjacent coastal waters, particularly if enriched by large terrestrial inputs. Conversely stagnant boreal fjords appear to be nutrient sinks.

There appears to be little information on nutrient cycling through polar fjords, but one or two interesting speculations may be suggested. Thus Burrell (in press) suggests that terrigenous nutrient input to high latitude fjords is low when compared to that of boreal fjords. However, Sugai and Burrell (1984) describe an interesting loop in the nutrient cycle of some Alaskan fjords. They suggest that high nitrogen and phosphate levels in the river outflow to Wilson Arm, a S.E. Alaskan fjord, in the late summer are attributable to the death and decay of large numbers of spawned pink and chum salmon in the head waters of the rivers. How common or generally important such episodic nutrient inputs are to high latitude fjord systems is a matter of speculation, but the influence of migratory populations on energy flow through these systems may be large.

AREA	KILLIARY HARBOUR		LOCH EWE		FANA-FJORD		PUGET SOUND		BOCA DE QUADRA		RESURRECTION BAY	
LATITUDE (N)	53°		58°		60°		48°		55°		60°	
ENERGY INPUTS gC m^{-2}yr^{-1} — PHYTOP.	179		111		416		465		145		230	
IMPORTS	157				25				41		35	
%of INPUTS	%TC	%PP	%TC	%PP	%TC	%PP	%TC	%PP	%TC	%PP	%TC	%PP
PRODU-CTION — ZOOP.									25	2		
CTION — BENTHIC					8	7			2	3		
RESPIR- — PELAGIC	62	117									32	36
-ATION — BENTHIC				31			8			66	27	30
SEDIMENTATION	19	35		27	23	22	5					
EXPORT	19	35			8	8						
BURIAL					1	1		13	20	25	3	3

Table 2. Summary of carbon budget information from six differing but well flushed fjords in boreal areas of Europe and North America.

4.2. Carbon budgets

As with nutrient budget studies more information is available on carbon flows through boreal fjordic systems than for polar systems. Recently a number of attempts have been made to produce carbon budgets for fjords in both Europe and North America. Table 2 summarizes the data available for six systems in a comparative manner. The exact comparability of the data must be open to question because of the wide variety of collection and analytical methods used in the various studies. To allow reasonable comparability the various budget compartments have been recalculated as proportions of the total primary productivity and total carbon input figures (when carbon input figures were available). The various systems compared have been chosen to represent a range of contrasting fjordic environments, although in terms of the quadripartite division of fjord types alluded to above they can all be categorised as boreal well-mixed systems. The three European systems included are Killiary Harbour, Western Ireland described by McMahon and Patching (1984); Loch Ewe in western Scotland, described by Davies (1975); and Fanafjord in western Norway, described by Wassmann (1984). Killiary Harbour is the most southerly of these (53 degrees North), is relatively shallow and has a high fresh water input from two rivers. Input of dissolved and particulate carbon from these sources was roughly equivalent to phytoplankton production. Phytoplankton respiration accounted for about 60% of the total carbon input with 20% being lost to the system over the sill and 20% being sedimented to the benthic systems. No estimates of benthic utilization or burial were made. Pelagic respiration accounted for more than the total primary production in this system, indicating that overall productivity was enhanced by the fluvial inputs. In Loch Ewe a carbon budget was attempted for a small subsidiary embayment of the main Loch. Fresh water inputs to this area are low, thus imports from terrestrial sources, though not measured, are probably slight. It was found that benthic production accounted for 30% of the primary produced carbon, a figure that was broadly equivalent to the measurable amount of material sedimenting from the water column. The Fanafjord is a branch of a complex fjordic system south of Bergen. A detailed carbon budget has been essayed in this system which has a fresh water input equivalent, on an annual basis, to only 20% of the total fjord volume. Carbon imports (sedimenting particulate organic carbon (POC)) were only 6% of the total primary production, but were 45% of the fraction of planktonic production estimated as sedimenting out (22%). It was estimated that 8% of the total carbon input was exported from the system, and only about 1% was accumulated in the sediments.

The three North American fjordic areas selected are Puget Sound, Washington, (Baker et al., 1985) and Boca da Quadra in S.E. Alaska and Resurrection Bay in south-central Alaska, both described by Burrell (1983). Puget Sound is an extensive and complex fjordic system many parts of which have been subjected to intensive study. Baker et al. provide a carbon budget for an area in the central basin of the Sound not far from the city of Seattle. The basin is over 200 m deep and is influenced by fresh water inputs thoughout its length. No estimates of carbon inputs to the system were made. The system is well-mixed through vigorous tidal exchange at the relatively deep sill leading to the Juan de Fuca Straits. Total primary production in the area is high but the bulk of this is cycled within the pelagic system and it was estimated that only about 5% sedimented out. Nevertheless, estimates of benthic respiration (8%) and burial of carbon (13%) considerably exceeded the sedimentation flux and it was suggested that this imbalance might be explained by alternative pathways of benthic carbon flow, involving sediment slumping, animal migration or surface water refluxing.

Burrell presented carbon budgets for two contrasting Alaskan fjords. The most southerly of these, Boca de Quadra lies just north of the Canadian border and is a multi-silled system. Fresh water volume flux amounts to less than 5% of the tidal prism and deep basin flushing takes place seasonally in summer through incursions of coastal water over the sill. The basin water is isolated over the winter months. Total sedimentation was not estimated but benthic production was calculated at about 2% of total carbon; benthic respiration accounted for 60-70%. In this fjord, which receives a considerable input of glacial melt water during the summer the proportion of carbon buried in the sediments was as much as 25% of the total inputs. This is attributable partly to the highly recalcitrant nature of much of the carbon input from terrestrial cources, and the high level of aluminosilicates in the sedimenting material, but may also be a consequence of the instability of the deep basin sediments. Sediment slumping and bed-load transport may rapidly remove even labile carbon from the biogenic zone in the upper sediments. The second Alaskan system studied, Resurrection Bay, is a single-silled fjord with a maximum basin depth of 290 m. Mean volume of fresh water influx via the Resurrection River is only 2% of the tidal prism, and is mainly concentrated in the late summer period when glacial melt water flows are greatest. Deep-water renewal patterns are similar to those found in Boca de Quadra, with winter isolation and intermittant summer renewals. In this fjord benthic respiration was about 60% of the phytoplankton carbon uptake rate, implying a rapid transfer of autochthonous carbon to the benthos. However, in contrast to the situation in Boca de Quadra, only a small proportion of the sedimentary carbon (3%) was accumulated. It is suggested that the greater lability of allochthonous inputs to this basin, largely comprising fish processing wastes, would account for this difference.

The effect of varying salinity on carbon flow in these types of fjord systems can be demonstrated by reference to the various studies undertaken in the Baltic Sea. Whilst not being a fjord in the true topographical sense, the Baltic may be regarded as such in its hydrographic characteristics (Jansson et al., 1984). Moreover, a number of fairly complete carbon budgeting exercises have been undertaken in various Baltic areas which, when compared, allow some assessment to be made of the effects of progressively decreasing salinities on carbon flow within the system. Table 3 details the carbon inputs and sinks in four areas of the Baltic ranging in salinity from 19% in Kiel Bay in the south west to less than 3% in the Bothnian Bay in the north east. Details of carbon flow at a 21 m deep station in Kiel Bight in the Belt Sea area (see Elmgren (1984) for Baltic subareas), are given by Graf et al. (1984), who measured benthic energy requirements directly using heat flow calorimetry. This was estimated at between 19 and 40% of primary production and somewhat greater than the estimate for carbon sedimentation of between 14 and 38%. Within the benthos macrofaunal production was assessed at between 3 and 9% of total primary production, whereas meiofauna accounted for 1% and heterotrophic bacteria up to 30%. The excess of benthic production over

FJORD TYPE	BOREAL STAGNANT		BOREAL FLUSHED								ARCTIC FLUSHED	
AREA	BYFJORD %TC	%PP	KIEL BAY %TC	%PP	ASKÖ %TC	%PP	TVÄRMINNE %TC	%PP	BOTH. BAY %TC	%PP	DISKO BUGT %TC	%PP
LATITUDE (N)	58°		55°		59°		60°		65°		69°	
SALINITY ‰	22-30		19		8		6		1-3		34	
ENERGY INPUTS (gC m⁻²yr⁻¹) PHYTOP.	232		160		160		100		28		50-90	
DETRITUS	104								21			
PRODUCTION / ENERGY INPUTS — ZOOPLANKTON	0·6	0·9		5-13		9		12	6	11		6-12
BACTERIA H.	5	7		15-30		24		9	10	19		
MEIOFAUNA	0·4	0·5		1		3		3	2	3		
MACROFAUNA	0·2	0·25		3-9		1		7	0·04	0·07		8-14
EPIFAUNA	2·5	4										
FISH	0·02	0·07					0·07		0·06	1		0·4-1
RESPIR-ATION — PELAGIC							†122					
BENTHIC			>100		25-38		45-76					
SEDIMENTATION	19	27	14-38		21-38		37-48		41	71		
EXPORT	32	47	import									
BURIAL					3		5		9			

Table 3. Summary of carbon budget information from a stagnant boreal fjord in Norway; from various areas of the Baltic having differing salinites; and from a well flushed area of a sub-artic fjord in Western Greenland.

water column inputs was ascribed to downward bed load transport of organic material from shallower areas, particularily in spring. Elmgren (1984) provides a comparative synopsis of energy flow estimates for a number of areas in the Baltic. The area chosen as representative of the Baltic proper is based largely on the extensive studies carried out in the Askö area of the western Gothland Sea, but is considered to be representative of much of the shallow central Baltic areas where salinities average about 8%. Primary production in this area is very similar to the levels recorded in the Belt Sea but secondary production is both lower and channeled somewhat differently. Thus, whereas zooplankton production falls within the same range proportionately to that in the more saline southern areas, macrobenthic production is lower and meiobenthic production higher. Total respiratory requirements and inputs through sedimentation are approximately in balance, but it is estimated that some 3% of total primary production is accumulated through burial in the sediments.

An area of the Gulf of Finland, Tvärminne, has been intensively studied by Kuparinen et al. (1984) who provide a detailed carbon budget for a 46 m deep basin area where the salinity is 6°/oo. Primary production here is only 60% of that found in the more saline areas. The proportion of secondary production attributable to the zooplankton is estimated at 12%, much the same as in other areas. Macrofaunal production is equivalent to the levels found in Kiel Bight, but meiofaunal production is also high. Total benthic and pelagic respiratory demands exceed the calculated inputs of fixed and sedimentary carbon, suggesting net import of carbon to this area, probably from terrestrial sources, even though sedimentation rates are higher than in the two more saline areas. In this area proportionately more carbon (5%) is accumulated in the sediments. The most brackish area of the Baltic is the Bothnian Bay, where salinities are only 1-3%. Elmgren (1984) has calculated energy flow for this area and this has been supplemented by information from Kankaala et al. (1984). Primary production in the area is considerably lower than in other more saline areas and almost equalled by allochthonous inputs. Pelagic secondary production is proportionately similar to that recorded in other areas, and sedimentation rate as a proportion of total energy inputs is similar to that recorded in the Gulf of Finland. However, benthic secondary production is considerably lower in general than in the other areas, with the exception of the meiofaunal fraction. Fish production is proportionately similar to the level estimated for more saline areas, but is concentrated overwhelmingly in pelagic fish populations (principally the herring). Carbon loss by burial in this area is proportionately three times that in the more saline southern areas. Elmgren points out that the low benthic production is largely attributable to the loss of benthic suspension feeders, thus removing a major link from the food web in these areas. In general then the effect of lowered salinities on this (large) fjordic system is to reduce primary production progressively, but only below an apparent threshold salinity level of perhaps 7°/oo. Pelagic secondary production remains stable proportionately but benthic production is reduced and emphasis is switched to meiofaunal components at the expense of large macrofaunal filter feeders. At the same time the organic carbon sequestered in the sediments increases progressively.

Table 3 also includes two other carbon budgets, one for a stagnant boreal fjord and another, less detailed, for a well-flushed arctic fjordic area. Rosenberg et al. (1977) provide a detailed study of energy flow through the Byfjord, a stagnant fjordic basin in western Sweden. The area is enriched by allochthonous inputs, and the lower waters of the 45 m deep basin are permanently anoxic. Primary production in the aerobic surface waters is relatively high, but secondary production is generally lower than in well-mixed systems. Pelagic secondary production attributable to zooplankton is <1% of the carbon inputs, as is macro- and meiofaunal production which is confined to those areas on the edges of the basin shallower than the 15 m oxycline. However, an additional loop in the food web, termed here epifauna, assumes considerable importance in the aerobic upper waters. This comprises the large benthic suspension feeders i.e. bivalve molluscs, particularly Mytilus and ascidians (Ciona) which dominate the aerobic communities and whose production represents 4% of the primary produc tion. Production by chemoautotrophic bacteria in the anoxic zone is also of some significance in carbon cycling in this system. However, a major feature of the budget estimates is that nearly 20% of the total carbon inputs to the basin are sedimented out below the oxycline and only an estimated 2% returned via the sulphur cycle. Moreover, an estimated 32% of total inputs is exported beyond the sill.

Although there is little detailed information available about energy flow through polar fjordic systems, Petersen (1984a and b and Petersen and Curtis, 1980) provide some details about the Disko Bugt, a fjord in west central Greenland. In that area, primary production varies between 50-90 $gC \cdot m^{-2} \cdot yr^{-1}$ i.e. somewhat lower than the figures recorded for many boreal fjords (see Matthews and Heimdal, 1980). Zooplankton and benthic macrofaunal production account for proportionately similar amounts of primary production and 1-7% of secondary production appears as fish production (yield). Petersen emphasizes the relative importance of benthic secondary production in arctic ecosystems when compared with temperate and tropical areas. The estimates quoted in Table 3 refer to a 26 m deep well-mixed area of the fjord. Petersen and Curtis (1980) provide some comparative information on secondary production in other parts of the fjord system with contrasting environmental characteristics (Table 4). Thus, in shallow exposed, i.e. well-mixed areas, benthic production is about 30 gC m^{-2} yr^{-1} i.e. three times that in the deeper mixed area. This high production is very largely attributable to high populations of bivalve suspension feeders. In sheltered, poorly flushed areas, benthic production is low (approx. 2 gC m^{-2} yr^{-1}) at all depths. The high production in shallow exposed areas must be dependent on carbon inputs from offshore areas, as is the case in the shallow communities in Swedish fjords described by Möller et al. (1985). No estimates of imports or exports are available for these systems.

CHARACTER OF LOCALITY	SHELTERED (Stagnation)	EXPOSED (Mixing)
SHALLOW (<18m)	2·25	28·94
DEEP (<26m)	2·08	7·19

gC m^{-2} y^{-1}

Table 4. Comparative benthic production in different localities of a sub-arctic fjord in Western Greenland (from Petersen and Curtis, 1984).

5. DISCUSSION

This brief overview of some of the information available on energy flows through fjordic systems serves to emphasize the fragmentary and limited nature of that information, despite the increased attention paid to fjordic studies in recent years. This is particularly true of polar fjords where the very few energy flow studies attempted have assessed only compartments of the system, and no general budgeting exercise has been attempted. Nevertheless, some general conclusions about fjordic energy flows and the consequences of their export/import budgets for the adjacent systems may be drawn. The importance of basin morphometry in regulating exchange with coastal waters as emphasized by Mann (1975) and Naiman and Sibert (1978) is again highlighted by the evidence presented here. Fjords retain allochtonous material and can be internally productive when hydrodynamically active. They can thus generate energy which is subsequently exported in one form or another. Unfortunately, few budgeting exercises have attempted to quantify the export/import balance in any exact way. Moreover, one of the most obvious export pathways, i.e. that through migration of higher carnivores into and out of the system, has never been adequately quantified holistically in any fjord system. Since fjords act as nursery areas for many coastal fish populations (e.g. Gordon, 1981; Cooper, 1985) or as passage and feeding areas for migrating salmonids, and many mammalian and avian species, then there may well be a net transfer of energy through these populations of some considerable significance for coastal waters. This may be of particular importance in high latitude fjords where the highly seasonal productivity is exploited by migratory populations which overwinter in lower latitudes (Pearson, in press; Petersen, 1984). The reverse case where fjordic productivity is stimulated by imports of energy through this pathway is exemplified by the observations of Sugai and Burrell (1984) on the contribution to the nutrient pool in Wilson Arm from decaying salmon following their spawning migration. The description of such pathways and an analytical quantification of their relative importance is an essential aim for future research.

The generalised energy flows considered here have been based largely on annual budgets and have necessarily ignored the highly seasonal nature of fjordic energy flows. Although reference has been made to the overwhelming importance of seasonality, comparative seasonal budgets have not been discussed, although some information is available. In many fjords, lower basin waters are isolated over winter and subsequently flushed in spring and early summer (c.f. Boca de Quadra, Burrell (1983)). This annual overturn fuels the spring bloom in the euphotic zone but may also contribute significantly to the coastal nutrient and carbon pool beyond the fjord (Parsons et al. 1983;

Burrell, in press). Such exports from the stored winter nutrient pool may be more significant in higher latitude fjords where herbivore populations are less closely coupled to the phytoplankton bloom. Numerous recent observations have suggested that much of the initial vernal bloom is transferred directly to the benthos through a mismatch of algal and herbivore population peaks (e.g. Valderhaug and Gray, 1984; Smetacek,1984). The extreme seasonality of high latitude fjords may exacerbate such mismatching and result in the emphasis of benthic over pelagic production in such areas, noted by Petersen (1984). During the later parts of the productive season fjords may often become net importers of carbon and perhaps nutrients from the surrounding coastal waters. An opposite example of this is documented by Therriault et al. (1984) in the Sagueny Fjord, where considerable net carbon inputs to the fjord from the St. Lawrence estuary were recorded in the late summer. In general, stratified fjords appear to be sinks for nutrients and carbon throughout the year, although no detailed seasonal budgets have been described for such systems. Following this resume, some emendations must be made to the summary of generalised basin characteristics presented by Pearson (1980) (Table 1). Thus stratified fjords are probably nutrient and carbon sinks at all latitudes, although the magnitude of their enrichment is inversely proportional to latitude. Well-flushed boreal systems are probably nutrient limited for much of the year, but may be net nutrient exporters following the spring overturn. Well-flushed polar fjords are probably net importers of carbon in the early part of the productive season through the phytoplankton and net exporters later in the season through the higher carnivores. These generalisations do, however, need much further quantification.

The caveat pronounced by Postma et al. (1984) concerning the difficulties inherent in attempting mass balances in any system is reinforced by the information reviewed here. Nevertheless, the approach does allow the identification of the important pathways fuelling the productivity of such systems, despite the obvious inconsistencies and contradictions in the various data bases. The budgets reviewed here also serve to emphasize the utility of such studies in fjordic systems. In these restricted systems the range of variability allows a comparative approach to identify and quantify the important components contributing to general productivity in a way impossible in more open systems.

206

REFERENCES

Baker, E.T., R.A. Feeley, M.R. Landry & M. Lamb, 1985. Temporal variations in the concentration and settling flux of carbon and phytoplankton pigments in a deep fjord like estuary. - Estuarine, Coastal and Shelf Science 21: 859-877.
Burrell, D.C., 1983. Patterns of carbon supply and distribution and oxygen renewal in two Alaskan fjords. - Sediment. Geol. 36: 93-115.
Burrell, D.C., (in press). Interaction between silled fjords and coastal regions of the Gulf of Alaska. - In D.W. Wood & S. Zimmerman (eds.): The Gulf of Alaska: physical environment and biological resources. NOAA, U,S, Gov. Print Offic, Washington D.C.
Cooper, A., 1980. Gadoid populations of western Scottish sea-lochs and their exchanges with west coast stocks. - In H.J. Freeland, D.M. Farmer & C.D. Levings (eds.): Fjord Oceanography, pp. 415-422. Plenum Press, New York and London.
Davies, J.M., 1975. Energy flow through the benthos in a Scottish sea loch. - Mar. Biol. 31: 353-362.
Duff, L., 1981. The Loch Eil project: effect of organic matter input on interstitial water chemistry of Loch Eil sediments. - J. Exper. Mar. Biol. Ecol. 55: 315-328.
Dugdale, R.C. & J.J. Goering, 1967. Uptake of new and regenerated forms of nitrogen in primary production. - Limnol. Oceanogr. 12: 196-206.
Elmgren, R., 1984. Trophic dynamics in the enclosed brackish Baltic Sea. - Rapp. Proc.-v. Reun. Con. Internat. l'Explor. Mer 183: 152-169.
Freeland, H.J., D.M. Farmer & C.D. Levings 1980 (eds). Fjord Oceanography. Plenum Press, New York and London. 715 pp.
Gade, H.G. & A. Edwards, 1980. Deep-water renewal in fjords. - In H.J. Freeland, D.M. Farmer & C.D. Levings (eds.): Fjord oceanography. Plenum Press, New York and London. pp. 453-490.
Gordon, J.D.M., 1981. The fish populations of the west of Scotland shelf. Part II. - Oceanogr. Mar. Biol. Ann. Rev. 10: 405-441.
Graf, G., W. Bengtsson, A. Faubel, L.-A. Meyer-Riel, R. Schulz, H. Theede & H. Thiel, 1984. The importance of the spring phytoplankton bloom for the benthic system of Kiel Bight. - Rapp. Proc.-v. Reun. Con. Internat. l'Explor. Mer 183: 136-143.
Holm, N.G., 1978. Phosphorous exchange through the sediment-water interface. Mechanism studies of dynamic processes in the Baltic Sea. - Contr. Microb. Geochem., Dept. Geol., Univ. Stockholm 3: 149 pp.
Jansson, B.-O., W. Wilmot & F. Wulff, 1984. Coupling the sub-systems - the Baltic Sea as a case study. - In. M.J.R. Fasham (ed.): Flows of energy and materials in marine ecosystems, pp. 549-595. Plenum Press, New York and London.
Kankaala, P., E. Alasaareia & A. Sundberg, 1984. Phytoplankton and zooplankton production in the northeastern and central Bothnian Bay - a review of studies carried out in 1968-1978. - Ophelia Suppl. 3: 69-88.
Kuparinen, J., J.-M. Leppänen, J. Sarvala, A. Sundberg & A. Virtanen, 1984. Production and utilization of organic matter in a Baltic ecosystem off Tvärminne, southwest coast of Finland. - Rapp. Proc.-v. Reun. Con. Internat. l'Explor. Mer 183: 180-192.
Mann, K.H., 1975. Relationship between morphometry and biological functioning in three coastal inlets of Nova Scotia. - In L.E. Cronin (ed.): Estuarine Research, pp. 377-398. Plenum Press, New York and London.
Matthews, J.B.L. & B.R. Heimdal, 1980. Pelagic productivity and food chains in fjord systems. - In H.J. Freeland, D.M. Farmer & C.D. Levings (eds.): Fjord oceanography, pp. 377-398. Plenum Press, New York and London.

McMahon, T.G. & J.W. Patching, 1984. Fluxes of organic carbon in a fjord on the west coast of Ireland. - Estuar. Coastal Shelf Science 19: 205-215.

Möller, P., L. Pihl & R. Rosenberg, 1985. Benthic faunal energy flow and biological interactions in some shallow marine soft bottom habitats. - Mar. Ecol. Prog. Ser. 27: 109-121.

Naiman, R.J. & J.R. Sibert, 1978. Transports of nutrients and carbon from the Nanaimo River to its estuary. - Limnol. Oceanogr. 23: 1183- 1193.

Parsons, T.R., R.I. Perry, E.D. Nutbrown, W. Hsieh & C.M. Lalli, 1983. Frontal zone analysis at the mouth of Saanish Inlet, British Columbia, Canada. - Mar. Biol. 73: 1-5.

Pearson, T.H., 1980. Macrobenthos of fjords. - In H.J. Freeland, D.M. Farmer & C.D. Levings (eds.): Fjord oceanography, pp. 569-602. Plenum Press, New York and London.

Pearson, T.H., 1982. The Loch Eil project: assessment and synthesis with a discussion of certain biological questions arising from a study of the organic. pollution of sediments. - J. Exper. Mar. Biol. Ecol. 57: 93-124.

Pearson, T.H. (in press). Fjordic ecosystems. - In Marine living systems of the far north.

Pearson, T.H., D.C. Burrell & H.M. Feder (in press). Fjords. - In R.E. Turner & W.J. Wolff (eds.): Coastal ecology source book.

Petersen, G.H., 1984a. Energy flow in comparable aquatic ecosystems from different climatic zones. - Rapp. Proc.-v. Reun. Con. Internat. l'Explor. Mer 183: 119-125.

Petersen, G.H., 1984b. Energy flow budgets in aquatic ecosystems and the conflict between biology and geophysics about earth-axis tilt. - In N.A. Morner & W. Karlen (eds.): Climate changes on a yearly to millennial basis, pp. 621-633. D. Reidel Publ. Com.

Petersen, G.H. & M.A. Curtis, 1980. Differences in energy flow through major components of sub-arctic, temperate and tropical marine shelf ecosystems. - Dana 1: 53-64.

Postma, H., W.M. Kemp, J.M. Colebrook, J. Horwood, I.R. Joint, R. Lampitt, S.W. Nixon, M.E.Q. Pilson & F. Wulff, 1984. - In M.J.R. Fasham (ed.): Flows of energy and materials in marine ecosystems, pp. 651-662. Plenum Press, New York.

Rosenberg, R., I. Olsson & E. Ölundh, 1977. Energy flow model of an oxygen-deficient estuary on the Swedish west coast. - Mar. Biol. 42: 99-107.

Skei, J.M. & S. Melsom, 1982. Seasonal and vertical variations in the chemical composition of suspended particulate matter in an oxygen deficient fjord. - Estuar. Coastal Shelf Science 14: 61-78.

Smetacek, C., 1984. The supply of food to the benthos. - In M.J.R. Fasham (ed.): Flows of energy and materials in marine ecosystems, pp. 517-548. Plenum Press, New York.

Sugai, S.S. & D.C. Burrell, 1984. Transport of dissolved organic carbon, nutrients and trace metals from the Wilson and Blossom Rivers to Smeaton Bay, southeast Alaska. - Can. J. Fish. Aquat. Sci. 41: 180-190.

Syvitski, J.P.M., D.C. Burrell & J.M. Skei, 1987. Fjords: processes and products. Springer-Verlag, New York. 379 pp.

Takahashi, M., D.L. Seibert & W.H. Thomas, 1977. Occasional blooms of phytoplankton during summer in Saanich Inlet, B.C. Canada. - Deep-Sea Res. 24: 775-780.

Therriault, J.C., R. de Ladurantaye & R.G: Ingram, 1984. Particulate matter exchange across a fjord sill. - Estuar. Coastal Shelf Science 18: 51-64.

Valderhaug, V.A. & J.S. Gray, 1984. Stable macrofauna community structure despite fluctuating food supply in subtidal soft sediments of Oslofjord, Norway. - Mar. Biol. 82: 307-322.

Wassmann, P., 1984. Sedimentation and benthic mineralisation of organic detritus in a Norwegian fjord. - Mar. Biol. 83: 83-94.

Wassmann, P. & A. Aadnesen, 1984. Hydrography, nutrients, suspended
   organic matter, and primary production in a shallow fjord system on
   the west coast of Norway. - Sarsia 69: 139-153.
Wilmot, W., P. Toll & B. Kjerfve, 1985. Nutrient transports in a
   Swedish estuary. - Estuar. Coastal Shelf Science 21: 161-184.

MASS BALANCE IN CORAL REEF-DOMINATED AREAS

S.V. Smith
Hawaii Institute of Marine Biology, 1000 Pope Road
University of Hawaii, Honolulu, Hawaii 96822

Coral reef community metabolic rates can be high. Autotrophic and heterotrophic communities apparently maintain high rates by juxta-position of "patches" or zones which intercept and process metabolic products swept from upstream within the system. The net result is that the new carbon production of entire coral reef systems appears no more than slightly elevated above the new production of plankton communities in the surrounding surface ocean.

Coral reef systems probably receive a slightly lower hydrographic supply of new nutrients than do the plankton communities of the surface ocean. Two biogeochemical considerations offset the effect of decreased nutrient supply. In the first place, benthic plants produce more carbon, relative to either N or P, than do plankton. Second, coral-reef systems are effective at both fixing atmospheric N and retaining it, thus allowing reefs to use residual dissolved inorganic P not immediately available to the surface-ocean plankton.

I conclude, from both data collected by me and others and considerations which I offer here, that shoal-water ecosystems called coral reefs have very limited metabolic interaction with the surrounding ocean. Important interactions which clearly do exist are probably related to aspects of ecosystems other than material balance.

I suggest that reefs are not metabolically different from other shoal-water systems, but that the lack of strong land-to-sea gradients makes reef mass balances and transports both easy and useful to evaluate.

1. INTRODUCTION

Coral reefs are perhaps unique in the insight which they provide to ecosystem function, but not for the reasons they tend to be studied. The plethora of meetings, journals, and researchers dealing exclusively with coral reefs has isolated reef studies, rather than encouraging comparison between coral reefs and other shoal-water ecosystems.

Lecture Notes on Coastal and Estuarine Studies, Vol. 22
B.-O. Jansson (Ed.), Coastal-Offshore Ecosystem Interactions.
© Springer-Verlag Berlin Heidelberg 1988

One interesting point of comparison has emerged in a few studies. Kaneohe Bay, Hawaii, a coral reef/estuary system with which I have had considerable experience, has appeared in a few comparisons among aspects of ecosystem metabolism (e.g. Nixon, 1981; Kemp et al., 1982). Kaneohe Bay tends to emerge from such comparisons of ecosystem metabolism as a low-nutrient end-member (even though the reported data were collected at a time when those of us working on the system considered it to be severely impacted by sewage); and Kaneohe Bay also does not particularly fall off general metabolic trends seen among more temperate-climate, eutrophic sites. The notion of the "extreme productivity" of coral reefs surrounded by an oceanic desert, and the search for the mystery nutrients to sustain this extreme productivity have been somewhat off the mark, I think.

The uniqueness which coral reefs can offer is very much in the context of the rationale for a workshop dealing with coastal-offshore couplings. Coral reefs are, in general, shoalwater ecosystems surrounded by an oligotrophic ocean. Many coral reefs have little (in the extreme cases, no) connectivity with an immediately adjacent land mass. Nor, in most instances, is there evidence of coral-reef coupling with major oceanic nutrient sources (e.g. in the form of either upwelling or organic import or export). There are obvious, important exceptions to both of these claimed "non-couplings", but I think the general absence of such ties makes for the unique value of coral reefs in ecosystem mass-balance studies.

For the most part, my discussion here will involve coral reefs which are in even more oligotrophic settings than Kaneohe Bay. I will couch my analysis in terms of the nutrient elements C, N, and P, which must obviously be linked (somehow) by metabolism. Despite what one might infer from the phrase "mass balance" in the title of this presentation, I do not intend to summarize the literature on coral-reef mass balance in any quantitative fashion. Nor will I delve into the major inorganic C mass balance pathways of $CaCO_3$ precipitation or gas flux. The former pathway does not seem especially relevant to a comparison with other ecosystems (but see Smith, 1978, 1983a, and Kinsey, 1985 for summaries of reef calcification rates), and the latter can be considered as a physically implicit response to biogeochemical forcing functions (Smith, 1985).

I have deliberately not made this paper into a review of coral-reef metabolism. I recommend the paper by Kinsey (1985) as a recent,

211

excellent review of coral-reef metabolism. I have, instead, attempted
to interpret available data on coral-reef metabolism within the context
of this workshop.

2. THE CORAL-REEF ECOSYSTEM

Let us first consider some generalities about the ecosystem of interest
to this presentation. A coral atoll is perhaps the "purest" form of
coral reef (Fig. 1). A roughly circular rim of reef separates an
internal lagoonal area (itself usually containing patch reefs) from the
surrounding deep ocean. Small islets composed of reef limestone debris
are scattered around the reef perimeter and are the only land
associated with most coral atolls. One or more passes breach the reef
rim of most atolls. A coral reef ecosystem consists of all these
interactive parts, not just the reef flat communities which have proven
so amenable to metabolic studies.

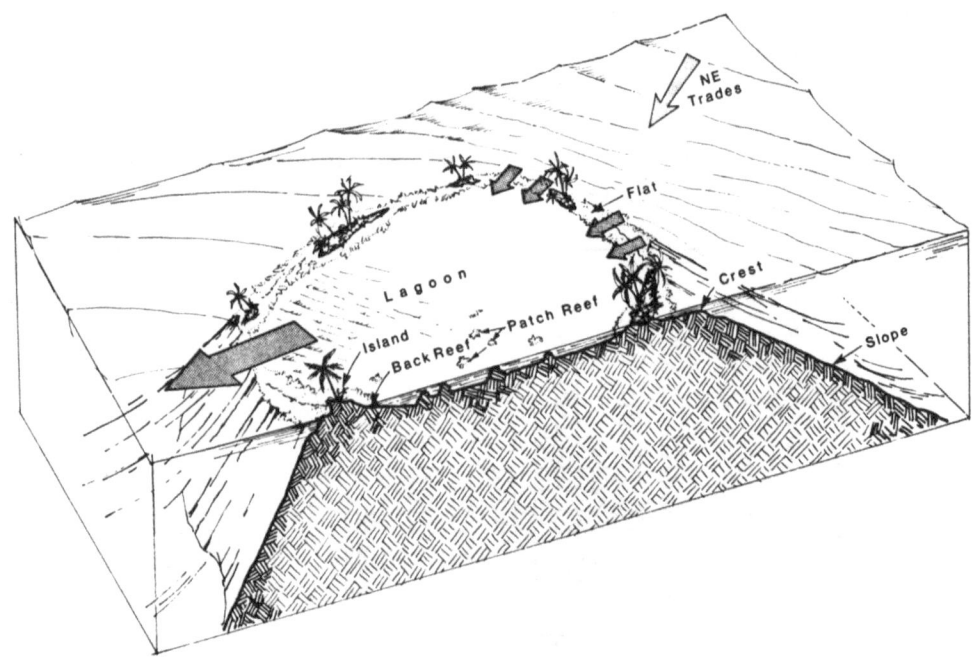

Fig. 1. Schematic diagram of a coral atoll, showing major physiographic
zones and water flow.

It is also important to consider the hydrographic regime of most coral reefs. Most atolls are embedded in a generally east-to-west surface ocean current regime imposed by prevailing tradewinds. Typically, water and dissolved materials enter atolls across the reef flat from the ocean to the east, and flow into the lagoon, (e.g. Atkinson et al., 1981). Plankton and other suspended material in this water are usually present in only low concentration. The water inflow is driven by waves which break along the windward reef crest. Some plankton may accumulate in the relatively confined lagoonal waters, but plankton biomass and metabolism ordinarily remain low in comparison to benthic biomass and metabolism. There is usually no clear unidirectional net transport across the leeward side of coral reefs. Instead, water tends to exit the atoll through the passes.

3. GROSS PRODUCTION OF CORAL-REEF COMMUNITIES

Coral reefs have long been noted for their high rate of organic production in the midst of a nutrient-poor ocean. This high production rate attributed to coral reefs is a feature of the gross, or total, primary production, and is largely a characteristic of reef flats and other shallow reef communities. The median gross production rate for these environments is about 7 $gC \cdot m^{-2} \cdot d^{-1}$, with some extreme values (Fig. 2). It is also obvious from this figure that the respiration rate

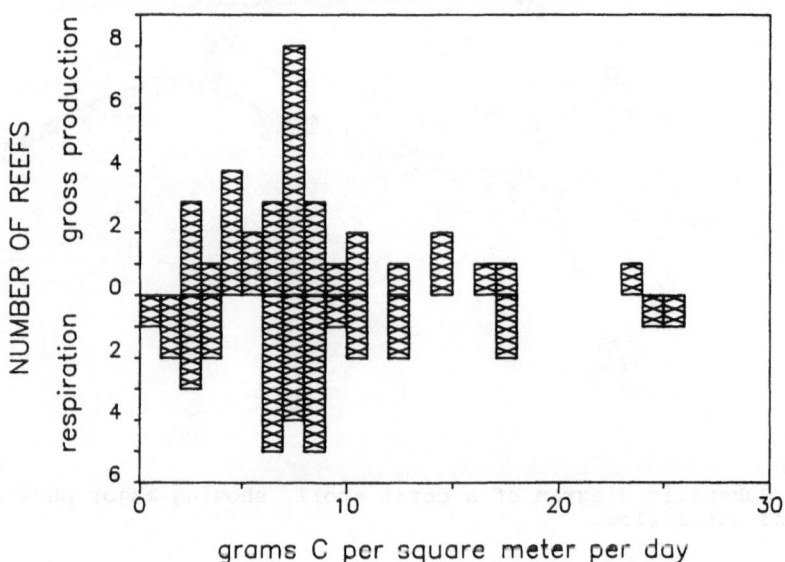

Fig. 2. Gross primary production and respiration of shallow reef environments. Data from Kinsey (1985).

of these shallow reef communities is very similar in both central tendency and dispersion to gross production. A point which is repeatedly overlooked is the fact that coral reef flats occupy only a small portion of most coral reefs. Aspects of reef-flat metabolism do give us insight into total-reef metabolism and will be used repeatedly during the rest of this presentation. However, reef flats and other shallow reef environments should be viewed in the context of the physiography and hydrography of entire coral-reef systems (Fig. 1).

4. NET PRODUCTION OF CORAL-REEF SYSTEMS

The net production of coral-reef systems (or of any other system) is of particular relevance in the study of metabolic coupling between eco-systems or between components within ecosystems. Net production, inte-grated over some period of time, is a measure of the maximum amount of organic matter that the system can export without losing biomass or having an external organic matter subsidy. This net production must be supported by the net supply of new inorganic reactants (nutrients).

Before considering the coupling between coral reefs and the surrounding ocean, it is useful, briefly, to review the use of the terms "new" and "net" production. The term "new production" originates from the recognition (by Dugdale and Goering, 1967) that the nitrogen to support primary production of plankton in the photic zone of the ocean includes both nitrogen brought up from deep water or delivered from land (both of these sources of "new nitrogen" being primarily NO_3^-) and nitrogen recycled within the photic zone (primarily NH_4^+).

Phytoplankton ecologists operationally define "new production" on the basis of new nitrogen from hydrographic sources but recognize that there can be a biochemical new nitrogen source (fixation) and a biochemical sink other than primary production (denitrification). The more obvious the importance of these alternative sources and sinks for new N becomes, the more I prefer to define new production on the basis of P, rather than N, uptake. Net P reservoirs in addition to dissolved P and organic matter are usually insignificant, and hydrographic considerations can often be used to evaluate the supply of "new P".

"Net production", as used by most community ecologists, is equivalent to "new production". Net community or system production is a measure of the rate of storage or export of organic matter produced by that community or system. At least on reef flats, net production is usually defined on the basis of either O_2 or CO_2 changes, rather than with nutrients. O_2 and CO_2 flux measurements, particularly O_2 flux measurements, are subject to severe interference from gas exchange, and P flux becomes a useful measure of new, or net production at the scale of the ecosystem. Obviously, new production derived from P uptake must be scaled by the appropriate C:P ratio in order for net C production to be derived (e.g. Smith and Jokiel, 1978; Atkinson, 1981; Atkinson and Smith, 1983).

Examples of net autotrophic and of net heterotrophic biological communities can be identified on individual reef flats, but the average net reef-flat metabolic performance has a strong 0 mode and is rather evenly distributed about 0 (Fig. 3). Piecewise metabolic studies extending from a coral reef crest, across the reef flat and back-reef area, and into the adjacent lagoon give the strong sense of a juxtaposition of autotrophic and heterotrophic components, adding towards near balance (Kinsey, 1979, 1985). The classical high-production reef flats which have been studied tend to be near the windward reef crest, upstream of a heterotrophic backreef area and a lagoon with autotrophic and heterotrophic components. Clear physiographic zonation evident on many windward reef flats tends to be replaced by a less zonal "patchwork" on leeward reef flats. There is apparently a similar patchwork of community metabolism.

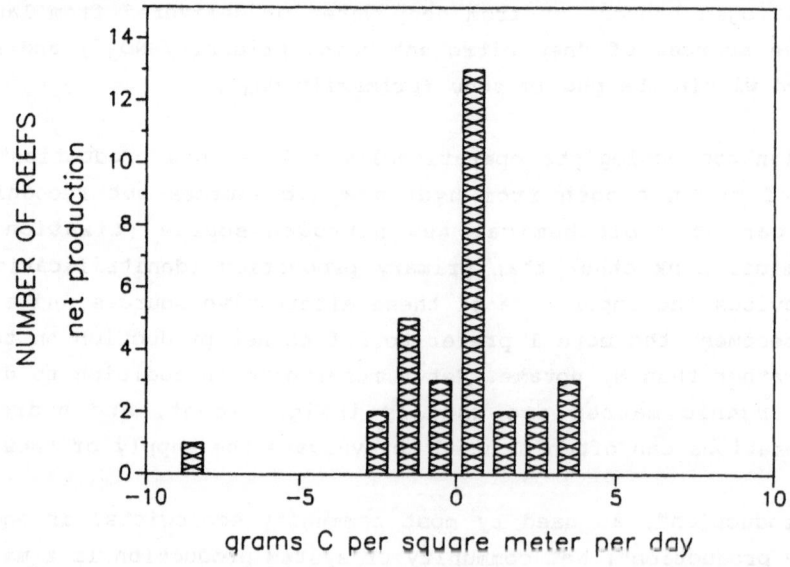

Fig. 3. Net production of shallow reef environments. Data from Kinsey (1985).

There is, in the coral-reef literature, an increasing amount of discussion about enhanced productivity in response to nutrient regeneration in coral heads, sediments, fish feces, and so forth. I will not go over that literature. The point which seems not to be emphasized is that this nutrient regeneration may contribute to the enhanced performance of local communities, but cannot support new production of the coral-reef system as a whole. The measured regeneration and inferred production enhancement are simply finer scale manifestations of patchiness than the community-scale patchiness I have already mentioned. Hatcher (1985) has recently presented a thoughtful scaling analysis of the processes affecting the flux of fixed N through coral-reef systems and the turnover of N within reef systems.

My own bias in evaluating total-system metabolism is towards strategies of total-system analysis, rather than piecewise summation at the scale of communities or smaller ecological units on the reef. Specifically, I have favored salt and water budgets to evaluate hydrographic residence time in reef systems, coupled with non-conservative deviations of C, N, and P from salinity as measures of net system biogeochemical fluxes.

There are only a few comprehensive studies of total-reef system net metabolism using this approach (Table 1). Studies of three unusual "enclosed atoll systems " (lagoonal reef complexes almost entirely bounded by low coral islands, rather than by oceanic reef flats) reenforce the idea of a metabolic unit with low net metabolism. Smith (1983b) summarized and "fine-tuned" earlier results for Fanning, Canton, and Christmas atolls by concluding that the net production of these systems lay between −5 and +170 $mgC \cdot m^{-2} \cdot d^{-1}$.

Table 1. Net productivity of four total reef systems.

SITE	DESCRIPTION	NET PROD. $mgC \cdot m^{-2} \cdot d^{-1}$	REFERENCE
Fanning Atoll	enclosed atoll total-system net	−5	Smith (1983b)
Christmas Atoll	enclosed atoll total-system net	70	Smith (1983b)
Canton Atoll	enclosed atoll total-system net	170	Smith (1983b)
French Frigate Shoals	open atoll piecewise sum	285	Atkinson and Grigg (1984)

Because these systems are hydrographically constrained, their net ecosystem metabolism may be biased somewhat. For comparison, Atkinson and Grigg (1984) used piecewise summation of community metabolism at French Frigate Shoals to derive a systemwide net rate of about 285 mgC $\cdot m^{-2} \cdot d^{-1}$. That system is very open on its leeward side, and it may well export more material than more circumscribed atolls.

The average among all these sites (including that studied by Atkinson and Grigg) is 130 mgC$\cdot m^{-2} \cdot d^{-1}$. This value, as a somewhat arbitrary central tendency for coral-reef net production, is about 2% of the median value which I quoted for the gross production of coral reefs. System-level net production is only a small fraction of the gross production of many reef communities. Piecewise summations for a few other reef systems (reviewed by Kinsey, 1985) yield rates indistinguishable from 0. I attribute these 0 values to uncertainties inherent in piecewise summations, not to proven characteristics that these systems have no net production. Low net production, relative to gross production, is a measure of the efficiency of a system at steady state in internally cycling metabolic products, rather than inefficiently losing them and making up the deficiency by import of new nutrients.

By comparison to the data in Table 1, the "new" production of oligotrophic oceanic plankton has been estimated to lie in the range 40-400 mgC$\cdot m^{-2} \cdot d^{-1}$ (Eppley and Peterson, 1979; Shulenberger and Reid, 1981; Jenkins, 1982). Platt et al. (1984) have argued from various theoretical grounds that 60 mgC$\cdot m^{-2} \cdot d^{-1}$ is probably the best global estimate. The point which emerges is that net coral reef organic carbon production may be higher than oceanic production, but only slightly.

From the empirical formula of Eppley and Peterson (1979), it can be calculated that new production is about 25% of gross production in the surface waters of the oligotrophic ocean. The striking production difference between reef and plankton systems lies in gross production, not net production. I attribute this difference to the obvious physiographic and hydrographic boundaries which retain reef products, in contrast to the density and photic boundary through which surface plankton readily sink.

I will offer some simple hydrographic and biogeochemical reasoning consistent with both the general similarity between reef and surface ocean new organic production rates, and the possible tendency of reefs to be slightly more productive than the surrounding ocean. Leaving aside, for the moment, a discussion of either N versus P limitation or N fixation as a nutrient to sustain "new" production, the nutrient source for an oceanic coral reef in the middle of an oligotrophic ocean is upward transport of nutrient-rich deep ocean water. This is exactly the source which supplies the reactants for oceanic plankton production.

If a coral reef system relies on upward mixing to supply its nutrients (as does the lower part of the photic zone in the oligotrophic ocean), then reef productivity (expressed in terms of nutrient uptake) should be of the same order as that of the adjacent ocean. The seamount forming the base of the reef may accelerate upward mixing somewhat (although there is no evidence that this phenomenon is of general significance). In any event (again, with some possible exceptions), upward mixing per unit area of reef will be impeded by the presence of the barrier. Thus, except for perhaps the very deep seaward fringe of a coral reef, supply of nutrients by upward mixing will be less for a reef than for the adjacent ocean.

Some reefs exist in a milieu of upwelling. For example, upwelling has been found along parts of the Australian Great Barrier Reef (Andrews and Gentien, 1982). The mechanism appears to be the result of the East Australia Current impinging upon the Australian continental shelf; it is apparently not reef-induced upwelling. Canton Atoll is close enough to the region of equatorial upwelling to receive somewhat elevated nutrients (Smith and Jokiel, 1978). Net system productivity there (170 $mgC \cdot m^{-2} \cdot d^{-1}$; Table 1) does appear to be elevated somewhat (Smith, 1983b).

However, even the presence of local upwelling (whether barrier-induced or inherent to the site) does not guarantee higher net production of coral reef systems. It is clear that high nutrient concentrations will stimulate primary production; this is seen in the case of plankton production associated with upwelling. However, coral reefs are ecosystems usually dominated by benthic, rather than by planktonic, organisms. While reef benthos productivity can be stimulated by elevated nutrient levels (Kinsey and Davies, 1979; Henderson and Smith, 1980; Smith et al., 1981), benthic organisms are likely to exhibit lower specific growth rates than plankton. Phytoplankton biomass will therefore be more likely to respond rapidly to elevated nutrients than will benthic plant biomass, so the reef response may be something other than elevated net production.

For example, the net result of upwelling in the vicinity of coral reefs may be stimulated plankton production and biomass (perhaps outside the boundaries of the reef system itself), increased transport of planktonic detritus to the reef, sedimentation, and stimulated heterotrophic activity of the reef benthos. There is evidence, from reef systems receiving sewage discharge, of elevated plankton production accompanied by elevated benthic respiration (Smith et al., 1981). If this elevated plankton production occurs within the reef system (as was the case in the cited study), then it must be considered as part of the total-reef net production. However, the plankton response to upwelling may be outside the reef, with the products swept in and deposited there. As an example of response to exogenous organic input, coral reefs which receive an exogenous supply of particulate organic matter show a response of net heterotrophy (Crossland et al., 1984).

Fast-growing benthic algae which are locally favored by high nutrient concentrations (often from pollution sources) on coral reefs tend to be taxa like Ulva, genera which are not usually common among reef algae. Such algae should be more common on reefs, if reef production were generally enhanced by upwelling.

How about nutrient supply by horizontal advection of low-nutrient water across a reef, together with the "benthic advantage" of being fixed in space, as stimuli for enhanced reef ecosystem production? Metabolic stimulation clearly occurs in upstream communities on a coral reef flat, although it is by no means evident that higher upstream nutrient concentration is the major mechanism at work. Photosynthetic biomass tends to build up and high net production tends to occur near reef crests where surface ocean water impinges upon the reefs. Downstream communities may become nutrient depleted, support lower plant biomass, and exhibit net heterotrophy, for little or no net difference in total nutrient availability across the reef system. The "benthic advantage" of being fixed in space relative to the moving water enhances patchiness, hence local production; it does not enhance the production of the total system.

In general, reef flats are of insufficient extent for the effect of local nutrient drawdown to be obvious; it becomes apparent only if a reef system which exhibits longer water residence time than a typical reef flat is examined. Figure 4 shows dissolved inorganic P (DIP) for two enclosed atolls normalized against water residence time. Note especially that it took several days in these systems for DIP to be depleted measurably (see also Atkinson, 1981).

Water flow across reef flats may be fast because of the force of breaking waves. Volume transport through entire reef systems will be impeded by the physical barriers of the reef topography, hence will tend to be less than wind drift in the adjacent open ocean. Therefore delivery of nutrients to a reef system by horizontal flow should be less than the horizontal advective nutrient delivery to an eulerian (i.e. spatially fixed) volume of the adjacent ocean.

I am forced to the conclusion that oceanic coral reef systems usually have somewhat less nutrients supplied to them than does the surrounding ocean. If all else were equal, coral reefs might therefore have somewhat lower net production than the surrounding ocean. Yet from the data from three hydrographically restricted atolls and one somewhat more open reef system (Table 1), we see suggestions that reefs are not less productive than the surrounding ocean. There are at least two reasons, based on biogeochemical considerations, why coral reefs may enjoy some productivity enhancement over plankton communities receiving an equivalent nutrient supply.

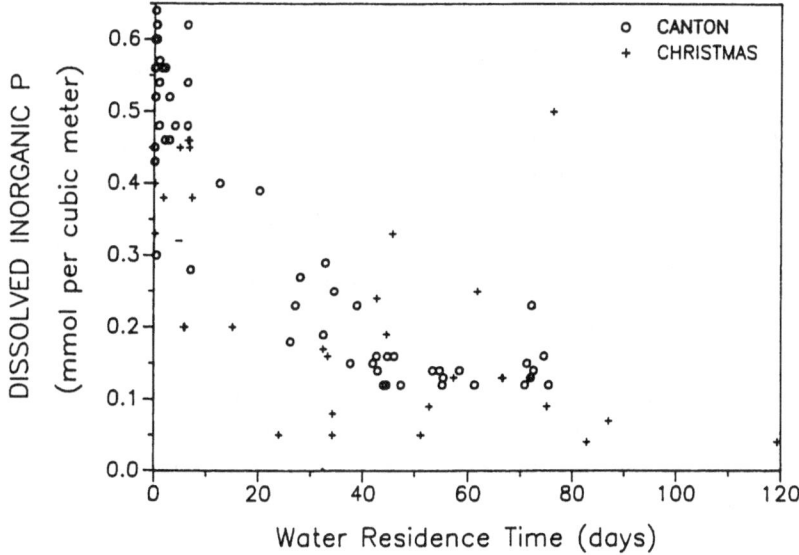

Fig. 4. DIP versus water residence time in two enclosed reef systems. Modified from Smith and Jokiel (1978) and Smith et al. (1984).

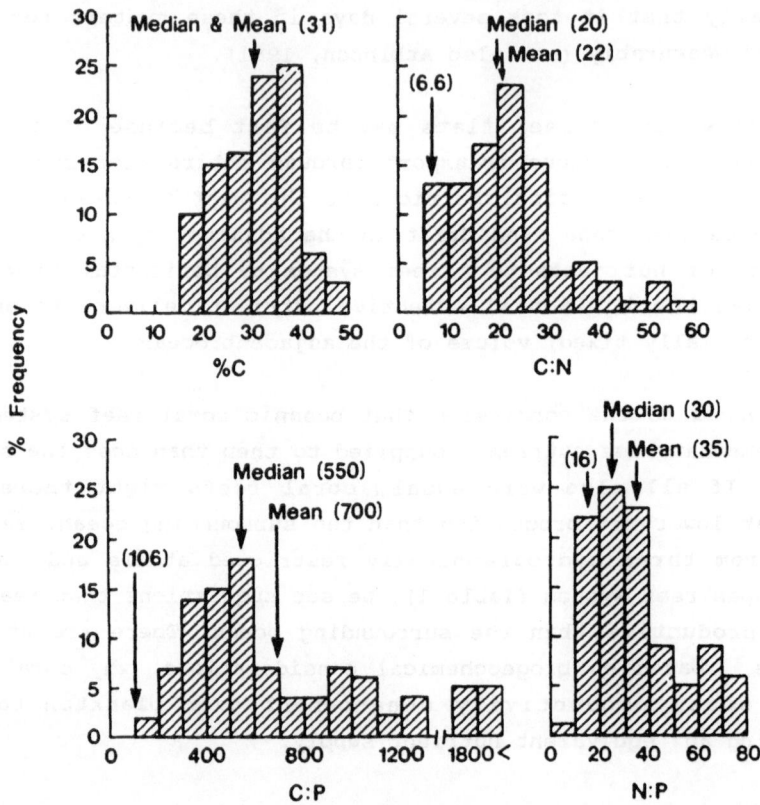

Fig. 5. Atomic C:N:P ratios of benthic marine plants: also shown are the "Redfield Ratios" for plankton. Data from Atkinson and Smith (1983)

Perhaps the most obvious reason is the nutrient composition of primary producers in benthic systems. Plankton typically have a C:N:P molar ratio of about 106:16:1, the so-called Redfield ratio. Benthic plants are depleted in N and especially in P, relative to C; the median C:N:P ratio of benthic plants (with considerable scatter) is about 550:30:1 (Atkinson and Smith, 1983) (Fig. 5).

With respect to the argument I am presenting here, the new production supported by a unit of P delivery to a benthic-based system will be about 5 times (550/106) as rapid as new production supported by that same unit P delivery to a plankton-based system. The "enhancement ratio" based on N, rather than P, is about 3-fold. This analysis of "production advantage" for a nutrient-limited benthic system over a nutrient limited planktonic system may bestow artificial importance to carbon, as the currency for comparison of organic production. These conclusions would have been quite different (in fact, simpler) if I had defined production directly on the basis of P or N, rather than converting nutrient uptake to C uptake; I choose not to belabor this point here.

There is a second reason why coral reefs enjoy some production enhancement relative to plankton in the surface ocean: There may be a shift from N to P-limitation as water leaves the open ocean and enters a reef system. Surface ocean waters tend to show residual dissolved inorganic P (DIP $\approx$ 0.1 μM) once dissolved fixed inorganic N (DIN) has become exhausted. This phenomenon has been observed by many investigators. Smith et al. (1986) recently summarized and discussed the phenomenon and derived the "oceanic regression line" illustrated in Figure 6 for a site in the oligotrophic central North Pacific Ocean. This DIN deficit, relative to residual DIP, is one line of evidence that the surface ocean (planktonic) system may be metabolically limited by the availability of N, not P (e.g. Thomas, 1966; Redfield et al., 1963).

Coral reef systems are recognized to have high rates of nitrogen fixation (e.g. Wiebe et al., 1975), hence to have the ability to continue taking up DIP at very low DIN concentrations. The characteristic of residual DIP, relative to DIN, largely disappears (reef data and regression line in Fig. 6). I have argued that confined systems, such as some coral reefs, may be metabolically limited by P while they supplement N internally (Smith, 1984). The "reef water" DIN:DIP slope of Figure 6 is about 7:1, or about one fourth of the N:P ratio of benthic plants (Fig. 5). If the "reef water" DIN:DIP slope of Figure 6 typifies net uptake of DIN and DIP from water by coral reef systems, then we might expect that reefs derive about three fourths of their N by N-fixation rather than by hydrographic processes.

If we use a C:N (molar) ratio of 18 for material being produced and use the net organic carbon production rate I have inferred for coral-reef systems, we can derive a general estimate of the N fixation required to supply three quarters of the reef N requirements; about 0.5 mmol·m^{-2} d^{-1}. This rate is modest in comparison to measured reef flat N fixation rates (e.g. Wiebe and Webb, 1975), although the reef flat rates include no adjustment for the proportional area over which they might occur.

I have suggested P-limitation to be a general characteristic of confined marine ecosystems which have plenty of time to exhaust available nutrients (see also Redfield, 1958), not a peculiarity of coral reefs; this suggestion requires further testing. Regardless of the ability to extrapolate my model to other systems, coral reefs gain advantage over the adjacent ocean by not losing their accumulated, atmospherically fixed N to either fallout or downward mixing.

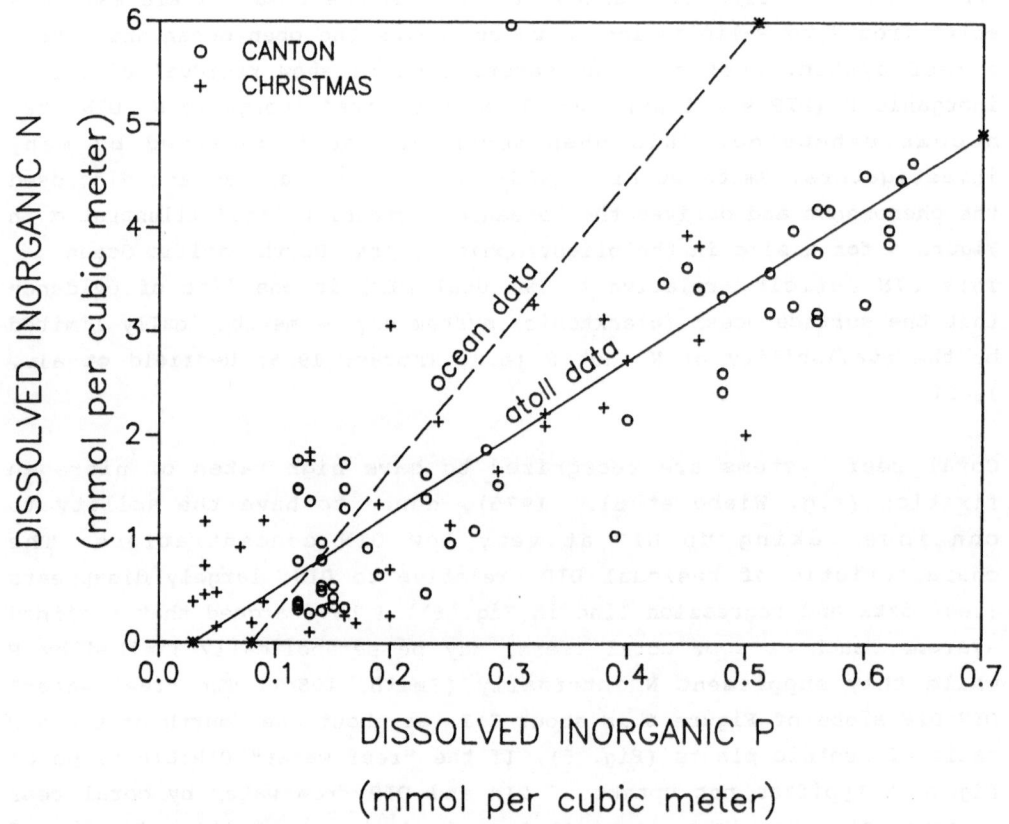

Fig. 6. DIN versus DIP in two enclosed atoll lagoons in the North Pacific Ocean. The geometric mean regression line through the data points (with one outlier removed) is
DIN = -0.24 + 7.35 x DIP (n = 87; r^2 = 0.804).
95% confidence limits on the intercept and slope are ± 0.17, ± 0.38 respectively. Data from Smith and Jokiel (1978) and Smith et al. (1984). Also shown is the geometric mean regression line for DIN versus DIP at a typical oceanic station in the oligotrophic North Pacific Ocean:
DIN = -1.07 + 13.98 x DIP (n = 460; r^2 = 0.996).
95% confidence limits on the intercept and slope are ± 0.14, ± 0.09 respectively. Data from Smith et al. (1986).

Thus, reef systems apparently can accumulate sufficient N to utilize DIP not immediately available to the adjacent surface ocean plankton community. It is noteworthy, as a side remark, that coral reef flats alone are <u>not</u> effective at retaining newly fixed N (Webb et al., 1975). However, water flow across these environments is usually sufficiently rapid to prevent any significant alterations of dissolved nutrient concentrations. These observations are a reminder that the metabolism of reef flats does not adequately characterize the metabolism of reef systems.

The conclusion from this biogeochemical analysis is that coral reefs may produce up to about five times as much new organic carbon as does the surface ocean, based strictly on P-limitation of new production. The organic production advantage for reefs over plankton will be even greater if we further argue for a switch from planktonic N-limitation to benthic P-limitation because of the accumulated products of N fixation. Hydrographic considerations will tend to temper this reef production advantage to some extent.

5. CONCLUSIONS

I have made a number of points about the net metabolism of coral reef ecosystems, relative to the adjacent surface ocean. I do not believe than any of these points is specific to coral reef systems. Rather, I suggest that the general lack of strong land-to-sea gradients calls attention to the metabolic characteristics of coral reefs. We "expect" high production in estuaries, which receive a large input of both dissolved and particulate material from land. Coral reefs do not generally receive this large input, so we are "surprised" to see high production there. We look for explanations for this paradox.

Does the paradox exist? With increased knowledge of estuarine ecosystem metabolism, we begin to suspect that there may be less organic "outwelling" than previously thought, and indeed there may sometimes be significant organic "inwelling". It has become popular to characterize estuaries as a filter for material carried from land to the sea (e.g. Kennedy, 1984). The filter is obviously ineffective for some materials. Leakage of terrigenous organic and inorganic nutrients through the land-sea filter tends to complicate interpretation of the biogeochemical processes accounting for the filtration. The mass balance of estuarine metabolism may not require land input of dissolved nutrients, so the net filtration may be relatively small. The new production of estuarine systems is not demonstrably high, despite locally high primary production. To what extent is this an example of the same production patchiness which I claim to characterize coral reefs?

Increased understanding of coral reef total-system metabolism points to low system net production, even though production of some communities is high. There appear to be sound oceanographic reasons for this low system production. Supply of oceanic materials to be biogeochemical reactants within reef systems is small, and export of organic products cannot exceed that supply. While important linkages clearly do exist between coral reefs and the adjacent ocean, I suggest these linkages are not well manifested by mass balance.

Ecologists seeking biotic, rather than chemical, interactions between systems might suggest that active migration patterns or passive dispersal processes are the logical alternatives to material exchange. Systems analysts might express such processes in terms of information exchange between reefs and the ocean. In either case, the evidence I have presented suggests that coral reef systems, and probably other systems with distinct physiographic and/or hydrographic boundaries, have very limited material coupling with adjacent systems. Large within-system material-flow couplings manifested by juxtaposed communities with high gross production and respiration suggest efficient cycling among those components, rather than significant external coupling.

REFERENCES

Andrews, J.C. & P. Gentien, 1982. Upwelling as a source of nutrients for the Great Barrier Reef ecosystems: a solution to Darwin's question? - Mar. Ecol. Prog. Ser. 8: 257-269.
Atkinson, M.J., 1981. Phosphate flux as a measure of net coral reef flat productivity. - Proc. Fourth Internat. Coral Reef Symp., Manila. v. 1, pp. 417-418.
Atkinson, M.J. & R.W. Grigg, 1984. Model of a coral reef ecosystem. II. Gross and net primary production at French Frigate Shoals, Hawaii. - Coral Reefs 3: 13-22.
Atkinson, M.J. & S.V. Smith, 1983. C:N:P ratios of benthic marine plants. - Limnol. Oceanogr. 28: 568-574.
Atkinson, M., S.V. Smith & E.D. Stroup, 1981. Circulation in Enewetak Atoll lagoon. - Limnol. Oceanogr. 26: 1074-1083.
Crossland, C.J., B.G. Hatcher, M.J. Atkinson & S.V. Smith, 1984. Dissolved nutrients of a high latitude coral reef, Houtman Abrolhos Islands, Western Australia. - Mar. Ecol. Prog. Ser. 14: 159-163.

Dugdale, R.C. & J.J. Goering, 1967. Uptake of new and regenerated froms of nitrogen in primary productivity. - Limnol. Oceanogr. 12: 196-206.

Eppley, R.W. & B.J. Peterson, 1979. Particulate organic matter flux and planktonic new production in the deep ocean. - Nature 282: 677-680.

Hatcher, A.I., 1985. The relationship between coral reef structure and nitrogen dynamics. - Proc. Fifth Internat. Coral Reef Symp., Tahiti, v. 3, p. 407-413.

Henderson, R.S. & S.V. Smith, 1980. Semi-tropical marine microcosms: facility design and an elevated-nutrient-effects experiment. - In J.P. Giesy Jr. (ed.): Microcosms in ecological research, pp. 869-910. DOE Symp. Ser. 52. U.S. DOE CONF-781101.

Jenkins, W.J., 1982. Oxygen utilization rates in the North Atlantic subtropical gyre and primary production in oligotrophic systems. - Nature 300: 246-248.

Kemp, W.M., R.L. Wetzel, W.R. Boynton, C.F. D'Elia & J.C. Stevenson, 1982. Nitrogen cycling and estuarine interfaces: some current concepts and research directions. - In V.S. Kennedy (ed.): Estuarine comparisons, pp. 209-230. Academic Press, New York.

Kennedy, V.S., 1979 (ed.). The estuary as a filter. Academic Press, New York.

Kinsey, D.W., 1979. Carbon turnover and accumulation by coral reefs. - Ph.D. Thesis, Univ. Hawaii, 248 pp.

Kinsey, D.W., 1985. Metabolism, calcification and carbon production. I System level studies. - Proc. Fifth Internat. Coral Reef Symp., Tahiti, v. 4, pp. 505-526.

Kinsey, D.W. & P.J. Davies, 1979. Effects of elevated nitrogen and phosphorus on coral reef growth. - Limnol. Oceanogr. 24: 935-940.

Nixon, S.W., 1981. Remineralization and nutrient cycling in coastal marine ecosystems. - In B.J. Neilson & L.E. Cronin (eds.): Estuaries and nutrients, pp. 111-138. Humana Press, Clifton, N.J.

Odum, H.T., 1956. Primary production in flowing waters. - Limnol. Oceanogr. 1: 102-117.

Platt, T., M. Lewis & R. Geider, 1984. Thermodynamics of the pelagic ecosystem: elementary closure conditions for biological production in the open ocean. - In M.J.R. Fasham (ed.): Flows of energy and materials in marine ecosystems, pp. 49-84. Plenum Press,

Redfield, A.C., B.H. Ketchum & F.A. Richards, 1963. The influence of organisms on the composition of sea water. - In M.N. Hill (ed.): The sea, v. 2, pp. 26-77. Wiley and Sons,

Shulenberger, E. & J.L. Reid, 1981. The Pacific shallow oxygen maximum, deep chlorophyll maximum and primary productivity, reconsidered. - Deep-Sea Res. 28: 901-919.

Smith, S.V., 1978. Coral reef area and the contributions of reefs to processes and resources of the world's oceans. - Nature 273: 225-226.

Smith, S.V., 1983a. Coral reef calcification. - In D.J. Barnes (ed.): Perspectives on coral reefs, pp. 240-247. Australian Inst. Mar. Sci.

Smith, S.V., 1983b. Net production of coral reef ecosystems. - In M.L. Reaka (ed.): The ecology of deep and shallow coral reefs, pp. 127-131. NOAA Symp. Ser. Undersea Res. v. 1.

Smith, S.V., 1984. Phosphorus versus nitrogen limitation in the marine environment. - Limnol. Oceanogr. 29: 1149-1160.

Smith, S.V., 1985. Physical, chemical and biological characteristics of CO_2 gas flux across the air-water interface. - Plant. Cell Environ. 8: 387-398.

Smith, S.V. & P.L. Jokiel, 1978. Water composition and biogeochemical gradients in the Canton Atoll lagoon. - Atoll Res. Bull. 221: 15-53.

Smith, S.V., S. Chandra, L. Kwitko, R.C. Schneider, J. Schoonmaker, J. Seeto, T. Tebano & G.W. Tribble, 1984. Chemical stoichiometry of lagoonal metabolism. - U. Hawaii/U. South Pacific Internat. Sea Grant Prog. Tech. Rep. UNIHI-SEAGRANT-CR-84-02, 30 pp.

Smith, S.V., W.J. Kimmerer, E.A. Laws, R.E. Brock & T.W. Walsh, 1981. Kaneohe Bay sewage diversion experiment: perspectives on ecosystem responses to nutritional perturbation. - Pac. Sci. 35: 279-396.

Smith, S.V., W.J. Kimmerer & T.W. Walsh, 1986. Vertical flux and biochemical turnover regulate nutrient limitation of net organic production in the North Pacific Gyre. - Limnol. Oceanogr. 31: 161-167.

Thomas, W.H., 1966. Surface nitrogenous nutrients and phytoplankton in the northeastern tropical Pacific Ocean. - Limnol. Oceanogr. 11: 393-400.

Webb, K.L., W.D. DuPaul, W. Wiebe, W. Sottille & R.E. Johannes, 1975. Enewetak (Eniwetok) Atoll: aspects of the nitrogen cycle on a coral reef. - Limnol. Oceanogr. 20: 198-210.

Wiebe, W.J., R.E. Johannes & K.L. Webb, 1975. Nitrogen fixation in a coral reef community. - Science 188: 257-259.

RIVERINE C, N, Si AND P TRANSPORT TO THE COASTAL OCEAN: AN OVERVIEW

David H. Peterson, Stephen W. Hager and Laurence E. Schemel
U.S. Geological Survey, 345 Middlefield Road - MS 496
Menlo Park, CA 94025

and

Daniel R. Cayan
Scripps Institution of Oceanography
La Jolla, CA 92093

1. INTRODUCTION

Terrestrial ecosystems cycle and recyle inorganic nutrients including a feedback to atmospheric dry deposition and precipitation (cf. Lewis et al., 1985). Each year, however, a small fraction per unit area of the atmosphere/plant/soil flux leaks from these land-based cycles via precipitation/runoff (Meybeck, 1982). These losses are, in general, unpreventable. Moreover, such nutrient "losses" have increased with increasing human population (Wollast, 1983); although to some extent this anthropogenic component can be controlled. Most rivers eventually flow into estuaries and the coastal ocean where their natural and anthropogenic nutrient loads continue to recycle, are lost to the atmosphere, or are buried in sediment. In one extreme, when riverine nutrient concentrations are exceedingly low, as in southwestern Canadian streams (Naiman and Sibert, 1978; Stockner and Shortreed, 1978, 1985), downstream plant biomass can be nutrient limited. In the other extreme, when these nutrient concentrations are very high such as in highly populated European river basins, downstream plant biomass can increase, perhaps intensifying natural anoxia cycles within the receiving estuarine/coastal ocean waters if these waters are stratified (Rosenberg, 1985).

To understand these nutrient cycles, researchers estimate and refine mass budgets of varying spatiotemporal detail and subsystem complexity. This paper uses such methods in an overview of the organic carbon, silica, nitrogen and phosphorus (C, Si, N, P) mass transport from rivers to estuarine/coastal ocean systems. Riverine C, Si, N, P transport is reviewed on a coarse (annual, regional, global) recycling scale and put into perspective with our best estimates of major

Lecture Notes on Coastal and Estuarine Studies, Vol. 22
B.-O. Jansson (Ed.), Coastal-Offshore Ecosystem Interactions.
© Springer-Verlag Berlin Heidelberg 1988

estuarine recycling processes: phytoplankton productivity and aphotic
and benthic mineralization.

Over an annual cycle, production and mineralization processes have
finite maximum rates per unit area. These finite rates can exceed or be
exceeded by riverine sources. In a generalized and very simple model,
when the area of the estuary is very small relative to the area of the
river basin (e.g. the river basin is hundreds of times larger), the
within-estuary nutrient cycling processes can be overwhelmed by a
strong riverine source. In this instance the within-estuary plant
nutrient recycling processes are probably relatively ineffective before
the river water reaches the coastal ocean. Factors which could modify
this general relation between river basin size and estuarine capacity
to recycle include increased human population density and thus nutrient
load in the river basin and increased riverine/estuarine turbidity,
decreasing the availability of light and, in turn, the photosynthetic
rates. When the area of the estuary increases relative to the area of
the river basin (e.g. the river basin is only about ten times larger),
estuarine cycling processes can dominate the relatively weak riverine
sources. In this instance nutrients can be recycled many times before
transit to the coastal ocean, atmosphere or bottom sediment. If these
estuarine/coastal systems where recycling processes are especially
important also remain stratified over long periods, then nutrient loss
from the upper photic layer is inevitable and low nutrient concentrations could limit biomass production. In this latter case the
magnitude of the anthropogenic nutrient factor is expected to be more
important than in systems where nutrient recycling processes are
relatively less impressive.

2. THE PROBLEM

Chemical budgets of riverine/estuarine sources of carbon, nitrogen,
silicon and phosphorous to the coastal ocean are reviewed with a broad
global or regional view. Before attempting this, several general
aspects of the problem are noted.

2.1 The last 10,000 years

Estuarine researchers generally do not think about variability over low
frequency or long time scales. The region of the eastern rim of the

north Pacific Ocean and western North America illustrates a long term perspective to be derived from large-scale considerations.

Fossil pollen records from Alaska to northwestern United States show a broad and consistent pattern of climate change. A warm and dry pattern persisted before 5,000 y.b.p. and a cool and humid pattern after 5,000 y.b.p. (Heusser et al., 1985). Heusser and coworkers suggested that during the warm and dry regime, the subtropical north Pacific high (anticyclonic) pressure center regulated climate at higher latitudes and for a greater part of the annual cycle than today. Presumably, about 5,000 years ago, the pressure system over the Pacific shifted more closely to its present behavior. Today winter is dominated by the Aleutian low pressure center (cyclonic), and winters are cooler and more humid (Heusser et al., 1985). It is very interesting that this inferred timing in dominance of cyclonic vs anticyclonic atmospheric pressure centers appears to be close to the timing of changes in marine sediment record. Before 4,500 y.b.p. laminated sediments were being deposited and preserved along the continental slope of California. This is evidence that the oxygen minimum was more intensely developed before 4,500 y.b.p. probably in response to a period of increased ocean productivity. Beginning 4,500 years ago oceanic conditions apparently changed (decreased upwelling?) as evidenced by the more recent sediments being bioturbated by macro-infauna that flourished with increased oxygenation along the continental slope (Gardner and Hemphill, 1986).

At least in this part of the world, then, both atmospheric and oceanic conditions apparently shifted around 4,000 to 5,000 y.b.p. and it seems that the present system has been operating in similar modes and annual patterns over several thousand years. We expect that the nature of estuarine and coastal ocean processes (in this region) was also changing over the last 10,000 years; but perhaps less so over the last several thousand years or until the beginning of man's intensive agricultural and urban activities.

2.2 Estuarine variability and circulation

External factors forcing estuarine variability may have changed little over the last several thousand years (in northwestern North America) or until man's activities increased. Nevertheless high variability is the rule in estuaries in contrast to classical oceanographic experience.

Some blue-water oceanographers probably have been interpreting nutrient variability in terms of results from only a few closely repeated estuarine cruises. Compare the relatively small variability in the proportions of nutrient concentrations observed during a transect along the entire axis of the Pacific Ocean (Fig. 1) with estuarine variability (Fig. 2) (one estuarine cruise probably cannot tell us as much about the estuary as one ocean cruise can tell us about the ocean). For purposes here, assume a travel time of 1,000 years for water from one end of the Pacific to the other and a sampling time of about one year (Craig et al., 1981). The ratio of sampling to transit time is 0.001. For an estuary (northern San Francisco Bay) similarly assume a river to ocean transit time of roughly 1 month and sampling time of about 10 years. Now the ratio of sampling to transit time is 120. Thus about 10 years of estuarine variability might be "scaled" to compare with more than 100,000 years of Pacific Ocean observations (1,000 years times 120). Of course we do not know in detail how the Pacific Ocean chemistry might have varied over such a long time scale (Broecker et al., 1985). But the inverse of this question, how estuarine chemistry might vary over a time scale of 40 minutes (1 month times 0.001) seems reasonable to consider. Compare, for instance, Figures 2 and 3 (note the estuarine observations are few and made over a 3-day not a 40-minute period). Because high estuarine variability is ever present, a first major caution is that estuarine variability is extremely important in confounding empirical estimates of estuarine chemical budgets.

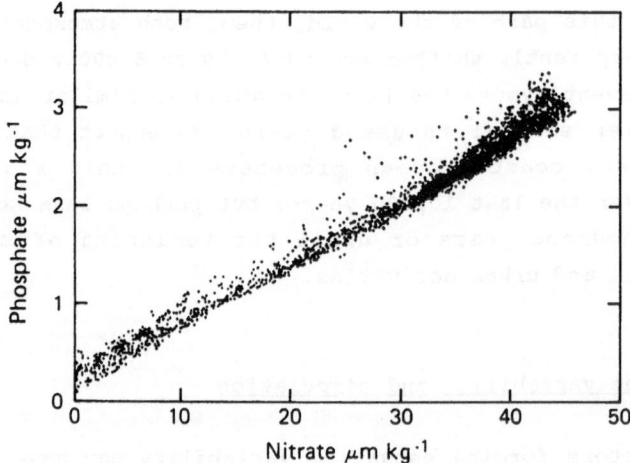

Fig. 1. Scatter of dissolved inorganic phosphate versus nitrate concentrations, GEOSECS Pacific Ocean Expedition, August 1973 to June 1974 (adapted from Craig et al., 1981).

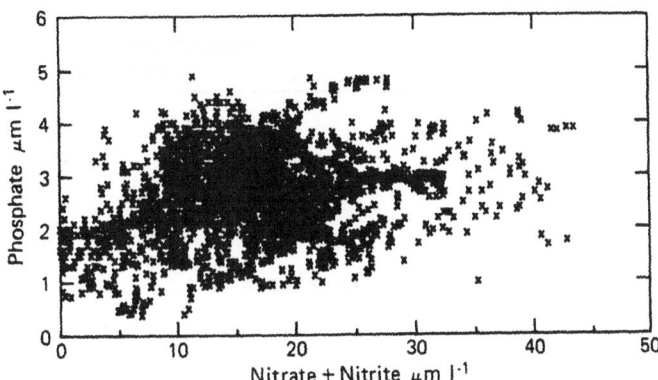

Fig. 2. Scatter of dissolved inorganic phosphate versus nitrate concentrations, northern San Francisco Bay estuary, 1969 to 1981.

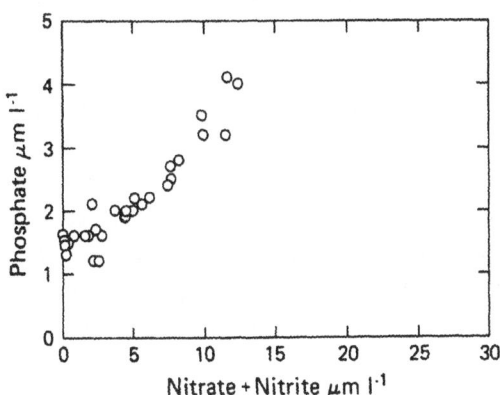

Fig. 3. Scatter of dissolved inorganic phosphate versus nitrate plus nitrite concentrations, norther San Francisco Bay estuary, September 13, 14, 15, 1972.

Table 1. Dissolved inorganic nutrient concentrations in aphotic
(> 500M) waters of the eastern Mediterranean Sea.

Nutrient	Approximate Concentration[1] Range (μg at 1-[1])
Phosphate	< 0.1 to 0.3
Nitrate	1 to 5
Silicate	3 to 9

[1] From McGill, 1965, Figures 2, 3 and 4.

Estuarine circulation can further complicate observation and analysis
of estuarine chemical budgets. Circulation plays a major role in
modifying or maintaining dissolved inorganic nutrient distributions in
estuaries as in the world's oceans and inland seas. Because the
dissolved inorganic nutrient concentrations in the deep ocean have not
been detectably influenced by man's activities, let us consider here
the North Pacific as an estuary (Stigebrandt, 1984). Its sea surface is
higher than the Atlantic (Reid, 1961) and its water near the bottom,
with relatively high concentrations of dissolved nutrients moves slowly
northward (cf. Fig. 4). A reverse flow pattern is maintained in the
Atlantic and even more so in the Mediterranean (Hopkins, 1978). Hence
natural processes cause nutrient concentrations in the Atlantic to be
lower than in the Pacific and to be even lower in the Mediterranean
(Table 1 and Redfield et al., 1963). We assume that the same or
similar circulation effects often apply to estuaries.

With regard to the north Pacific being estuarine-like in its
circulation and nutrient concentrations, the higher proportion of
silica relative to nitrate in its lower salinity surface waters to the
north is also consistent with an estuarine-like behavior (Zentara and
Kamykowski, 1977). Higher concentrations of silica relative to nitrate
are probably typical of estuaries relatively unaffected by
anthropogenic chemicals (Peterson et al., 1986). Further, lower ratios
of silica to nitrate may result from the reverse flow pattern (e.g.
Table 1).

On a smaller scale, estuarine circulation can also maintain relatively
high dissolved inorganic nutrient concentrations by trapping particles
which serve as nutrient sources. It is not easy to determine if the

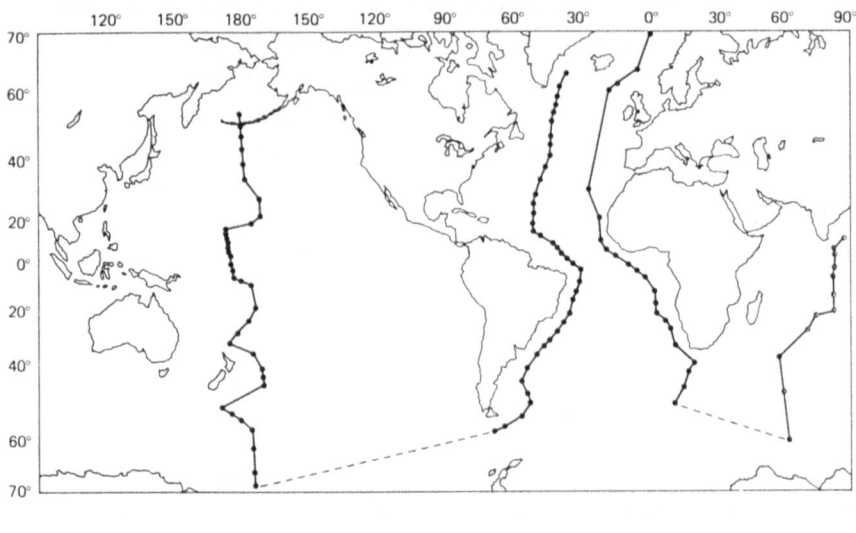

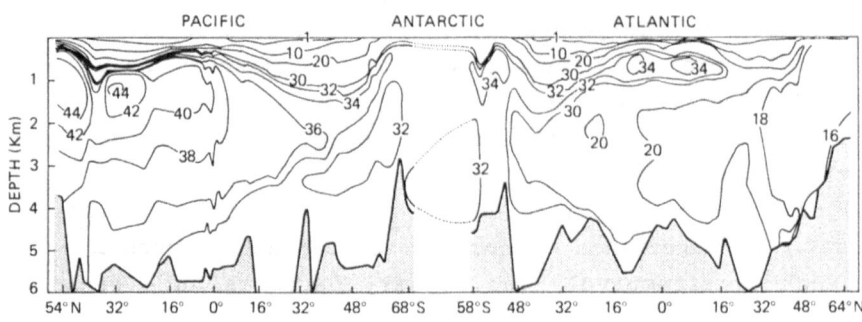

Fig. 4. Dissolved inorganic nitrate concentration (μmole kg^{-1}) along north-south transects (see upper panel) of the Pacific and Atlantic Oceans (adapted from Sharp, 1983).

interior estuarine "null zone" decomposition processes are largely maintained by; 1) organic fluxes from landward sources such as nonviable freshwater phytoplankton and riverborne detritus (cf. Cadee, 1978; Van Bennekom et al., 1978); 2) organic fluxes seaward of the null zone (cf. Morris et al., 1982); or, in some instances, 3) in situ organic sources as the nearby shoals (Cloern et al., 1983). All of these factors can obscure estuarine/coastal ocean chemical budgets. For example, even after several decades of surveys, identifying the dominant sources maintaining dissolved silica concentration maxima in the freshwater tidal regime of estuaries remains controversial (Anderson, 1986).

2.3 Direct and indirect estimates

Measurement of estuarine/coastal ocean fluxes is a fundamental and difficult problem that usually requires high space-time resolution (Kjerve and Seim, 1984; Wilmot et al., 1985; Uncles et al., 1985). Understandably such large data sets are not available for C, N, Si and P and other chemicals in large rivers. Thus indirect flux estimates are the usual recourse. One method, often used by geochemists, is to assume that the coastal ocean receives an amount equal to the river mass transport. This estimate can also be modified by subtracting possible losses of riverine materials due to estuarine sedimentation or other processes. The importance of such losses can depend upon many factors including how effectively the estuary performs as a sediment trap (Schubel and Carter, 1984; Biggs and Howell, 1984), what form the chemical substances are in (i.e. particulate or dissolved), the role of solid surface activity (e.g. as in phosphate versus silica, Mayer and Glass, 1980) and estuarine biological processes (e.g. photosynthesis, nitrification/denitrification). Considerably more is known about riverine transport than estuarine processes and the rates at which they can modify this transport to the sea. In some instances, then, a riverine estimate of transport to the coastal ocean can be too high.

3. A LARGE-SCALE PERSPECTIVE

With the above cautions and controversies as a backdrop we boldly characterize the magnitude of riverine/estuarine C, N, Si, and P transport to the coastal ocean. First we review global estimates of their riverine sources. Then these estimates are compared to very simple global estimates of general estuarine processes: phytoplankton productivity, water-column mineralization and benthic exchange. These are not, of course, all of the pertinent processes, but we have some understanding of their biochemical effects on riverine sources.

3.1 Global runoff and chemistry

Global land area, 149 x 10^6 km^2 has a mean specific water yield of 1 m^3 s^{-1} per 100 km^2 (UNESCO, 1978). As discussed in Meybeck (1982) about two thirds of this global land size, or 100 x 10^6 km^2, is drained by rivers to the sea with a specific water yield of 1.18 m^3 s^{-1} per 100 km^2 river basin (Baumgartner and Reichel, 1975, cited in Meybeck, 1982). For comparison, the ten largest world rivers have a water yield of 1.4 m^3 s^{-1} per 100 km^2 whereas the next in size (20 to 60) have a yield of only 0.66 m^3 s^{-1} per 100 km^2 (Peterson et al., 1985; note decimal error, p. 43, Table 4). Thus the world's largest rivers tend to have a higher specific yield than intermediate sized rivers (Leopold, 1962).

A synthesis of diverse literature on riverine chemistry has provided estimates of carbon, nitrogen, silica and phosphorus concentrations in global runoff (Table 2, 3). These estimates include annual anthropogenic N and P inputs as measures of eutrophication effects on the coastal zone (assuming Redfield's ratios of C:N:P, Peterson and Melillo, 1985). Our purpose here is to put this global riverine source into some perspective via crude global estimates of estuarine sources and sinks and some coastal ocean fluxes. A base for such comparisons is the global estuarine area.

Global estuarine area is 1.06 x 10^{12} m^2, or 1.36 x 10^{12} including the Baltic Sea (Woodwell et al., 1973). Using this value the global ratio of land area to estuarine area is 94 to 1 without the Baltic Sea and 74 to 1 with it. In order to simplify the estuarine mass balance problem and be able to relate to any global estimate of estuarine area, first consider the following series of assumptions: 1) the mean replacement time of the fresh water entering the world's estuaries is, more or less, one month; 2) the mean estuarine depth is 5 m; 3) the mean salinity is, approximately, 16 parts per thousand (e.g. the fresh water volume is half of the total volume using a salinity of 32 for ocean water and 0 for river water); and 4) global runoff is 3.7 x 10^{13} m^3 y^{-1} (from Meybeck, 1982). Based on these largely hypothetical assumptions,

Table 2. Global riverine dissolved silica transport to the ocean.

TRANSPORT (moles y^{-1})	REFERENCE
7.0 x 10^{12}	Livingstone, 1963[1]
6.5 x 10^{12}	Meybeck, 1979[2]

[1] Cited in DeMaster, 1981

[2] Based on an estimated world mean riverine silica concentration, 173 μmole l^{-1} (Meybeck, 1979) and river flow, 3.74 x 10^{13} m^3 y^{-1} (Baumgartner and Reichel, 1975 cited in Meybeck, 1982).

Table 3. Global geostatics of riverine C, Si, N and P composition.

CHEMICAL COMPOSITION	RIVERINE CONCENTRATION (umole l^{-}1)	MASS EMISSIONS PER LAND AREA (mmole m^{-2} d^{-1})	MASS TRANSPORT (10^{12} moles y^{-1})	RATIO
POC + DOC[1]	880	0.88	33	110
DSi[2]	170	0.18	6.5	22
PON + DON+[3] DIN + AN	80	0.08	3	10
POP + DOP+[4] DIP + AP	8	0.008	0.3	1

[1] From Meybeck, 1982, Figure 6, p. 435, POC is particulate organic carbon DOC is dissolved organic carbon.

[2] See Table 1, DSi is dissolved silica

[3] From Meybeck, 1982, Figure 7, p. 436, PON is particulate organic nitrogen (PON + DON = 72% of Total), DIN is dissolved inorganic nitrogen (DIN = 12% of Total), AN is 1970 anthropogenic nitrogen (AN = 16% of the Total).

[4] From Meybeck, 1982, Figure 8, p. 437, POP is particulate organic phosphorus, (POP + DOP = 92% of the Total), DIP is dissolved inorganic phosphate phosphorous (DIP = 4% of the Total), AP is 1970 anthropogenic phosphorous (AP = 4% of the Total).

global estuarine area is estimated to be 1.2 x 10^{13} m^2. Using this
value the global ratio of continental area drained by rivers to the
ocean (1 x 10^{12} m^2, Meybeck, 1982) to global estuarine area is 83 to 1.
Thus we assume that the "global estuarine area" is, more or less, 1.2 x
10^{12} m^2.

3.2 Estuarine modifications

Because of the serious complications of estuarine variability and
circulation noted earlier, it seems premature to consider here all of
the processes that can modify estuarine and coastal ocean fluxes. Thus
the estuarine source/sink processes are considerably simplified to
include mean-annual values of phytoplankton production, water column
respiration and benthic exchange. Phytoplankton productivity, a major
estuarine nutrient sink and a major estuarine organic matter source, is
considered first.

It is noteworthy that a review of estuarine phytoplankton production of
only about a decade ago (Woodwell et al., 1973) found virtually no
estimates of productivity from estuaries of the world published in the
open literature and, therefore, was limited to a few observations from
the adjacent coastal ocean including river plumes. In a more recent
review, 27 annual estimates of estuarine phytoplankton productivity
were available ranging from about 30 to 600 gC m^2 y^{-1} (Boynton et al.,
1983). These annual values can reflect a variety of factors including
the availability of light due to high concentrations of suspended
sediment (Fig. 5). Even in turbid coastal plumes the productivity can
be suppressed until water clarity increases due to sedimentation and
dilution with ocean waters (Edmond et al., 1981). A major uncertaintly
then, in arriving at a reasonable estimate of phytoplankton
productivity in the world's estuaries is some notion of their typical
range in estuarine suspended sediment concentrations. Such values are
reported in geologic and engineering literature but are frequently
absent in biological studies (which, unfortunately, are often conducted
in estuaries or locations different from those of geologic and
engineering studies). Nevertheless, for the purpose of serving as a
point of reference, we assume global mean net photic phytoplankton
productivity is 250 gC m^{-2} y^{-1}, (with a probable range of $\pm$ 100 gC m^{-2}
y^{-1}.

238

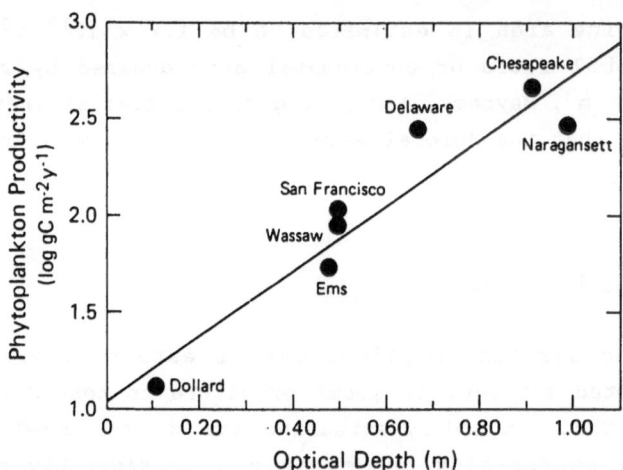

Fig. 5. Mean annual phytoplankton productivity in estuaries in relation
to optical depth, the reciprocal of the diffuse light extinction
coefficient (from Peterson et al., 1985).

Aphotic estuarine water column organic mineralization rates are very
poorly known. Preliminary estimates from a variety of estuaries
indicates 12 μmole l^{-1} d^{-1} is reasonable, apparently depending on
organic input and water depth or dilution (Fig. 6, upper panel).

There appears to be more information regarding rates of estuarine
benthic nutrient exchange than water column mineralization. Benthic N
and P fluxes in Table 4 are from a literature review of Callender and
Bennett (cited in Peterson et al., 1985). Selection of a global average
estuarine benthic oxygen consumption (Table 4) was guided by the work
of Nixon (Fig. 6, lower panel). Silica flux is typical of observed
rates as per Hammond et al. (1985) and D'Elia et al. (1983).

Of course there is considerable latitude for adjustment of net
phytoplankton productivity, aphotic mineralization and benthic exchange
estimates (Table 4). Also, other possible sources and sinks or other
methods of making such estimates might be considered. For instance
DeMaster (1981) estimated estuarine biogenic silica removal to be 20
percent or less of the riverine transport, or, about 1.3×10^{12} moles
SiO_2 y^{-1} (from Table 2). From this, an estimate of silicious plankton
carbon assimilation is 0.7×10^{13} moles C y^{-1}, assuming a C:Si ratio of
106:20 (Kaul and Froelich, 1984). As expected, this is less than our
estimate for total assimilation, about one third of the total value in
Table 4.

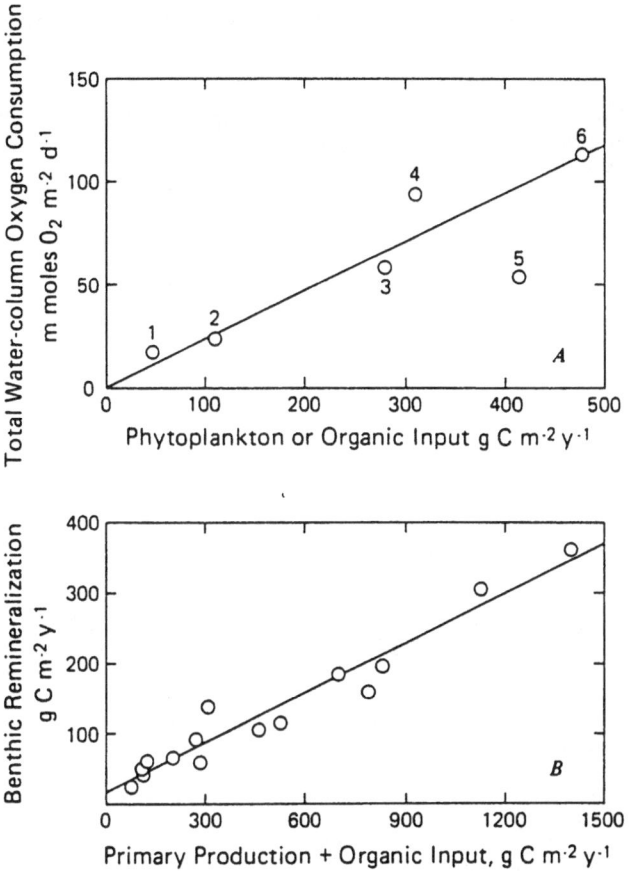

Fig. 6. Panel A: Water-column oxygen consumption versus organic input:
1) Newport Estuary (Williams & Murdoch, 1966; Williams, 1966) assuming
mean-annual oxygen consumption is 13 μmole l^{-1} d^{-1} and mean depth is
1.2 m; 2) northern San Franciso Bay estuary (Peterson, 1979; Cole &
Cloern 1984) assuming mean-annual oxygen consumption is 6 μmole l^{-1} d^{-1}
and mean depth is 4 m; 3) Wassaw estuary (Turner, 1978; Turner et al.,
1979) assuming mean-annual phytoplankton productivity is 90 and
macroalgae organic exudate is 190 gC m^{-2} y^{-1}, mean annual oxygen
consumption is 14 μmole l^{-1} d^{-1}, and mean depth is 4 m; 4) Narragansett
Bay (Furnas et al., 1976; Oviatt et al., 1981) assuming mean annual
oxygen consumption is 10 μmole l^{-1} and mean depth is 9m; 5) Dollard
estuary (Van Es and Ruardij, 1982) assuming mean annual phytoplankton
productivity is 16 μmole l^{-1}, and mean depth is 3.3 m; 6) Chesapeake
Bay (Boynton et al., 1982; Kemp & Boynton, 1981) assuming mean-annual
oxygen consumption is 16 μmole l^{-1} d^{-1} and mean depth is 7 m. Panel B:
Benthic oxygen consumption (assuming $CO_2:O_2$ = 1:1) versus organic input
(adapted from Nixon, 1981).

Table 4. Global river source of C, Si, N and P and estimates of estuarine sources and sinks

Chemical Species	River[1] Source (moles y^{-1})	Photic Net[2] Productivity (moles y^{-1})	Aphotic[3] Mineralization (moles y^{-1})	Benthic[4] Exchange (moles y^{-1})
C	3.3×10^{13}	2.5×10^{13}	-1.3×10^{13}	-1.3×10^{13}
Si	6.5×10^{12}	-4.7×10^{12}	2.5×10^{12}	2.6×10^{12}
N	3.0×10^{12}	-3.8×10^{12}	2.0×10^{12}	1.3×10^{12}
P	3.0×10^{11}	-2.4×10^{11}	1.2×10^{11}	0.9×10^{11}

[1] From Table 3.

[2] Assuming global estuarine area is 1.2×10^{12} m^2, net productivity is 250 gC m^{-2} d^{-1} and Redfield ratios are C:Si:N:P = 106:20:16:1. A negative value means loss from the dissolved phase to particulate organic matter.

[3] Assuming global estuarine area is 1.2×10^{12} m^2, an average estuarine depth is 5 m, average photic depth is 2.5 m, average rate of oxygen consumption is 12 μmole 1^{-1} d^{-1}, a ratio of oxygen consumption to carbon mineralization is 1:1 and Redfield ratios are C:Si:N:P = 106:20:16:1. A negative value means mineralization of organic matter.

[4] Assuming global estuarine area is 1.2×10^{12} m^2, average benthic oxygen consumption is 30 mmoles m^{-2} d^{-1}, a ratio of oxygen consumption to carbon mineralization 1:1, Si flux = 6 mmole m^{-2} d^{-1} NH_4 flux = 3 mmole m^{-2} d^{-1}, PO_4 flux = 0.2 μmole m^{-2} d^{-1}. A negative value means mineralization of organic matter.

Budget estimates are satisfying to make when they help us identify or understand major elements of a system. It is in this frame of mind that we refer to results of annual estuarine sources and sinks, global or otherwise. In an overview of the results in Table 4 the values for net photic productivity and benthic exchange are similar largely because both estimates are based on their estimated global area (considered within $\pm$ 50 percent by Woodwell et al., 1973). Benthic oxygen consumption, a measure of organic matter mineralization, is 22 percent of the total riverine organic input plus net estuarine photic productivity. It is instructive to note that the river inputs and benthic sources appear to be of similar magnitude. To clearly understand why this is so let us compare two divergent estuaries, Chesapeake Bay and San Francisco Bay, using annual estimates of their riverine and benthic dissolved silica sources.

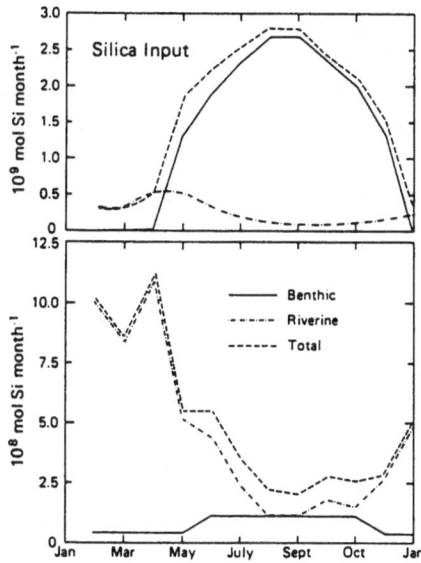

Fig. 7. Upper panel: Riverine versus benthic dissolved silica input to Chesapeake Bay (adapted from D'Elia et al., 1983). Lower Panel: Riverine versus benthic dissolved silica input to northern San Franscisco Bay. Riverine silica from Peterson et al., 1986; benthic silica from assuming an area of 6×10^8 m^2, a winter flux of 2 mmoles S_1O_2 m^2 d^{-1} and a summer flux of 6 mmoles S_1O_2 m^2 d^{-1} (see Hammond et al., 1985). Note the reverse roles of benthic/riverine input.

Benthic inputs of dissolved silica in Chesapeake Bay are larger than the riverine inputs principally because the ratio of river basin area to estuarine area is about 15:1 (Fig. 7, upper panel). In contrast, benthic inputs of dissolved silica in San Francisco Bay are smaller than the riverine inputs (Fig. 7, lower panel) principally because the ratio of river basin area to estuarine area is nearly 100:1. The ratio for our global estuary is closer to the ratio for San Francisco Bay than Chesapeake Bay. Accordingly, under the assumptions used in the estimates, riverine sources are nearly equivalent to or stronger than benthic sources on a global scale.

On a global scale average P fluxes appear to be sufficient to support annual net estuarine phytoplankton productivity but the situation is less certain for N (Table 4). Intuitively, the role of elemental recycling appears to be less critical in our global estuary with a ratio of river basin area to estuarine area of 83 to 1 than if it were an order of magnitude lower. Generally as the ratio of river basin area to estuarine area decreases, the ratio of river input to estuarine net photic productivity also probably decreases. We assume as this ratio decreases, for instance as the proportion of estuarine area increases, elemental recycling becomes more and more important and some elements are used many times before their ultimate fates such as transport to the sea or estuarine burial are decided.

In our global-mean estuary, net photic productivity is almost completely mineralized either in the aphotic water column or on the bottom. Presumably only a small fraction is permanently buried in sediment or available for transport to the coastal ocean. It seems that as organic sources increase, such as net photic productivity, so do water column and benthic mineralization rates (Fig. 6). It seems that after net photic phytoplankton productivity supports aphotic and benthic mineralization, any leftovers for coastal exchange are extremely difficult to quantify by empirical methods (with the possible exception of certain riverine organic substances or tracers such as lignins).

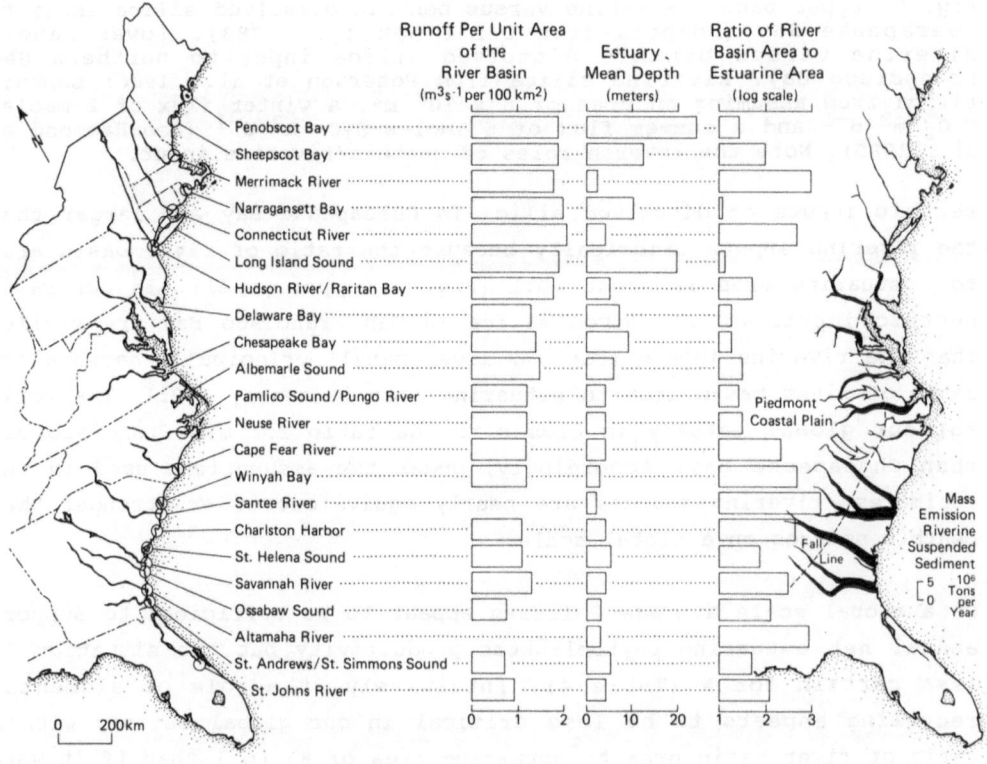

Fig. 8. River basin-estuary statistics for eastern seaboard, United States (data from National Oceanic and Atmospheric Administration, 1985). Riverine suspended sediment adapted from Meade, 1969).

A generalization to be made here is that leakage of riverine organic matter from an estuary to the coastal ocean probably increases as the ratio of river basin to estuarine area increases. For example, the southern United States continental shelf appears to be an excellent candidate for such transfers because the ratio of river basin to estuarine area is clearly greater than 100 for many of its estuaries (Fig. 8). Furthermore, the probability of net transfer is also high because these estuaries are relatively shallow and the strength of estuarine circulation (or return flow from the coastal ocean) decreases strongly with decreasing water depth (Festa and Hansen, 1976). These rivers are also characterized by extremely high suspended sediment concentrations and, therefore, probably lower Si, N and P losses to photosynthesis. Interestingly, however, remote sensing images (Atkinson and Menzel, 1985, p.4 Fig. 2) and other surveys (Bingham, 1973) indicate turbid river plumes in this region are restricted to a relatively narrow band along the inner shelf. But even here, where we have an obvious candidate for major estuarine-coastal ocean exchange, the larger scale onshore fluxes from the open ocean still dominate the coastal nutrient budgets. In the Georgia Bight alone, ocean transport of dissolved inorganic nitrogen is 1.6×10^{11} moles N y^{-1} (Atkinson, 1985). This, from a tiny segment of the world's coastal ocean, is equivalent to 26 percent of the <u>total</u> global riverine dissolved inorganic plus anthropogenic input!

From the above example it seems reasonable to expect that on an annual basis nutrient transport into and along the world's coastal waters is dominated by oceanic processes. Until detailed observations can demonstrate otherwise, riverine/estuarine nutrient sources to coastal waters are considered secondary to the ocean except locally. Our basis for this conclusion was drawn from the eastern seaboard of the United States which has a relatively wide shelf and probably reduced cross-shelf circulation with resulting low ambient dissolved inorganic nitrogen concentrations (Walsh, 1983; Walsh et al., 1985). The shelf along western United States is relatively narrow. Oceanic mechanisms of nutrient transport into this coastal region includes eastern boundary currents (Chelton et al., 1983) and upwelling (Bakun and Parrish, 1980). Because of the narrow shelf, oceanic processes are expected to be even more important than for a broad shelf (the eastern seaboard), even adjacent to a major river source such as the Columbia River (Carpenter, in press). In fact plankton biomass in this narrow shelf region is linked to the California Current's strength and position which in turn depends on the cyclonic/anticyclonic atmospheric circulation on a larger basin scale (Chelton et al., 1983) and coastal upwelling (Bakun and Parrish, 1980). In summary, the above suggests that on an annual time scale riverine nutrient fluxes are probably important to the coastal ocean only locally (with the possible exception of the world's largest rivers).

In a more local analysis, based specifically on individual rivers, the Hudson River/New York Bight region has been identified as a coastal region with significant anthropogenic nutrient sources (Garside et al., 1976). But even here anoxia biocycles appear to be dominated by anomalous large-scale atmospheric/coastal ocean circulation patterns (Falkowski et al., 1980).

4. SYNOPSIS

Practical considerations generally constrain annual estuarine experiments to minimal sampling densities in time or space or both. Estuarine variability is a major complication in any attempt at developing relatively tight chemical budgets. Another topic, not considered here, is experimental error inherent in methods. In many instances the estuarine processes of interest do not lend themselves to highly quantitative estimates, for example, net photic phytoplankton productivity per unit area. These limitations make estimates of system-wide annual budget errors tenuous. What, for example, is the error in annual summaries of net photic phytoplankton productivity over an entire estuary? To our knowledge interannual or other experimental designs have not yet identified the magnitude of such error for system-wide annual budgets. Because we are lacking this knowledge we envision differences in annual source/sink estimates less than $\pm$ 10 percent to be insignificant. In fact, for most annual estimates of total estuarine system sources and sinks including river inputs, net photic phytoplankton productivity, aphotic mineralization and benthic exchange, it would be optimistic to be able to statistically support differences less than $\pm$ 25 percent as meaningful, and perhaps more commonly, differences less than $\pm$ 50 percent or even larger. Fortunately this is not always the case; sometimes differences in replicate mass budget estimates of large space-time averages are on the order of $\pm$25 percent or less. For example, Meybeck's (1979) riverine global silica estimate is only 8 percent less than an earlier estimate (Table 2); Walling and Webb (1983) considered small a 17 percent difference between their independent estimates of the global mean riverine total dissolved solids load to the ocean (33.3t km^2 y^{-1}) and Meybeck's (1979) estimate (38.8t km^2 y^{-1}); Woodwell et al. (1973) reported the Baltic Sea surface area to be 3 x 10^{11} m^2, a number which is 30 percent less than more recent figures (4 x 10^{11} m^2 (Jansson et al., 1984), etc. This level of differences averaged over large areas is often negligible (for instance, in our very simple example based on large differences in silica flux caused by large differences in area, Fig. 7). The differences, however, in specific rates (e.g. per unit area) are considerably more difficult to identify and interpret. For example, D'Elia et al. (1983) consider the Chesapeake Bay mean-annual benthic dissolved silica flux, 7.5 mmole m^{-2} d^{-1}, as a conservative estimate for this system. Helder et al. (1983) estimate a mean-annual flux of 2.3 μmole $m^{-2}d^{-1}$ for the Ems-Dollard estuary, an estimate which is one third that of the Chesapeake Bay. Such differences might be caused by a combination of factors including differences in species composition and productivity, water temperature, and, possibly, methodology. Clearly identifying the causes of such differences is an important topic of ongoing and future research.

From a broad and preliminary global budget perspective riverine/estuarine sources of C, Si, N and P to the coastal zone appear to be of secondary importance compared to oceanic sources except locally. Even within estuaries, where riverine nutrient sources are more important, the role of riverine sources to estuarine biogeochemistry is difficult to sort out (Officer et al., 1984; Seliger et al., 1985). Within estuaries vertical Si and N transport and exchange processes are often on the order of mmoles m^{-2} d^{-1} and P fluxes 0.1 mmoles m^{-2} d^{-1}. Such vertical rates are typically the same or even higher for soil/plant exchange process in the upland river basin whereas the horizontal transport or leakage from a river basin is often only a few percent of its vertical fluxes when averages over its entire basin. Therefore, riverine sources can rival vertical estuarine fluxes, such as benthic fluxes, when the river basin is roughly two orders of magnitude greater than the estuarine area. This is close to the ratio of global river basin area to estuarine area (ca. 83). Thus it seems useful to identify one extreme where river inputs are direct to the sea (Fig. 9, coastline - river); an example is the Amazon River where the null zone is positioned over the continental shelf (Gibbs, 1977). Another extreme would be where estuarine river inputs are inland from the coastline (Fig. 9, coastline - ocean); an example is Chesapeake Bay and its tributary estuaries. Also, some systems are probably in close balance between these extremes (e.g. the Columbia

River estuary, Hansen, 1965), where the null zone is close to the mouth of the estuary depending on river flow (Fig. 9, coastline - null zone). Furthermore, some estuaries are more like one extreme during peak flows and the other during droughts.

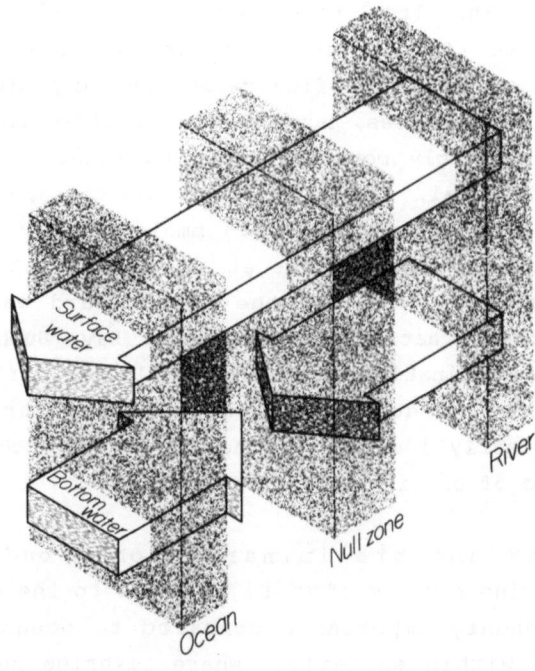

Fig. 9. A highly idealized ocean/estuarine/riverine circulation pattern

To some extent the above general characteristics of estuaries are revealed simply from a listing of publications because scientists study processes which are measurable and important. For example, two estuaries, Narragansett Bay and Connecticut River, are geographically adjacent (Fig. 8) but their respective ratios of river basin to estuarine area are 10.8 and 555. Accordingly, Narragansett Bay has been the site of studies of recycling processes and benthic fluxes (cf. Nixon et al., 1976, 1980; Nixon, 1981; Seitzinger et al., 1980; Oviatt et al., 1981; Garber, 1984; Kelley and Nixon, 1984; Nowicki and Nixon, 1985). In contrast, the Connecticut River is known for its studies of extra-estuarine river plumes (cf. Garvine, 1874; Garvine and Monk, 1974; Garvine, 1977, 1982). In the Baltic Sea, with a 4.3 ratio of river basin to sea area (Table 5) a strong anthropogenic signal has not yet been identified (Jansson et al., 1984). Apparently the long term effects of anthropogenic increases in nutrient loads are obscured by natural nutrient recycling variability.

Table 5. Geonutrient statistics of the Baltic Sea Region[1]

CHEMICAL COMPOSITION	RIVERINE CONCENTRATION (μmole 1^{-1})	MASS EMISSIONS PER LAND AREA (mmole m^{-2} d^{-1})	MASS TRANSPORT (10^9 moles y^{-1})	RATIO
N	67	0.22	32	25
P	2.7	0.0087	1.3	1

RIVER FLOW (m^3 s^{-1})	RIVER FLOW PER UNIT AREA OF THE RIVER BASIN (m^3 s^{-1} per 100 km^2)	BALTIC SEA AREA (km^2)	RATIO OF RIVER BASIN AREA TO BALTIC SEA
15,000	0.87	400,000	4.3

[1] Adapted from Jansson et al., 1984; Rosenberg, 1985.

To understand how nutrients control the biochemistry of estuarine-coastal systems we also need to understand how the nutrient concentrations themselves are controlled. We generalized that for some estuaries riverine dissolved Si, N and P are recycled once or not at all before reaching the coastal zone. In other estuaries, most likely with even smaller riverine sources relative to estuarine sinks, dissolved Si, N and P are recycled many times before reaching the coastal zone (or sediment burial or loss to the atmosphere). Obviously, however, such an extremely broad-brush approach leaves many questions. Note for example, the "anomalously" low oxygen consumption associated with waste sources in Fig. 6 (panel A, the data point labelled 5). Does this anomaly represent 1) noise, 2) an artifact of the averaging method, or 3) a real effect (note that Sayler and Gilmore (1978) report a low waste organic carbon: oxygen consumption ratio, approximately 2:1 by moles). Also, it is well known that riverine nutrient concentrations are controlled by a series of factors in addition to natural and anthropogenic causes. What, for example, causes the unusually low dissolved inorganic nitrogen and phosphorous concentrations in the Umpqua estuary (Fig. 10, State of Oregon, western United States). Are such low values due to 1) the low human population density in the river basin or, in addition, 2) high regional precipitation? Low chemical concentrations are also reflected in the river, which indicated the lowest total organic carbon values observed in a survey of North

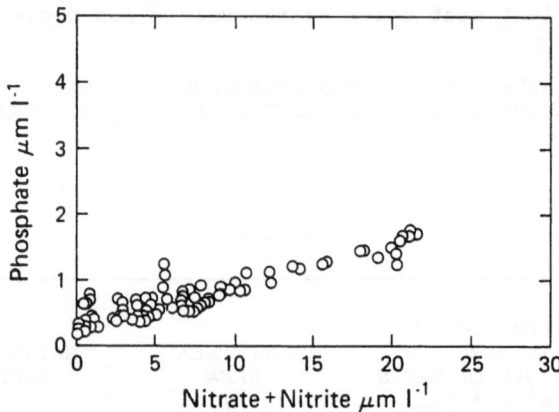

Fig. 10. Range in dissolved inorganic nitrate plus nitrite and phosphate concentrations Umpqua River estuary, Oregon, composite of August 7, 8 and October 3,4,5, 1974 and March 16, 1975.

American rivers Mulholland and Watts, 1982). Nutrients from a relatively undisturbed river basin can be effectively washed out early on in the precipitation-runoff cycle depending on the history (strength and timing) of precipitation events. But in a densely populated river basin such washout may be seldom achieved. Apparently this high-precipitation washout is so relentless in some regions of southwestern Canada and northwestern United States (presumably over the last thousands of years, Huesser et al., 1985) that the downstream estuarine receiving water concentrations can be higher than in their rivers (Naiman and Siebert, 1978).

Perhaps a major uncertainty in the budget approach used here is a lack of detailed knowledge of the role of riverine particulate and dissolved organic C, N and P in the recycling process. As estimated by Meybeck (1982) these forms are a substantial part of the budget. In a look to the future we expect more cross-discipline research to unify the details of the important source/sink processes on larger spatiotemporal scales than generally attempted in the past.

ACKNOWLEDGEMENTS

We thank B.-O. Jansson, F. Nichols, M. Pamatmat, J. Sharp and L. Smith for their reviews of the manuscript and J. Dileo-Stevens and M. Nichols for technical assistance.

REFERENCES

Anderson, G.F:, 1986. Silica, diatoms and a freshwater productivity maximum in Atlantic coastal plain estuaries, Chesapeake Bay. - Est. Coast. Shelf Sci. 22: 183-197.

Atkinson, L. P. & D.W. Menzel, 1985. Introduction; Oceanography of the southeast United States Continental Shelf. - In L.P. Atkinson, D.W. Menzel & K.A. Bush (eds.): Oceanography of the southeastern U.S. continental shelf, Coastal and estuarine sciences 2, pp. 1-9. American Geophysical Union, Washington, D.C.

Atkinson, L.P., 1985. Hydrography and nutrients of the southeastern U.S. continental shelf. - In L.P. Atkinson, D.W. Menzel & K.A. Bush (eds.): Oceanography of the southeastern U.S. continental shelf, Coastal and estuarine sciences 2, pp. 77-92. American Geophysical Union, Washington, D.C.

Bakun, A. & R.H. Parrish, 1980. Environmental inputs to fishery population models for eastern boundary current regions. - In G.D. Sharp (ed.): Workshop on the effects of environmental variation on the survival of larval pelagic fishes. IOC Workshop Rep. 38: 67-104.

Bascom, W., 1982. The effects of waste disposal on the coastal waters of Southern California. - Environ. Sci. Technol. 16: 226A-236A.

Baumgartner, A. & E. Reichel, 1975. The world water balance. Elsevier, Amsterdam. 179 pp.

Bigham, G.N., 1973. Zone of influence - inner continental shelf of Georgia. - J. Sed. Pet. 43: 207-214.

Biggs, R.B. & B.A. Howell, 1984. The estuary as a sediment trap; alternate approaches to estimating its filtering efficiency. - In V.S. Kennedy (ed.): The estuary as a filter, pp. 107-130. Academic Press, Inc., London.

Boynton, W.R., C.A. Hall, P.G. Falkowski, C.W. Keefe & W.M. Kemp, 1983. Phytoplankton productivity in aquatic ecosystems. - In O.L. Lange, P.S. Nobel, C.B. Osmond & H. Ziegler (eds.): Physiological plant ecology II. Encyclopedia of plant physiology New series vol. 12B, pp. 305-327. Springer-Verlag, New York.

Boynton, W.R. W.M. Kemp & C.W. Keefe, 1982. A comparative analysis of nutrients and other factors influencing estuarine phytoplankton production. - In V.S. Kennedy (ed.): Estuarine comparisons, pp. 69-90. Academic Press, New York.

Broecker, W.S., D.M. Peteet & D. Rind, 1985. Does the ocean-atmosphere system have more than one stable mode of operation? - Nature 315: 21-26.

Cadee, G.C., 1978. Primary production and chlorophyll in the Zaire River, estuary and plume. - Neth. J. Sea Res. 12: 368-381.

Carpenter, R. (in press). Has man altered the cycling of nutrients and organic C on the Washington continental shelf and slope? - Deep-Sea Res.

Chelton, D.B., P.A. Bernal & J.A. McGowan, 1982. Large-scale interannual physical and biological interaction in the California current. - J. Mar. Res. 40: 1095-1125.

Cloern, J.E., A.E. Alpine, B.E. Cole, R.L. Wong, J.F. Arthur & M.D. Ball, 1983. River discharge controls phytoplankton dynamics in the northern San Francisco Bay estuary. - Est. Coast. Shelf Sci. 16: 415-429.

Cole, B.E. & J.E. Cloern, 1984. Significance of biomass and light availability to phytoplankton productivity in San Francisco Bay.- Mar. Ecol. Prog. Ser. 17: 15-25.

Craig, H., W.S. Broecker & D. Spencer, 1981. Geosecs Pacific Expedition Vol. 4, Section and Profiles. IDOE and National Science Foundation, U.S. Government Printing Office, Washington, D.C. 251 pp.

De Master, D.J., 1981. The supply and accumulation of silica in the marine environment. - Geochim. Cosmochim. Acta 45: 1715-1732.

D'Elia, C.F., D.M. Nelson & W.R. Boynton, 1983. Chesapeake Bay nutrient and plankton dynamics: III. The annual cycle of dissolved silicon. - Geochim. Cosmochim. Acta 47: 1945-1955.

Ducklow, H.W., D.A. Purdie & P.J. LeB. Williams, 1986. Bacterioplankton: a sink for carbon in a coastal marine plankton community. - Science 232: 865-867.

Edmond, J.M., E.A. Boyle, B. Grant & R.F. Stallard, 1981. The chemical mass balance in the Amazon plume I: the nutrients. - Deep-Sea Res. 18A; 1339-1374.

Falkowski, P.G., T.S. Hopkins & J.J. Walsh, 1980. An analysis of factors affecting oxygen depletion in the New York Bight. - J. Mar. Res. 38: 479-500.

Festa, J.F. & D.V. Hansen, 1976. A two-dimensional numerical model of estuarine circulation: the effects of altering depth and river discharge. - Est. Coast. Mar. Sci. 4; 309-323.

Furnas, M.J., G.L. Hitchcock & T.J. Smayda, 1976. Nutrient-phytoplankton relationships in Narragansett Bay during the 1971 summer bloom. - In M. Wiley (ed.): Estuarine processes, pp. 118-133. Academic Press, Inc., New York.

Garber, J.H., 1984. Laboratory study of nitrogen and phosphorus remineralization during the decomposition of coastal plankton and seston. - Est. Coast. Shelf Sci. 18: 685-702.

Gardner, J.C. & E. Hemphill-Haley, 1986. Evidence for a stronger oxygen-minimum zone off central California during late Pleistocene to early Holocene. - Geol. 14: 691-684.

Garside, C., T.C. Malone, O.A. Roels & B.A. Shartstein, 1976. On evaluation of sewage-derived nutrients and their influence on the Hudson estuary and New York bight. - Est. Coast. Mar. Sci. 4: 281-289.

Garvine, R.W., 1974. Physical features of the Connecticut River outflow during high discharge. - J. Geophys. Res. 79: 831-846.

Garvine, R.W., 1977. River plumes and estuary fronts. - In Estuaries, geophysics and the environment, pp. 30-35. National Academy of Sciences, Washington, D.C.

Garvine, R.W., 1982. A steady state model for buoyant surface plume hydrodynamics in coastal waters. - Tellus 34: 293-306.

Gibbs, R.J., 1977. Suspended sediment transport and the turbity maximum. - In Estuaries geophysics and the environment, pp. 106-109. National Academy of Sciences, Washington, D.C.

Hammond, D.E. C. Fuller, D. Harmon, B. Harman, M. Korosec, L.G. Miller, R. Rea, S. Warren, W. Berelson & S.W. Hager, 1985. Benthic fluxes in San Francisco Bay. - Hydrobiologia 129: 69-90.

Hansen, D.V., 1965. Currents and mixing in the Columbia River estuary. Ocean science and ocean engineering, Transactions of Joint Conference, pp. 943-953. Oceanic Technol. Soc. Amer. Soc. Limnol. Oceanogr., Washington, D.C.

Helder, W., R.T.P deVries & M.M. Rutgers van der Loeff, 1983. Behavior of nitrogen nutrients and silica in the Ems-Dollard estuary. - Can. J. Fish. Aquat. Sci. 40(1): 188-200.

Hopkins, T.S., 1978. Physical processes in the Mediterranean Basins.-
In B. Kjerfve (ed.): Estuarine transport processes, pp. 269-310.
The Belle W. Baruch Library in Marine Science number 7, University
of South Carolina Press, Columbia, South Carolina.

Huesser, C.J., L.E. Huesser & D.M. Peteet, 1985. Late-Quaternary
climate change on the American North Pacific coast. - Nature 315:
485-487.

Jansson, B.-O., W. Wilmot & F. Wulff, 1984. Coupling the subsystems-
Baltic Sea as a case study. - In M.J.R. Fasham (ed.): Flows of
energy and materials in marine ecosystems, pp. 549-595. Plenum
Publ. Corp.

Kaul, L.W. & P.M. Froelich, Jr., 1984. Modeling estuarine nutrient
geochemistry in a simple system. - Geochim. Cosmochim. Acta 48:
1417-1433.

Kelly, J.R. & S.W. Nixon, 1984. Experimental studies of the effect of
organic deposition on the metabolism of a coastal marine bottom
community. - Mar. Ecol. Prog. Ser. 17: 157-169.

Kemp, W.M. & W.R. Boynton, 1984. Spatial and temporal coupling of
nutrient inputs to estuarine primary production: the role of
particulate transport and decomposition. - Bull. Mar. Sci. 35: 522-
535.

Kemp, W.M. & W.R. Boynton, 1981. External and internal factors
regulating metabolic rates of an estuarine benthic community.-
Oecologia 51: 19-27.

Kjerfve, B. & H.E. Seim, 1984. Construction of net isopleth plots in
cross-sections of tidal estuaries. - J. Mar. Sci. 42: 503-508.

Leopold, L., 1962. Rivers. - Am. Sci. 50: 511-537.

Lewis, Jr., W.M., M.C. Grant & S.K. Hamilton, 1985. Evidence that
filterable phosphorus is a significant atmospheric link in the
phosphorus cycle. - Oikos 45: 428-432.

Livingstone, D.A., 1963. Chemical composition of rivers and lakes.-
U.S. Geol. Surv. Prof. Paper 440-G. 61 pp.

Mayer, L.M. & S.P. Glass, 1980. Buffering of silica and phosphate in a
turbid river. - Limnol. Oceanogr. 25:12-22.

McGill, D.A., 1965. The relative supplies of phosphate, nitrate and
silicate in the Mediterranean Sea. - Rapp. P.-v. Reun. Comm. int.
Explor. Mer Mediterr. 18: 737-744.

Meade, R.H., 1969. Landward transport of bottom sediments in estuaries
of the Atlantic coastal plain. - J. Sed. Pet. 39: 222-234.

Meentemeyer, V., 1978. Macroclimate and lignin control of litter
decomposition rates. - Ecology 59: 465-472.

Meybeck, M., 1982. Carbon, nitrogen, and phosphorus transport by world
rivers. - Amer. J. Sci. 282: 401-450.

Meybeck, M., 1979. Concentration des eaux fluviales en elements mareurs
et apports en solution aux oceans. - Rev. Geol. Dyn. Geogr. Phys.
21: 215-246.

Morris, A.W., D.H. Loring, A.J. Bale, R.J. Howland R.F.C. Mantoura &
E.M.S. Woodward, 1982. Particle dynamics, particulate carbon and
the oxygen minimum in an estuary. - Oceanog. Acta 5: 349-353.

Mulholland, P.J. & J.A. Watts, 1982. Transport of organic carbon to the
oceans by rivers of North America: a synthesis of existing data.-
Tellus 34: 176-186.

Naiman, R.J. & J.R. Sibert, 1978. Transport of nutrients and carbon
from the Nanaimo River to its estuary. - Limnol. Oceanogr. 23:
1183-1193.

National Oceanic and Atmospheric Administration, 1985. National
Estuarine Inventory, Data Atlas, Vol. I: Physical and hydrologic
characteristics, U.S. Dept. Commerce, Washington, D.C.

Nixon, S.W., C.A. Oviatt & S.S. Hale, 1976. Nitrogen regeneration and
the metabolism of coastal marine bottom communities. - In J.M.
Anderson & A. MacFadyen (eds.): The role of terrestrial and aquatic
organisms in decomposition processes, pp. 269-283. Blackwell
Scient. Publ., Oxford.

Nixon, S.W., J.R. Kelley, B.N. Furnas, C.A. Oviatt & S.S. Hale, 1980. Phosphorus regeneration and the metabolism of coastal bottom communities. - In K.R. Tenore & B.C. Coull (eds.): Marine benthic dynamics, pp. 219-242. Univ. South Carolina Press, Columbia, S.C.

Nixon, S.W., 1981. Remineralization and nutrient cycling in coastal marine ecosystems. - In R.J. Neilson & L.E. Cronin (eds.): Estuaries and nutrients, pp. 111-138. The Humana Press.

Nowicki, B.L. & S.W. Nixon, 1985. Benthic community metabolism in a coastal lagoon ecosystem. - Mar. Ecol. Prog. Ser. 22: 21-30.

Officer, C.B., R.B. Biggs, J.L. Taft, L.E. Cronin, M.A. Tyler & W.R. Boynton, 1984. Chesapeake Bay anoxia: origin, development, and significance. - Science 223: 22-27.

Oviatt, C.A., B. Buckley & S.W. Nixon, 1981. Annual phytoplankton metabolism in Narragansett Bay calculated from survey field measurements and microcosm observation. - Estuaries 4: 167-175.

Peterson, B.J., 1984. Synthesis of carbon stocks and flows in the open ocean mixed layer. - In J.E. Hobbie & P.J. LeB. Williams (eds.): Heterotrophic activity in the sea, pp. 547-554. Plenum Press, N.Y.

Peterson, B.J. & J.M. Melillo, 1985. The potential storage of carbon caused by eutrophication of the biosphere. - Tellus 37B: 117-127.

Peterson, D.H., R.E. Smith, S.W. Hager, D.D. Harmon, R.E. Herndon & L.E. Schemel, 1985. Interannual variability in dissolved inorganic nutrients in northern San Francisco Bay estuary. Hydrobiologia 192: 37-58.

Peterson, D.H., D.R. Cayan & J.F. Festa, 1986. Interannual variability in biogeochemistry of partially mixed estuaries: dissolved silicate cycles in northern San Francisco Bay. - In D.A. Wolfe (ed.): Estuarine variability, pp. 123-138. Academic Press, London.

Peterson, D.H., 1979. Sources and sinks of biologically reactive oxygen, carbon, nitrogen and silica in northern San Francisco Bay. - In T.J. Conomos (ed.): San Francisco Bay: The urbanized estuary, pp. 175-193. Pacific Division, AAAS.

Redfield, A.C., B.H. Ketchum & R.A. Richards, 1963. The influence of organisms on the composition of sea water. - In M.N. Hill (ed.): The sea, 2, pp. 26-77. Wiley and Sons, N.Y.

Reid, J.L., 1961. On the temperature, salinity and density difference between the Atlantic and Pacific oceans in the upper kilometre.- Deep-Sea Res. 7: 265-275.

Rosenberg, R., 1985. Eutrophication - the future marine coastal nuisance? - Mar. Pollut. Bull. 16: 227-231.

Sayler, G.S. & C.M. Gilmor, 1978. Heterotrophic utilization of organic carbon in aquatic environments. - J. Environ. Qual. 7: 385-391.

Schubel, J.R. & H.H. Carter, 1984. The estuary as a filter for fine-grained suspended sediment. - In V.S. Kennedy (ed.): The estuary as a filter, pp. 81-106. Academic Press, London.

Seitzinger, S., S. Nixon, M.E.Q. Pilson & S. Burke, 1980. Denitrification and N_2O production in near-shore marine sediments. - Geochim. Cosmochim. Acta 44: 1853-1860.

Seliger, H.H., J.A. Boggs & W.H. Biggley, 1985. Catastrophic anoxia in the Chesapeake Bay in 1984. - Science 228: 70-73.

Sharp, J.H., 1983. The distributions of inorganic nitrogen and dissolved and particulate organic nitrogen in the sea. - In E.J. Carpenter & D.G. Capone (eds.): Nitrogen in the marine environment, pp. 1-35. Academic Press, N.Y.

Stigebrandt, A., 1984. The North Pacific: a global-scale estuary. - J. Phys. Ocean. 14:464-470.

Stockner, J.G. & K.R.S. Shortreed, 1978. Enhancement of autotrophic production by nutrient addition in a coastal rainforest stream on Vancouver Island. - J. Fish. Res. Bd. Can. 35: 28-34.

Stockner, J.G. & K.S. Shortreed, 1985. Whole-lake fertilization experiments in coastal British Columbia lakes: empirical relationships between nutrient inputs and phytoplankton biomass and production. - J. Fish. Res. Bd. Can. 42: 649-658.

Turner, R.E., 1978. Community plankton respiration in a salt marsh estuary and the importance of macrophytic leachates. - Limnol. Oceanogr. 23; 442-451.

Turner, R.E., S.W. Woo & H.R. Jitts, 1979. Phytoplankton production in a turbid, temperate salt marsh estuary. - Est. Coast. Mar. Sci. 9: 603-613.

Uncles, R.J., R.C.A. Elliott & S.A. Weston, 1985. Observed fluxes of water, salt and suspended sediment in a partly mixed estuary.- Est. Coast. Shelf Sci. 20:147-168.

United Nations Educational, Scientific and Cultural Organization, 1978. World water balance and water resources of the earth. - UNESCO Press, Paris. 663 pp.

Van Bennekom, A.J., G.W. Berger, W. Helder & R.T. DeVries, 1978. Nutrient distribution in the Zaire estuary and river plume. - Neth. J. sea Res. 12: 296-323.

Van Es, F.B. & P. Ruardij, 1982. The use of a model to assess factors affecting the oxygen balance in the water of the Dollard. - Neth. J. Sea Res. 15: 313-330.

Walling, D.E. & B.W. Webb, 1983. The dissolved loads of rivers: a global overview. Dissolved loads of rivers and surface water quantity/quality relationships. - IAHS 141: 3-20.

Walsh, J.J., E.T. Premuzic, J.S. Gaffney, G.T. Rowe, G. Harbottle, R.W. Stoenner, W.L. Balsam, P.R. Betzer & S.A. MacKoll, 1985. Organic storage of CO_2 on the continental slope off the mid-Atlantic bight, the southeastern Bering Sea, and the Peru coast. - Deep-Sea Res. 32: 853-883.

Walsh, J.J., 1983. Death in the sea: enigmatic phytoplankton losses.- Prog. Oceanog. 12: 1-86.

Williams, R.B., 1966. Annual phytoplankton production in a system of shallow temperate estuaries. - In H. Barnes (ed.): Some contemporary studies in marine science, pp. 689-716. Georg Allen and Unwin, London.

Williams, R.B. & M.B. Murdoch, 1966. Phytoplankton production and chlorophyll concentration in the Beaufort Channel, North Carolina. - Limnol. Oceanogr. 11: 73-82.

Wilmot, W., P. Toll & B. Kjerfve, 1985. Nutrient transports in a Swedish estuary. - Est. Coast. Shelf Sci. 21: 161-184.

Wollast, R., 1983. Interactions in estuaries and coastal waters. - In B. Bolin & R.B. Cook (eds.): The major biochemical cycles and their interactions, SCOPE 21, pp. 385-410. John Wiley and Sons, Chichester.

Woodwell, G.M., P.H. Rich & C.A.S. Hall, 1973. Carbon in estuaries. In G.W. Woodwell & E.V. Pecan (eds.): Carbon and the biosphere, pp. 221-240. U.S. Atomic Energy Commission.

Zentara, S.J. & D. Kamykowski, 1977. Latitudinal relationships among temperature and selected plant nutrients along the west coast of North and South America. - J. Mar. Res. 35: 321-337.

III. ACTIVE TRANSPORT

FISH MIGRATIONS BETWEEN COASTAL AND OFFSHORE AREAS

J.J. Zijlstra
Netherlands Institute for Sea Research, P.O. Box 59
1790 AB, Den Burg, Texel, The Netherlands

1. INTRODUCTION

A large part of the transport of matter and energy between coastal and offshore areas depends on water movements, induced by tides, wind or other physical factors. A form of transport which is largely or partly independent of physical processes concerns active migrations of animals with adequate means of locomotion or with a behavioural pattern, which allows a selective use of water movements. This form of transport may occur in several classes of the animal kingdom but is most highly developed in fishes and larger crustacea (e.g. shrimps, prawns, crabs), various species of which may depend on coastal-offshore migrations for survival.

This paper deals with fishes, of which most adult specimens with a length of over 20 cm are able to maintain a swimming speed of 50 to over 200 cm s^{-1} (Harden-Jones, 1968). They can outswim the currents or stem tidal currents in an estuary in wait for the turn of the tide. Larval and small juvenile fish, with too small swimming power to counter the currents, often show adaptive behaviour to cross the coastal-offshore boundary. For instance, North Sea plaice (Cushing, 1975) and herring larvae on the coast of Maine (Graham, 1972) select a land ward directed undercurrent in an estuarine system by moving to the bottom layers. On the other hand, elvers of the Atlantic eel (Creutzberg, 1961), select landward directed flood currents in estuarine regions while burying in the sediment during ebb tide. A similar behaviour has been described for (sub)-adult plaice by Harden Jones et al. (1978) and de Veen (1978) and may in fact be quite common. Therefore, fishes employ various means of crossing the coastal-offshore boundary, from "riding the tides" or selecting a particular current system to outswimming the currents, sometimes changing in the course of their development.

Migration involving inshore-offshore movements are well known, as for instance those of diadromous fishes like the salmons, sturgeons, shads and eels that migrate between fresh water and the offshore region as

Lecture Notes on Coastal and Estuarine Studies, Vol. 22
B.-O. Jansson (Ed.), Coastal-Offshore Ecosystem Interactions.
© Springer-Verlag Berlin Heidelberg 1988

part of their reproductive behaviour. Another form of large-scale, but less conspicuous, migration is undertaken by various fish species, of which the adults inhabit the offshore area while their juveniles occupy a coastal nursery (Table 1). In the offshore and the continental

TABLE 1

Presence of the most dominant species in the offshore North Sea and in the continental coastal area of the Netherlands, Germany and Denmark, according to commercial catches and bycatches (from HOLDEN, 1978 and TIEWS, 1978).

Dominant species	North Sea	Continental coastal area
Pollachius virens	+	
Melanogrammus aeglifinus	+	
Trisopterus esmarkii	+	
Scomber scrombus	+	
Ammodytes spec.	+	juv.
Limanda limanda	+	juv.
Pleuronectes platessa	+	juv.
Solea solea	+	juv.
Gadus morhua	+	juv.
Merlangius merlangus	+	juv.
Clupea harengus	+	juv.
Agonus cataphractus		+
Liparis liparis		+
Zoarces viviparus		+
Myoxocephalus scorpius		+
Osmerus eperlanus		+
Pomatoschistus spec.		+

coastal waters of Denmark, Germany and the Netherlands, 7 out of 17 common species have only juveniles inhabiting the coastal zone, which seems to serve as a nursery (Zijlstra, 1972). This distribution implies an offshore migration of the fish at the end of or during the juvenile stage. The resulting transport of living matter could be considerable, as the species concerned constitute about 60% of the annual North Sea fish landing of 2-3 million tons (Holden, 1978).

Similar distributions, with juvenile fish inhabiting the coastal zone, have been observed in other parts of the world: on the west coast of Britain (Claridge et al., 1986), along the east coast of the United States (Smith et al., 1966; Gunter, 1967; McHugh, 1967) and in the northwest Pacific (Pearcy and Myers, 1974; Misitano, 1977). Also in tropical areas a considerable part of the fish fauna in the coastal zone may consist of juveniles from offshore species: the Terminos Lagoon in Mexico (Yanes-Arancibia et al., 1980; Yanes-Arancibia and Day, 1982), an estuarine area in Indonesia (Hardenberg, 1931), mangrove swamps in Puerto Rico (Austin, 1971), and some lagoons in Guadeloupe (Louis and Lasserre, 1982). In most cases a large part, up to 70% of the fish species present in the estuary are claimed to be represented only by juveniles that can comprise about 30-75% of the fish fauna in terms of numbers or weight.

2. TYPES OF COASTAL OFFSHORE RELATIONS IN FISH STOCKS

The importance of coastal-offshore migrations in a fish stock can only be fully comprehended after a complete study of the distribution and population-dynamics of all developmental stages of the species involved. Such studies do exist as will be shown below, but they are rare and usually concern highly valued commercially important fish species in well-studied parts of the world oceans. More often, even in commercially important species our knowledge is restricted to the sub-adults and adults. The distribution of the postlarval juvenile stage are rarely known.

Oftentimes fish fauna are intensively studied in coastal areas, which are small in comparison to the bordering offshore region, and therefore carry relatively small fish populations. Because of the smaller stocks, the movements of fish in and out of coastal waters are most clearly reflected in the fish faunas of inshore areas as estuaries or lagoons. Based on such studies six components have been recognized in estuarine and coastal fish faunas, neglecting the occasional presence of fresh-water species, which are outside the scope of this paper. These six components are:

(1) Anadromous and catadromous species.
(2) Resident species, spending their entire lives in the coastal area
(3) Marine species, spawning in the coastal region and using it as a nursery ground.
(4) Marine species, which use the coastal area as a nursery ground, but spawning and spending most of their adult life in the offshore region.
(5) Marine species, which pay regular seasonal visits to the coastal zone, usually as adult.
(6) Adventitious visitors, which appear irregularly and have no apparent estuarine or coastal requirements.

This classification is based on those of McHugh (1967), Zijlstra (1978) and Louis and Lasserre (1983), made for different regions ranging from temperate to tropical climates and may therefore have a world-wide application.

1) The diadromous group includes the most striking examples of fishes migrating across a coastal boundary. The anadromous species, the salmons, sturgeons, seatrouts, lampreys, sticklebacks, smelts, shads

and some bass (<u>Morone saxatilis</u> and <u>M. americana</u>) occur in the sea as adults and sometimes as juveniles, but migrate into rivers for spawning. Most species tend to limit their seaward extension to the coastal area, and are therefore not strongly involved in linking coastal and offshore ecosystems. However, some make extensive migrations into the open ocean as the Atlantic and Pacific salmons, which spend a year or more in the arctic region, near Greenland, Iceland and Lofoten in the Atlantic and in the Bering Sea in the Pacific. The juveniles of these species, in particular those of the Pacific salmons, feed and grow in estuaries. In some of the salmons (<u>Salmo salar</u>, <u>Oncorhynchus kisutch</u>) part of the juvenile life occurs in the river. Anadromy seems to be limited mainly to temperate regions and is probably absent in tropical waters (Haedrich, 1983). At present the contribution of anadromous species to the coupling of coastal-offshore ecosystems is likely to be small as most species have declined strongly due to engineering works and pollution in the rivers. For instance, in Chesapeake Bay the contribution of anadromous fish to the ichthyofauna was about 14% of the numbers in the late 1950's (McHugh, 1967), in most European waters it is likely to be lower (Table 2). Only the Pacific seems still to carry a significant stock of salmon (annual yield appr. 300,000 tons).

TABLE 2

The distribution of the fish fauna of 4-5 estuaries into five components: residents, seasonal visitors, nursery type, diadromous fish and adventitious visitors. Classification as given by the authors, except for Rokan estuary where it was made on the information available on the stage of the fish (juvenile, mature). A. showing the numbers of species per component, together with the percentage distribution; B. percentage distribution according to number of individuals (in case Guadeloupe: weight). Data from: Wadden Sea (DANKERS & DE VEEN, 1978), Chesapeake Bay (MCHUGH, 1967), Guadeloupe (LOUIS & LASERRE, 1982), Terminos Lagoon (YÁNEZ-ARACIBIA & DAY, 1982), Rokan estuary (HARDENBERG, 1932).

	Residents	Seasonal visitors	Nursery type	Diadromous	Adventitious
A. Number of species					
Wadden Sea	7 (9%)	19 (25%)	12 (16%)	8 (11%)	30 (39%)
Chesapeake Bay	10 (12%)	53 (65%)	9 (11%)	10 (12%)	?
Guadeloupe	7 (20%)	1 (3%)	24 (70%)	–	3 (8%)
Terminos Lagoon	12 (9%)	54 (45%)		–	55 (45%)
Rokan estuary	18 (17%)	5? (5%)	34 (33%)	–	47?(45%)
B. Number of individuals (%)					
Wadden Sea	25%	22%	45%	4%	4%
Chesapeake Bay	<1%	4 %	82%	14%	?
Guadeloupe*	46%	0.3%	53%	–	0.3%
Terminos	34%	62%		–	4%

*distribution by weight

Catadromy, the migration from fresh water to the sea to spawn, is a rare phenomenon on a world scale. It is restricted mainly to fresh water eels (<u>Anguilla</u> spec.), although sometimes mullets (Mugilidae) and some other species may be involved. Eels migrate from fresh water to the open ocean (with a reverse migration of the offspring); however,

eels may grow up in coastal water (estuaries), as reported for
Chesapeake Bay (McHugh, 1967), the Wadden Sea (Zijlstra, 1978), and
other estuaries. Whatever net transport may occur thereby between
coastal and offshore waters remains obscure.

2) The resident species are of no direct importance for our problem as
no transport or relationship across the coastal-offshore boundary of
any significance is involved. The group usually constitutes a
significant but not dominant part of the fish fauna, 9-20% of the
species and up to 46% of the numbers (Table 2). The estimates indicate
that few species have adapted to a completely coastal life, and only
moderately successful.

The poor adaptation to coastal conditions by fishes and other fauna may
be related to the usually unpredictable and adverse environment, with
large variations in temperature, salinity, oxygen saturation, pH, etc.,
leading sometimes to catastrophic mortalities (Austin, 1971). In
addition, spawning in coastal regions seems to require the production
of non-buoyant eggs. The success of species with non-buoyant eggs in
the coastal area may be related to the usually strong and
unpredictable density fluctuations occurring in these waters
(Zijlstra, 1978). In addition, strong tidal currents tend to disperse
buoyant eggs out of the coastal zone (Haedrich, 1983).

3) These constraints may explain why few marine species inhabiting the
offshore region have successfully spawned in coastal environments
(Haedrich, 1983). Well known examples of species that do in temperate
regions are some herring stocks in the Atlantic (_Clupea harengus_), and
in the Pacific (_Clupea harengus pallasi_) (Nikolsky, 1963; Hourston,
1980), the winter flounder (_Pseudopleuronectes americanus_) in the
western North Atlantic (Haedrich, 1983), the capelin (_Mallotus
cillosus_) in the North Atlantic (Gjosaeter and Saetre, 1974) and the
garfish (_Belone belone_) in the eastern North Atlantic. Some
observations suggest a similar situation in tropical and subtropical
waters (Louis and Lasserre, 1982; Bozeman and Dean, 1980). Therefore
this group, although connecting coastal and offshore ecosystems by an
inshore migration of adults and an offshore movement of immature fish,
is in terms of quantity probably of limited importance, except the
Pacific herring and the capelin in some areas where they are the
subject of important fisheries.

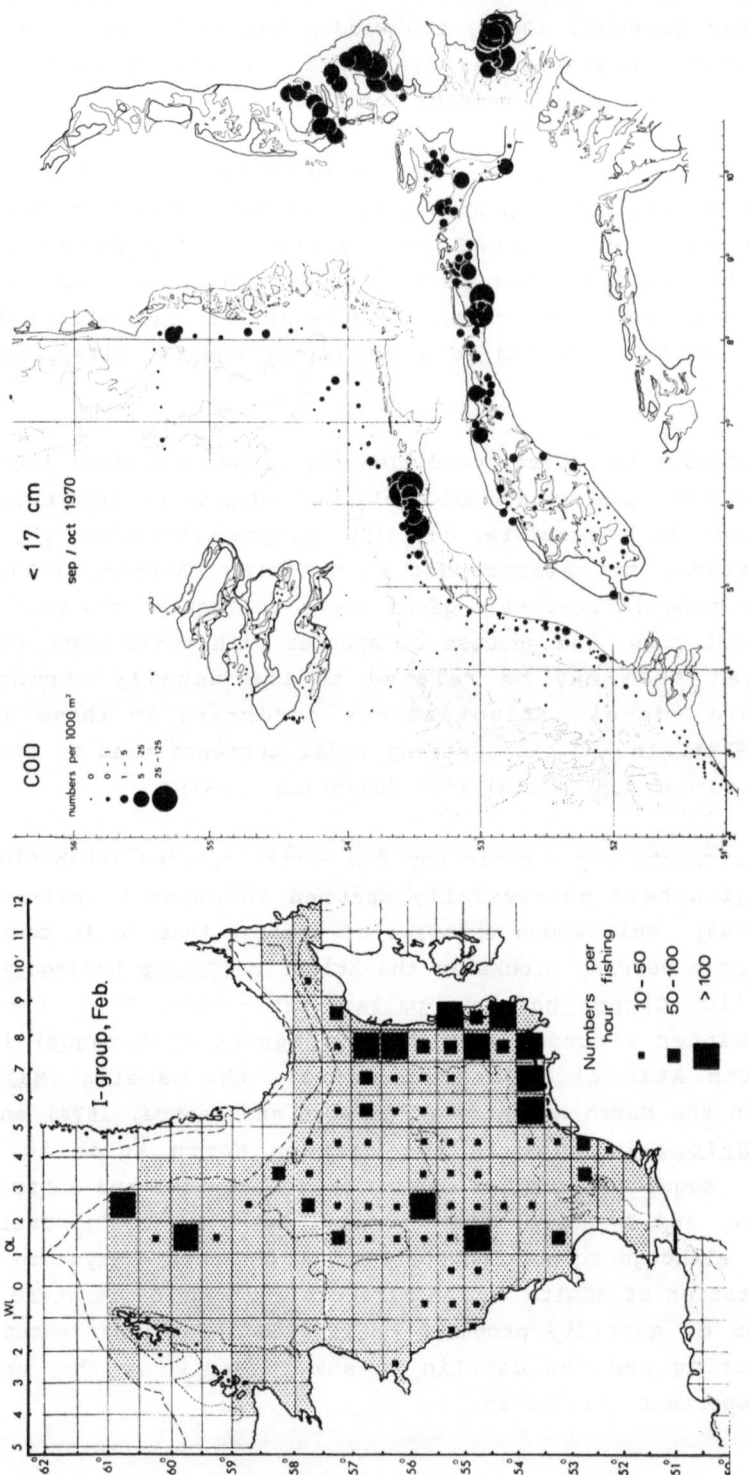

Fig 1. Distribution of juvenile cod in the offshore part (left)
and in the coastal waters of the southern part of the North Sea.
(from Daan, 1978).

4) The importance of the "nursery type group" in the fish fauna of coastal regions is demonstrated in Table 2, both for temperate and tropical regions, in terms of species numbers and the proportion of individuals or weight. In all areas the information is based on bottom- trawl catches (except in the Rokan estuary, where trap-nets were used) and may therefore underestimate pelagic fish. Among the abundant fish species of the Rokan estuary presumably half belonged to this group. Data on northern estuaries along the eastern U.S. seaboard indicate a strong share, as much as 75% in numbers, of "nursery type" representatives in the demersal fish fauna (Oviatt and Nixon, 1973). As the fish belonging to this component may enter the coastal area as larvae or early post-larvae, feed and grow there, and migrate to the open sea at some time during the juvenile stage, they may transport living matter across the coastal-offshore boundary. However, the presence of only juveniles of a fish species in a coastal area is no proof that indeed all juveniles of that species grow up in inshore waters. Most of the cod (Gadus morhua) found in the Wadden Sea are juveniles, but juvenile cod are also found, partly pelagically, in a large part of the open North Sea (Daan, 1978; Fig. 1). The presence of juvenile cod in coastal waters probably signifies that this life-stage tolerates the summer conditions of inshore waters, in particular the higher temperatures. On the other hand, juvenile plaice are restricted to the coastal area (Zijlstra, 1972; Fig. 2) indicating the importance of the area as a nursery. Of 12 North Sea offshore species populations having mainly juveniles in the Wadden Sea, about 8 require a coastal nursery. Before estuarine dependence is claimed for a fish species, a thorough study has to be made of the distribution of the juveniles, as also pointed out by Haedrich (1983).

A question to be considered is why some open sea fishes have adopted coastal nurseries, whereas the majority has obviously not. The high productivity of coastal areas has been mentioned together with the relative scarcity of resident species, offering a niche and good feeding conditions for those juvenile fish able to adapt to the other-wise adverse abiotic conditions (e.g. Haedrich, 1983). For the North Sea plaice, it could be shown that growth in the Wadden Sea was optimal under all densities encountered and was stimulated by the relatively high summer temperatures in the area (Zijlstra et al., 1982; Fig. 3).

The relatively high temperatures in late spring and summer in coastal areas as compared to the offshore region may be as important as the high food abundance. Higher temperatures enable the young fish to

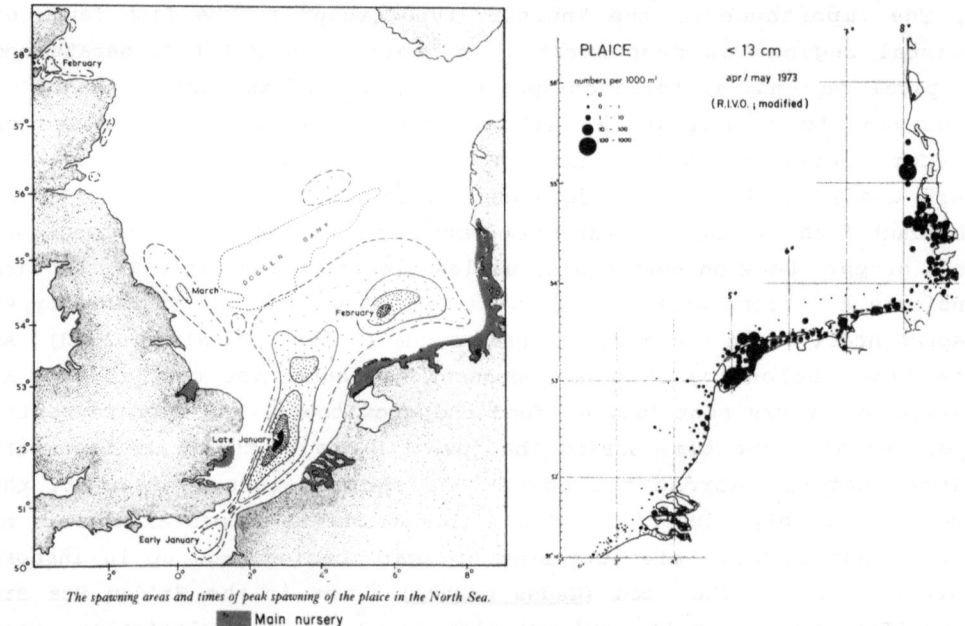

The spawning areas and times of peak spawning of the plaice in the North Sea.
▓ Main nursery

Fig. 2. The distribution of planktonic eggs of North Sea plaice (left) and the juvenile fish in the southern North Sea (right).

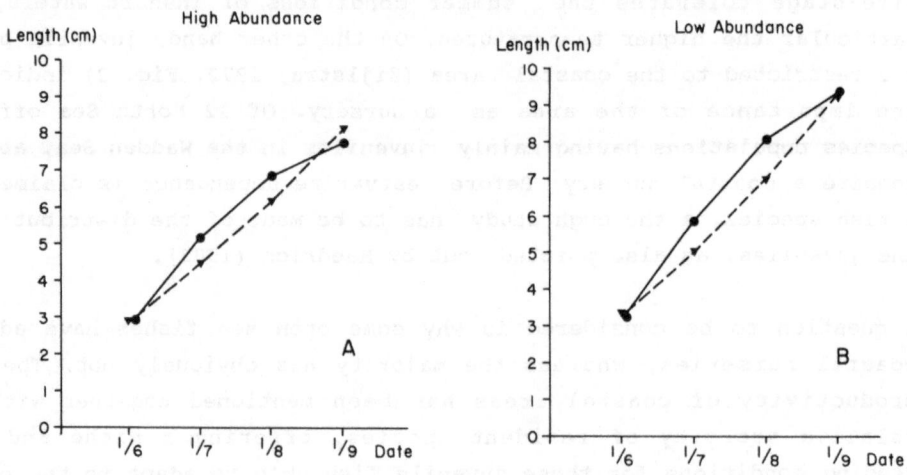

Fig. 3. Growth of 0-group plaice in the Wadden Sea during high and low abundance (full drawn line) compared to simulated growth (interrupted line), according to experiments with excess feeding under different temperature conditions (from Zijlstra and Witte, 1985).

quickly grow beyond their vulnerable size. The Wadden Sea plaice have exceptionally high mortalities up to a length of 30 mm due to predation of brown shrimp (van der Veer and Bergman, 1987). In open coastal areas along the British coast temperatures and growth rates were much lower (Zijlstra et al., 1982) and mortalities remained high all during summer (e.g. Lockwood, 1980). Moreover, there are indications that large predators, including adults of the same species, are scarce or absent in estuaries and lagoons. Therefore low predation pressure may be an important factor in coastal nurseries. This conclusion is supported by a recent study on predation by North Sea fish on their juveniles, showing predation to be low in plaice (P. platessa) and sole (Solea sole) with a prolonged juvenile stage in the coastal zone (Daan, 1983). 5 & 6) The last two groups of open sea fishes that visit the coastal zone seasonally or occasionally are hard to separate. There are only a few species of visitors but their relative numbers may be large. It seems questionable whether a true distinction exists between them as the evidence may simply reflect the abundance of the species in the nearby offshore region. The "visits" may be simple dispersion into an area at a time when conditions as temperature, salinity etc. are favourable rather than a directed migration. In any case, sometimes large quantities of adult fish may enter the coastal zone, to feed and grow for a shorter or longer period in the area. It is even possible that such excursions are made on a daily base, with invasions during the day or night period. It seems doubtful that the survival of the species depends on such visits, but the fish transport unknown quantities of living matter across the coastal-offshore boundary.

It is practically impossible to estimate these quantities in the absence of adequate data on number and growth of the species during their visits.

3. ESTIMATES OF COASTAL-OFFSHORE TRANSPORT BY FISHES

From the foregoing section, significant transports of living matter between coastal offshore waters can only be expected from fishes whose juveniles exploit the inshore zone as a nursery and from seasonal visitors. The group of offshore species spawning in the coastal zone and using it as a nursery is presumably small and can be considered as a special case of the "juvenile" type. Moreover, in some of these species, such as the capelin, the juveniles seem to leave the coastal zone already as larvae (Vilhjalmsson, 1983).

TABLE 3

Mean yearclass-size at 3 yars of age (a), mean numbers departing from coastal nursery annually (b), with their mean individual weight (c) and total weight (d) of three North Sea fish species, with a coastal nursery. Columns (e) and (f) give the estimated transport of fish biomass per m^{-2} (live weight) from the coastal zone to the offshore area, based on the size of the total continental ccoastal zone of Belgium, the Netherlands, Germany and Denmark, assuming a width of 15 km (e), and based on the size of the Wadden Sea (f). 1) Data from BANNISTER, 1978; 2) data from DE VEEN, 1978; 3) data from BURD, 1978.

Species	Yearclass size (at 3 years)	Estimated numbers departing coastal zone	Average weight at departure	Total weight at departure	Weight per m^2 total coastal zone 25 000 km^2)	Weight per m^2 Wadden Sea (8000 km^2)
Plaice 1)	360 × 10^6	360 × 10^6	150 g	54 000 tons	2.16 g	6.75 g
Sole 2)	120 × 10^6	120 × 10^6	120 g	14 000 tons	0.58 g	1.80 g
Herring 3)	6000 × 10^6	10 000 × 10^6	9 g	90 000 tons	3.60 g	11.25 g
Sum				158 400 tons	6.34 g	19.80 g

Most of the transport will be from the inshore area to the open sea, due to the offshore migration of immature fish. It is true that this offshore transport is partly compensated by an import of larvae (or eggs) each year, but that import is extremely small compared to the offshore transport. For instance, for North Sea plaice with an average recruitment of 360 million individuals at 3-years of age (Bannister, 1978), the live weight of the fish is estimated at about 54,000 tons at the time of offshore migration (Table 3). The number of larvae entering the coastal zone is estimated at 11 x 10^{10} using mortality rates (97%) determined by Zijlstra et al. (1982) for the period between larvae and age 3. With an individual weight of about 20 mg, total larval weight would have been about 200 tons, and therefore very small compared to the export. It seems likely that this example also applies to most other species. In addition, net transport of living matter resulting from visits of seasonal and adventitious migrants will also be in an offshore direction, as these fish come to feed on estuarine production.

These conclusions support a statement by Haedrich (1983) that "feeding by fish is an important pathway for the export from estuaries of the nutrients that might otherwise be trapped there". The question is how this export can be quantified.

A crude estimate could be derived from the annual variation in the weight of the fish fauna of a coastal area, estimated per unit area, assuming the difference between the lowest and highest estimate to be due to export. Oviatt and Nixon (1973) give data on the annual amplitude of fish biomass estimates for some New England coastal areas. The "export" estimated in this way would range from 4.4 g·m^{-2} live weight (approx. 0.4 gC·m^{-2}) in Narragansett Bay and 18.3 g·m^{-2} ($\pm$ 1.8 gC·m^{-2}) for Block Island Sound. The result will not be accurate because a decline in fish biomass will not only be caused by export but also by locally induced mortality. Moreover, the fish biomass is made up of various species, that are often not synchronized in their migrations (Haedrich, 1983). Finally, estimating fish biomass per unit area by fishing methods gives highly uncertain results, as all fishing gears take only a proportion of the fish present, that moreover may vary in the course of the year. The effect of gear selection on the method is therefore highly uncertain. Taking all these sources of error into consideration leads to a final estimate of uncertain accuracy.

An indirect method could be based on the knowledge of individual species using the coastal area as nursery exclusively, their distribution and abundance at various stages of life and growth. Such data are only available to me for three fish stocks of the North Sea with coastal nurseries. All three species are commercially important and data on numerical abundance and growth rate are available (Bannister, 1978; Burd, 1978; de Veen, 1978). Plaice and sole mainly grow in the coastal zone of Belgium, the Netherlands, Germany and Denmark, up to an age of about 2-3 years (Zijlstra, 1972; de Veen, 1978). Post-larval herring (_Clupea harengus_) smaller than about 10 cm are exclusively found in the coastal regions of the North Sea (e.g. Postuma et al., 1965) (Fig. 4) and may be distributed in the same area as plaice and sole in view of the fact, that they are found concentrated in the southeastern North Sea after leaving the coast (Cushing, 1962; Postuma et al., 1965) (Fig. 5). In contrast to plaice and sole the herring stay in the coastal zone for only a few months up to a year, depending on their growth rate. All three species enter the coastal nursery between early and late spring, as larvae. Table 3 gives the result of our calculations, based on data of yearclass size and average weight of each species at the time of departure from the nursery. Under the assumption that all North Sea plaice, sole and herring grow in the area indicated, the mean annual juvenile fish production of the 3 species would amount to approximately 160,000 metric tons fresh weight. The assumptions made find some justification in the fact, that in a shrimp-fishery in German coastal waters some 170 million 0-group plaice and some 60 million 0-group sole are destroyed annually, whereas these waters cover only about 1/3 of the total coastal area considered. If the fish were utilizing the whole coastal area of 25.000 km^2 between the Dover Straits and the tip of Denmark, that coastal area would export on average some 6.34 g m^{-2} live weight annually (approx. 0.6 gC·m^{-2} y^{-1}). The export would be 19.80 g·m^{-2} live weight per year (approx. 2 gC·m^{-2} y^{-1}) if only the 8.000 km^2 of

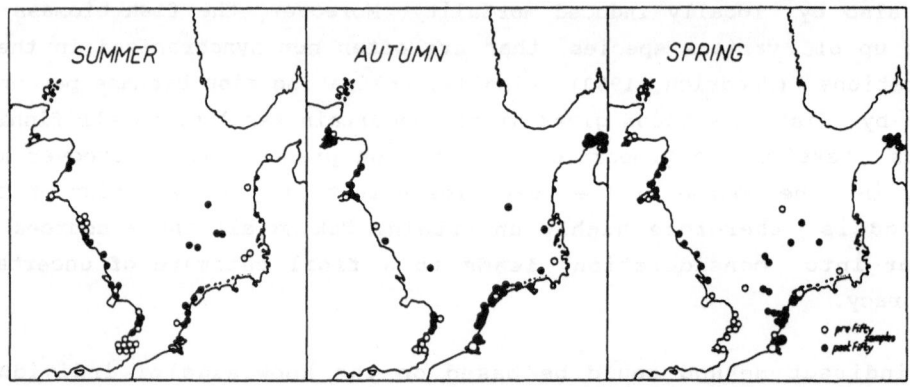

Fig. 4. Distribution of North Sea herring smaller than 11 cm. (from Postuma et al., 1965).

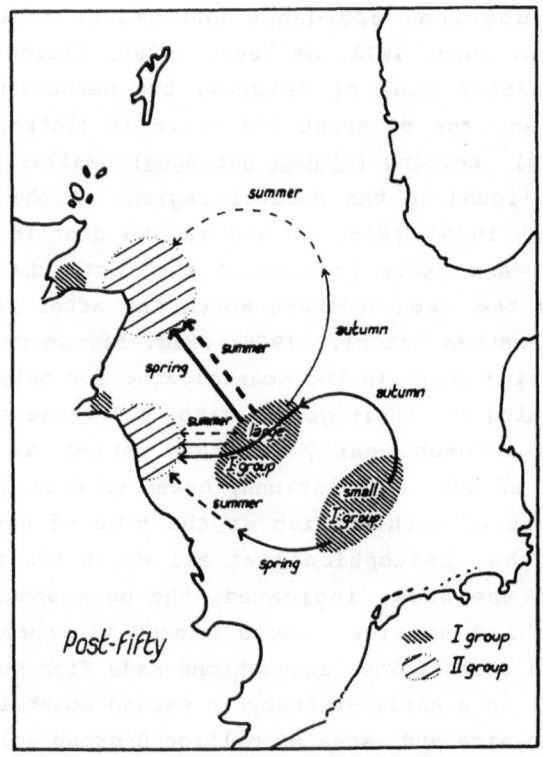

Fig. 5. Schematic presentation of the migration of North Sea herring with increasing age and size, indicating a general north-westerly direction (from Postuma et al., 1965).

the Wadden Sea were occupied as nursery, which seems most likely in the case of plaice and sole.

The estimates indicate the order of magnitude of the offshore export of recruits. The assumption that all juveniles of the three species grow up in the area indicated may not be quite true for the herring, of which part may utilize the east-coast of Denmark and the British coast. On the other hand, the numbers assumed to depart from the nursery are probably too low in herring, as the immatures of this species suffer heavily from predation after their migration to the open sea, so that mortality between 1-3 years of age and hence abundance at a length of 10 cm might be seriously underestimated (Daan, 1983). Moreover, our estimates include only 3 species, whereas probably more are using the area as a nursery. Also, no account has been given of export of fish due to seasonal and adventitious migrants, which are extremely hard to assess.

In view of our estimates, which are in the same order as those obtained from the U.S. east coast by other methods, it seems unlikely, that transport of living matter by fish is very important in comparison with quantities of organic matter or nutrients transported by physical processes. For instance, for the Wadden Sea Postma (1954) estimated the import from the North Sea at 80 $gC \cdot m^{-2} \cdot yr^{-1}$, an amount which may have been doubled or trebled since due to eutrophication (de Jonge and Postma, 1974). Therefore, the statement of Haedrich (1983) is probably to be questioned if considered quantitatively.

Nevertheless, transport processes due to fish migrations across the coastal-offshore boundary do exist and even if they are quantitatively of lesser importance, they do affect the ecosystems on both sides of the boundary significantly. On the offshore side large fish populations, like plaice, sole and herring in the North Sea, are presumably dependent on these migrations. In the coastal area the regular visits of juvenile fish, derived from large offshore stocks, affect both the planktonic and benthic food webs and trophic dynamics. For instance, the annual consumption of plaice alone takes about 20-30% of the production of macrobenthos in the Wadden Sea (Kuipers, 1977; de Vlas, 1979).

REFERENCES

Austin, H.M., 1971. A survey of the ichthyofauna of the mangroves of western Puerto Rico during December 1967-August 1968. - Caribb. J. Sci.11: 27-39.

Bannister, R.C.A., 1978. Changes in plaice stocks and plaice fisheries in the North Sea. - Rapp. P.-v. Reun. Cons. int. Explor. Mer 172: 86-101.

Bozeman, E.A. & J.M. Dean, 1980. The abundance of estuarine larval and juvenile fish in a South Carolina intertidal creek. - Estuaries 3: 89-97.

Burd, A.C. 1978. Long-term changes in North Sea herring stocks. - Rapp. P.-v. Reun. Cons. int. Explor. Mer 172: 117-153.

Claridge, P.N., I.C. Potter & M.W. Hardisty, 1986. Seasonal changes in movements, abundance, size composition and diversity of the fish fauna of the Severn estuary. - J. mar. biol. Ass. UK 66: 229-258.

Creutzberg, F., 1961. On the orientation of migrating elvers (Anguilla vulgaris Turt.) in a tidal area. - Neth. J. Sea Res. 1: 257-338.

Cushing, D.H., 1962. Recruitment to the North Sea herring stocks. - Fish. Invest. Lond. 27 pp.

Cushing, D.H., 1975. Marine ecology and fisheries. Cambridge University Press, Cambridge, London, New York, Melbourne. 278 pp.

Daan, N., 1978. Changes in cod stocks and cod fisheries in the North Sea. - Rapp. P.-v. Reun. Cons. int. Explor. Mer 172: 39-57.

Daan, N., 1983. The ICES stomach sampling project in 1981: aims, outline and some results. - Northwest Atlantic Fisheries Organisation 5th Annual Meeting, A2 (mimeo).

Dankers, N.M.J.A. & J.F. de Veen, 1978. Variations in relative abundance in a number of fish species in the Wadden Sea and the North Sea coastal area. - In N. Dankers, W. Wolff & J.J. Zijlstra (eds.): Fishes and fisheries of the Wadden Sea, Rep. 5 of the Wadden Sea Working Group.

Gjosaeter, J. & R. Saetre, 1974. The use of data on eggs and larvae for extimating spawning stock of fish populations with demersal eggs. - In J.H.S. Blaxter (ed.): The early life history of fish, pp. 139-149. Springer Verlag, Verling, Heidelberg, New York.

Graham, J.J., 1972. Retention of herring larvae within the Sheepscot estuary of Main. - Fish. Bull. US 70: 299-305.

Gunter, G., 1967. Some relationships of estuaries to the fisheries of the Gulf of Mexico. - In G.H. Lauff (ed.): Estuaries 83: 621-638. Am. Assoc. Adv. Sci. Publ.

Haedrich, R.L., 1983. Estuarine fishes. - In D.W. Goodall (ed.): Ecosystems of the world, vol. 26. B.H. Ketchum (ed.): Estuaries and enclosed seas, pp. 183-207. Elsevier Sci. Publ. Com., Amsterdam-Oxford-New York.

Harden-Jones, R., 1968. Fish migrations. Esward Arnold Publ. Ltd., London. 325 pp.

Harden-Jones, F.R., G.P. Arnold, M. Greer Walker & P. Scholes, 1978. Selective tidal stream transport and the migration of plaice (Pleuronectes platessa L.) in the southern North Sea. - J. Cons. perm. int. Explor. Mer 38: 331-337.

Hardenberg,J.D.F., 1931. The fish fauna of the Rokan mouth. - Treubia 13: 81-168.

Holden, M.J., 1978. Long-term changes of fish from the North Sea. - Rapp. P.-v. Cons. Explor. Mer 172:11-26.

Houston, A.S., 1980. The decline and recovery of Canada's Pacific herring stocks. - Rapp. P.-v. Reun. Cons. int. Explor. Mer 177: 143- 153.

Jonge de, V.N. & H. Postma, 1974. Phosphorus compounds in the Dutch Wadden Sea. - Neth. J. Sea Res. 8: 139-153.

Kuipers, B.R., 1977. On the ecology of juvenile plaice on a tidal flat in the Wadden Sea. - Neth. J. Sea Res. 11: 56-91.

Lockwood, S.J., 1980. Density-dependent mortality in O-group plaice
(<u>Pleuronected platessa</u> L.) populations. - J. Cons. int. Explor. Mer
39: 148-153.

Louis, M. & G. Laserre, 1982. Etude des peuplements des poissons dans
les lagunes des mangroves de Guadeloupe (Antilles francaises). -
Ocean. Acta, Spec. Vol. Coastal Lagoons, pp. 333-338.

McHugh, J.L., 1967. Estuarine nekton. - In G.H. Lauff (ed.): Estuaries.
Am. Assoc.   Adv. Sci. Pupl. 83: 581-620.

Misitano, D.A., 1977. Species composition and relative abundance of
larval and post-larval fishes in the Columbia River estuary, 1973.
- Fish. Bull. US 75: 218-221.

Nikolsky, G.V., 1963. The ecology of fishes. Academic Press, London-
New York. 352 pp.

Oviatt, C.A. & S.W. Nixon, 1973. The demersal fish of Narragnasett Bay:
an analysis of community structure, distribution and abundance. -
Est. Coast. mar. Sci. 1: 361-378.

Pearcy, W.G. & S.S. Myers, 1974. Larval fishes of Yaquina Bay, Oregon:
a nursery ground for marine fishes? - Fish. Bull. US 72: 201-213.

Postuma, K.H., J.J. Zijlstra & N. Das, 1965. On the unmature herring of
the North Sea. - J. Cons. int. Explor. Mer 29: 256-276.

Postma, H., 1954. Hydrography of the Dutch Wadden Sea. - Arch. néerl.
Zool. 10: 405-511.

Smith, R.F., A.H. Swartz & W.H. Massman, (eds.), 1966. A symposium on
estuarine fisheries. Trans. Am. Fish. Soc. Suppl. 95: 1-154.

Tiews, K., 1978. The German industrial fisheries in the North Sea and
their by-catches. - Rapp. P.-v. Reun. Cons. int. Explor. Mer 172:
230-238.

Tiews, K., 1978. Non-commercial fish species in the German Bight.
Records of by-catches of the brown shrimp fishery. - Rapp. P.-v.
Reun. Cons. int. Explor. Mer 172: 259-265.

Veen de, J.F., 1978. Changes in North Sea sole stocks (<u>Solea solea</u> L.).
- Rapp. P.-v. Reun. Cons. int. Explor. Mer 172: 124-136.

Veen de, J.F., 1978. On selective tidal transport in the migration of
north Sea plaice (<u>Pleuronectes platessa</u>) and other flatfish
species. - Neth. J. Sea Res. 12: 115-147.

Veer, H.W. van der & M.J.N. Bergman, 1987. Predation by crustaceans on
a newly settled O-group plaice <u>Pleuronectes platessa</u> population in
the western Wadden Sea. - Mar. Ecol. Prog. Ser. 35:203-215.

Vilhjalmsson, H., 1983. On the biology and changes in exploitation and
abundance of the Icelandic capelin. - FAO Fisheries Rep. 291, 2:
537-553.

Vlas, J. de, 1979. Annual food intake by plaice and flounder in a
tidal flat area in the Dutch Wadden Sea, with special reference to
consumption of regenerating parts of macrobenthic prey. - Neth. J.
Sea Res. 13: 117-153.

Yanez-Arancibia, A., F. Amezcua Linares & J.W. Day Jr., 1980. Fish
community structure and function in Terminos Lagoon, a tropical
estuary in the southern Gulf of Mexico. - In V. Kennedy (ed.):
Estuarine perspectives, pp. 465-482. Academic Press Inc., New York.

Yanez-Arancibia, A. & J.W. Day Jr., 1982. Ecological characterisation
of Terminos Lagoon, a tropical lagoon estuarine system in the
southern Gulf of Mexico. - Ocean. Acta Spec. Vol. on Coastal
Lagoons. pp. 431-440.

Zijlstra, J.J., 1972. On the importance of the Wadden Sea as a nursery
area in relation to the conservation of the southern North Sea fish
resources. - Symp. zool. Soc. Lond. 29: 233-258.

Zijlstra, J.J., 1978. The function of the Wadden Sea for the members of
its fish fauna. - In N. Dankers, W.J. Wolff & J.J. Zijlstra (eds.):
Fishes and fisheries of the Wadden Sea, pp. 20-25. Rep. 5   Wadden
Sea Working Group.

Zijlstra, J.J., R. Dapper & J. IJ. Witte, 1982. Settlement, growth and
mortality of post-larval plaice (<u>Pleuronectes platessa</u>) in the
western Wadden Sea. - Neth. J. Sea Res. 15: 250-272.

Zijlstra, J.J. & I.J. Witte, 1985. On the recruitment of O-group plaice in the North Sea. - Neth. J. Zool. 35: 360-376.

LARVAL TRANSPORT IN COASTAL CRUSTACEA: THREE CASE HISTORIES

Peter C. Rothlisberg
CSIRO Division of Fisheries Research
P.O. Box 120, Cleveland, Qld, 4163 Australia

Larval transport mechanisms of three coastal Crustacea are presented as three case histories: (1) Panulirus cygnus, the western rock lobster, found off western Australia, (2) Pandalus jordani, the pink shrimp, distributed off the west coast of North America, and (3) Penaeus merguiensis, the banana prawn, which occurs throughout the Indo-Pacific region.

Adult P. cygnus migrate offshore 40 to 60 km before the larvae hatch. Through the interaction of larval vertical migratory behavior and offshore surface wind drift the larvae are transported over 1000 km offshore in the southeastern Indian Ocean. Slight changes in the day-night vertical distribution of the older larvae move them back towards the coast, to the shelf edge where they metamorphose to the postlarval puerulus stage. The puerulus is capable of directed swimming across the shelf and settles in the coastal limestone reefs 7 to 14 months after hatching.

The center of the distribution of P. jordani is off the Oregon coast, an area of intense seasonal upwelling. The larval period of P. jordani is characterized by considerable variation in current regimes. Larvae hatch in early spring and are moved northward and shoreward by the Davidson Current. In mid-April, when upwelling usually begins, the current direction changes to southward flowing with an offshore component. During the three to four month larval life, water temperatures are usually maintained between 10 to 12°C in the coastal zone, which is optimal for larval growth and survival. When wind and current regimes are anomalous advective losses and temperature induced mortality are higher.

The Gulf of Carpentaria is the center of the distribution of P. merguiensis in Australia and an area where the species has been intensively studied. Postlarval recruitment in the Gulf is highly seasonal, with two peaks each year. Peak postlarval recruitment however, does not occur after peaks of larval production. Differential larval advection, controlled by the interaction of diurnal vertical migration of the

Lecture Notes on Coastal and Estuarine Studies, Vol. 22
B.-O. Jansson (Ed.), Coastal-Offshore Ecosystem Interactions.
© Springer-Verlag Berlin Heidelberg 1988

larvae and a seasonal change in the phase of the tidal currents that dominate the Gulf, are responsible for the postlarval recruitment patterns.

1. INTRODUCTION

Of the world's benthic invertebrate species, about 70% have pelagic larvae (Thorson, 1950). While having a pelagic larval phase may be advantageous for, for example dispersal, colonization, recolonization (for review see Mileikovsky, 1971), it also carries considerable risks or disadvantages. Biotic factors such as predation and starvation affect larval mortality (for reviews see May, 1974; Hunter, 1981). Abiotic factors such as currents are especially critical for species that can only complete the distinct phases of their life cycles in different habitats (Hjort, 1914; Gunter, 1980; Jackson and Strathmann, 1981; Norcross and Shaw, 1984).

The advantages and disadvantages of the pelagic phase are both related to the length of the planktonic phase. In decapod Crustacea, this varies from 2 days for a ocypodid crab, Uca subcylindrica (Rabalais and Cameron, 1983) up to almost 2 years for the palinurid lobster Jasus edwardsii (Lesser, 1978). The magnitude of the advantages or disadvantages, however, is not a simple relationship with the length of the planktonic phase. Complications arise with the superimposition of time and place of spawning, fecundity, and time of recruitment on the ambient hydrographic regime (Parrish et al., 1981). Habitat availability for postlarval and adult stages also affect the ultimate distribution and abundance of the species.

Larvae cannot be regarded as passive drifters in unidirectional currents, as often depicted in hydrographic atlases. A detailed knowledge of both larval ecology and the dynamics of the hydrographic milieu are essential for an understanding of the distance, direction and timing of larval transport (for more detailed discussions see Rothlisberg et al., 1983; Norcross and Shaw, 1984).

In this paper I shall summarize the larval transport case histories of three crustaceans: (1) the western rock lobster, Panulirus cygnus, off western Australia (2) the pink shrimp, Pandalus jordani, off western North America and (3) the banana prawn, Penaeus merguiensis, in the Gulf of Carpentaria, northern Australia. These species were chosen because they are all commercially important and their larval ecologies

are therefore reasonably well known; they are found in different
hydrographic regimes, ranging from oceanic, to coastal upwelling, to a
large, tidally active, embayment; each taxon has a different larval
form and life span. Using these examples I shall illustrate how the
interaction of larval behavior and hydrographic regime affects larval
transport, and the role the offshore pelagic phase plays in determining
both the patterns and magnitude of postlarval recruitment to the
nearshore zone.

2. CASE HISTORIES
2.1 Case 1: Panulirus cygnus

Phyllosoma larvae are unique to spiny (or rock), coral and slipper
lobsters in the families Palinuridae, Synaxidae and Scyllaridae
respectively. These larvae are very long-lived and therefore have the
potential for wide dispersal (Phillips and Sastry, 1980). Because they
are long-lived and widely distributed, studies of larval and
postlarval ecology are limited to a few species (Phillips and
McWilliam, 1986). The Australian western rock lobster, Panulirus
cygnus, has been by far the most extensively and intensively studied
(summarized in Kanciruk, 1980; Phillips and Sastry, 1980; Phillips,
1981; Phillips and McWilliam, 1986). Synoptic studies of the
oceanographic environment off western Australia have also been
undertaken (Wyrtki, 1962, 1963; Hamon, 1965; Rochford, 1969a, b;
Cresswell and Golding, 1980).

Adult P. cygnus move from their shallow coastal reef habitats into
deeper water, 40 to 60 km from the coast, during the time of egg
deposition and hatching (Phillips, 1981) (Fig. 1). Larvae are released
here in summer (Rimmer, 1980) These early larvae, migrating actively
into the surface waters at night, are transported offshore by surface
wind drift (<50m) at about 5.25 km/day (Rimmer and Phillips, 1979),
against the general direction of circulation of the upper 300 m of the
water column (Wyrtki, 1962; Hamon, 1965). The larvae become widely
distributed in the southeastern Indian Ocean, up to 1500 km offshore
(the extent of sampling), with the highest abundance 375 to 1030 km
offshore (Phillips et al., 1979). Mid- to late-stage phyllosoma larvae
are virtually absent from the continental shelf and coastal waters
(Phillips et al., 1978).

The return transport of the late phyllosoma larvae is not completely understood. It is thought that, because the older larvae spend proportionately less time in surface waters by night, and migrate to deeper waters by day, that they are carried coastward by the prevailing currents (Phillips, 1981). The larvae complete the planktonic period offshore, with concentrations of late phyllosoma larvae found in waters near the edge of the continental shelf (Phillips et al., 1978). It is here outside the southward flowing Leeuwin current that the metamorphosis to a unique postlarval stage, the puerulus, probably occurs, with salinity changes being the likely stimulus (Phillips and McWilliam,1986). The puerulus is a nektonic stage capable of directed swimming at 15 to 46 cm/sec (Phillips and Olsen, 1975).

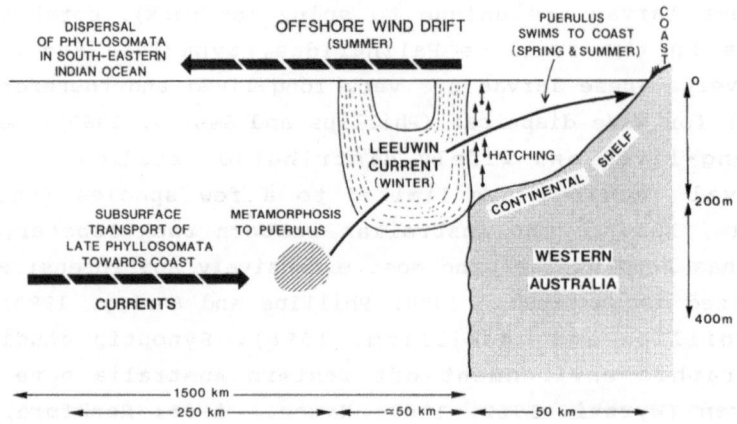

Fig. 1. Summary or the oceanic cycle of <u>Panulirus cygnus</u> (from Phillips and McWilliam, 1986).

The method by which puerulus larvae orient to their shallow coastal nursery grounds is not known. The ability to sense breaking waves, or pressure changes of passing wave trains or surface gravity waves have all been suggested (Phillips, per. comm.). The use of internal wave fronts, as has been demonstrated for other postlarval Crustacea (Shanks, 1983, 1985; Zeldis and Jillett, 1982) has been discounted, because the puerulus is not found in surface waters, except very close to shore (Phillips and McWilliam, 1986). Recruitment of pueruli occurs year round, with a peak of settlement between September and January (Phillips and Hall, 1978). The duration of the combined larval and postlarval stages is therefore between 7 and 14 months, which offers considerable potential for prolongation of either the phyllosoma or puerulus stage, dependent on environmental conditions.

2.2. Case 2: Pandalus jordani

The physical and biological oceanography off western North America, particularly the upwelling ecosystems off the Oregon coast, have been extensively studied (e.g. Smith, 1974; Huyer, 1977; Peterson and Miller, 1975; Peterson et al., 1979: Parrish et al., 1981). Studies of the larval ecology of pandalid shrimps, which form important fisheries in the northern Pacific, have been limited to a few species (Haynes, 1985). The larval ecology of P. jordani off the central Oregon coast has been the most intensively investigated, but only over a two year period (Rothlisberg and Pearcy, 1977; Rothlisberg, 1979, 1980; Rothlisberg and Miller, 1983).

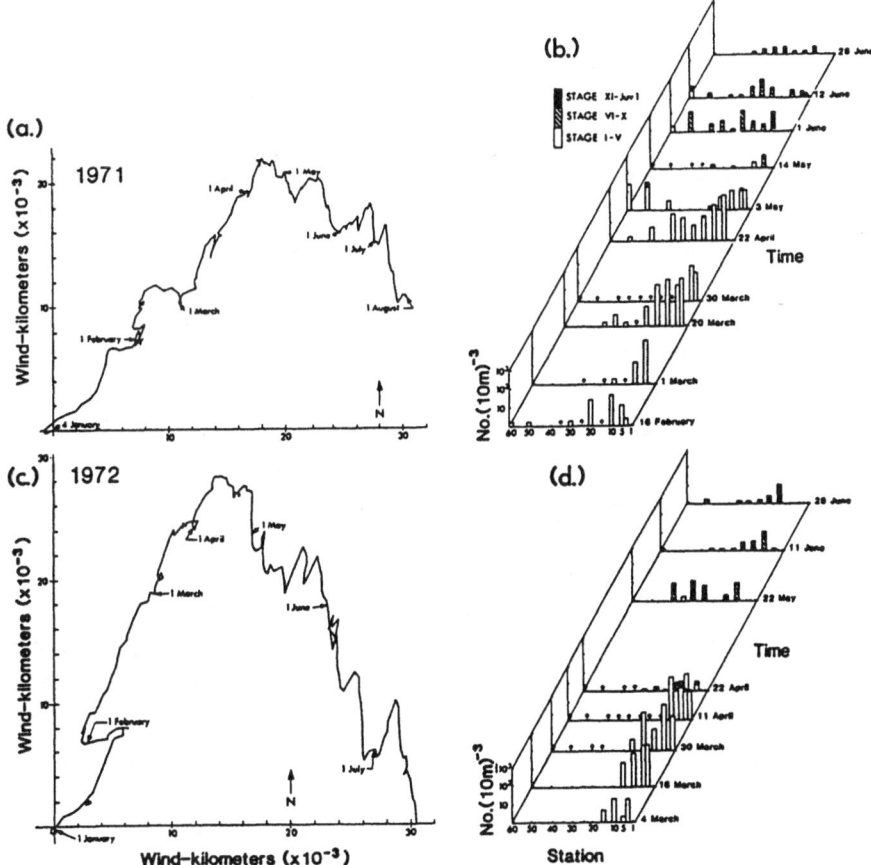

Fig. 2. Wind regimes and larval Pandalus jordani distribution off central Oregon. A. Progressive wind vectors 4 January to 1 August 1971. B. Larval distribution and abundance 1971. Station numbers are nautical miles from coast. C. Progressive wind vectors 1 January to 20 July 1972. D. Larval distribution and abundance 1972 (redrawn from Rothlisberg and Miller 1983).

The pink shrimp, P. jordani, is found on the west coast of North America from Unalaska to San Diego. Commercial quantities occur from Vancouver Island to Morro Bay, with the center of the distribution off Oregon in depth from 73 to 256 m on continental shelf areas characterized by green mud/sand substrates (Lukas, 1981). The shrimp are fished during the day, with benthic otter trawls. At night, the adults migrate vertically into the water column to feed (Pearcy, 1970). The females carry eggs from October to March; the larvae usually hatch off Oregon over two weeks in March (Robinson, 1971), when early larvae appear in the plankton (Rothlisberg and Miller, 1983) (Fig. 2d). Typically at this time of the year, the northerly flow of the Davidson Current is accompanied by strong southerly winds (Fig. 2c). Because of the north-south orientation of the Oregon coast and Coriolis deflection, early larvae in the surface waters are pushed shoreward from the midshelf. The seawater temperature remains relatively high (10 to 12°C). At these temperatures, larvae molt at almost weekly intervals (Rothlisberg, 1979). The larval life span covers 13 stages, over three to four months, and is the longest known for any pandalid (Rothlisberg, 1980; Haynes, 1985). By mid-April the wind field reverses and intermittent episodes of upwelling begin (Fig. 2c). Surface waters are moved offshore and temperatures remain below 12°C in the coastal zone. By this time mid-stage larvae are present, still largely confined to the coastal zone. With persistent northerly winds and upwelling, later larvae and early juveniles are moved offshore (Fig. 2d), over the parental grounds, although this offshore dispersal is limited because the later larval stages are distributed lower in the water column (Rothlisberg and Pearcy, 1977).

Anomalous wind and current regimes recorded in 1971 (Bakun, 1973), gave some insight into the role of the hydrographic regime and its effect on larval distribution and survival. In 1971 southerly and northerly winds and current velocities were all lower than normal (Fig. 2a). Early larvae were distributed more widely offshore and their growth rates were initially slower than usual because of the lower water temperatures (Fig. 2b). When weak upwelling began, they were at an earlier stage of development and their offshore dispersal beyond the shelf edge was increased because they were higher in the water column and moved offshore by the surface currents. The elevated water temperatures increased larval mortality, as demonstrated in laboratory experiments (Rothlisberg, 1979). Survival was an order of magnitude lower in 1971 than in 1972 because of the increased advective losses and increased temperature (Rothlisberg and Miller, 1983) (Fig. 3).

The strength of the relationship between hydrographic conditions and larval survival is reflected in a positive correlation between an upwelling index and larval survival, as measured by egg and recruitment surveys in the commercial fishery, using a 20-year data set. The correlation explains 55% of the variability in larval survival (Rothlisberg and Miller, 1983).

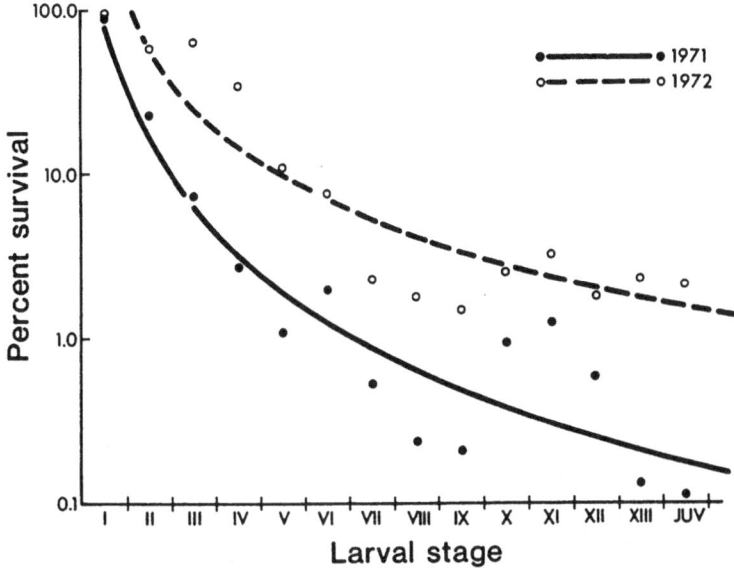

Fig. 3. Larval survival of <u>Pandalus jordani</u> 1971 and 1972 (redrawn from Rothlisberg and Miller 1983).

2.3. <u>Case 3: Penaeus merguiensis</u>

The Gulf of Carpentaria, in tropical northern Australia, is the center of the Australian distribution of the banana prawn, <u>Penaeus merguiensis</u>. This species is widespread throughout the entire Indo-Pacific, from the Persian Gulf in the west, to Australia and New Caledonia in the east. It has a complex life history, spawning offshore and spending its postlarval and juvenile stages in nursery grounds in mangrove-lined estuaries (Staples et al., 1985). Staples (1979) found, in a survey of the nursery grounds around the Gulf, that the rivers in which postlarval recruitment occurred fell into distinct zones, based on the timing of postlarval ingress (Fig. 4, areas 1 to 4). The peak of recruitment to the rivers in area 1 was 6 months out of phase with the peak recruitment in area 3, approximately 500 km to the south. Furthermore, the peak of egg production offshore (Crocos and Kerr, 1983) in both areas was not reflected later in the peak numbers of postlarvae recruiting to the adjacent rivers. Differential larval advection on a seasonal scale was examined to explain the zonation and asymmetry.

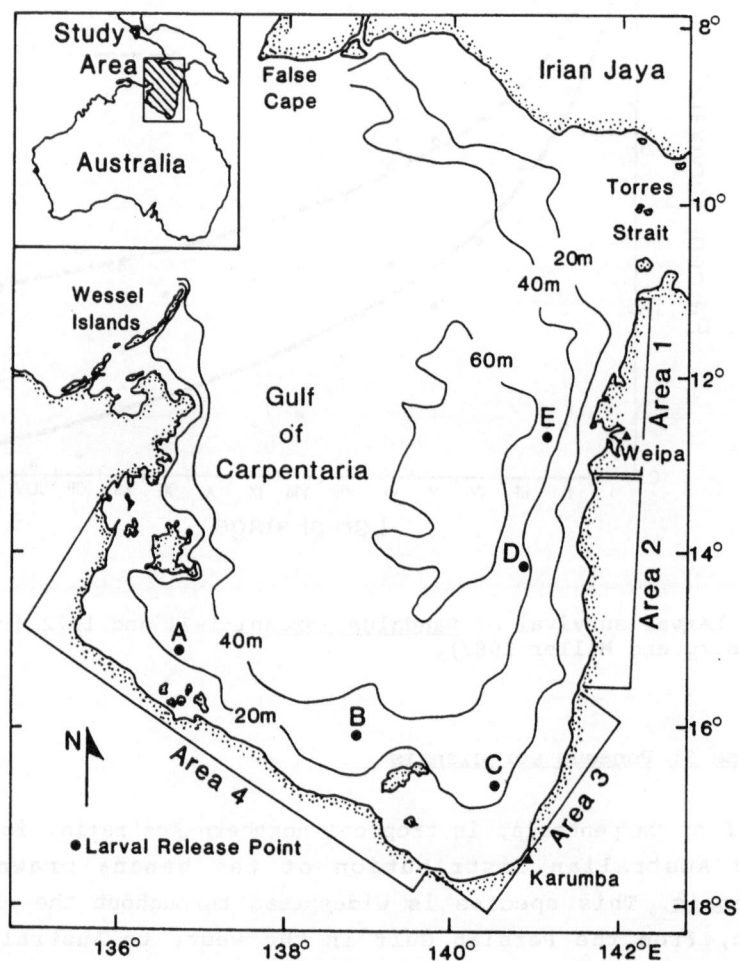

Fig. 4. The Gulf of Carpentaria, showing postlarval <u>Penaeus merguiensis</u> recruitment Areas 1 to 4 (after Staples 1979). Points A to E are larval release points in model (redrawn from Rothlisberg et al. 1983).

The Gulf of Carpentaria is a large (ca. $3.7 \ 10^5 \ km^2$), shallow (<70 m), and almost landlocked tropical embayment (Fig. 4). The hydrological data from a series of 10 Gulf-wide cruises (Rothlisberg and Jackson, 1982) were summarized by Forbes (1984), who found that most of the variation in temperature, salinity and density was in the coastal zone, due to seasonal wind regimes and heavy rainfall in the austral summer. Summer is also the season of maximum stratification of temperature but not of salinity. These seasonal changes are local processes, produced by for example, wind mixing, tidal stirring, solar heating and terrestrial runoff. The local hydrography and tidal wind data, were

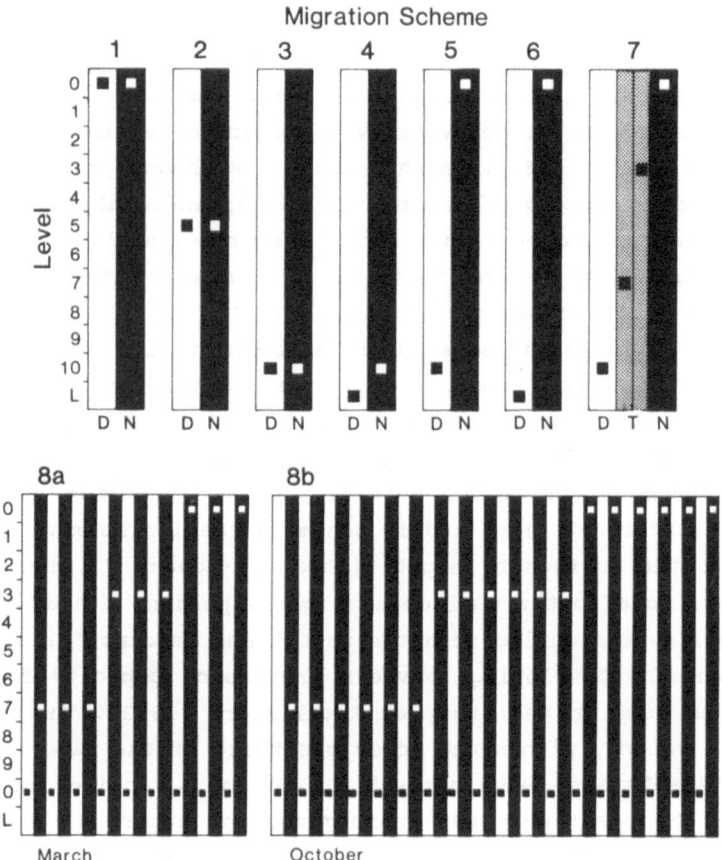

Fig. 5. Schematic representation of eight larval vertical migration schemes used in the two- and three-dimensional models. D = daytime depth, N = night-time depth, T = transitional depths at dawn and dusk (redrawn from Rothlisberg et al., 1983).

incorporated in a two-dimensional model of Gulf circulation (Church and Forbes, 1981, 1983; Forbes and Church, 1983). They found that the Gulf circulation is dominated by a large, clockwise gyre, modified or sub divided by seasonal winds.

The effect of vertical migratory behavior on dispersal of penaeid larvae was also investigated (Rothlisberg, 1982). Both migratory ability and vertical excursion increased with the age and stage of development of the larvae. Simultaneous current meter recordings showed that the length of time the larvae spent at various depths affected the distance and direction of larval transport. Extrapolations from this study were limited by the observations which were short-term and localized.

A three-dimensional numerical model of larval advection, which included both currents and larval behavior, was therefore developed for the Gulf (Rothlisberg et al., 1983). In this model the Ekman dynamics of surface and bottom drag were applied to the two-dimensional Gulf-wide tidal model to give three-dimensional current speeds and directions. At the same time eight larval behaviors (Fig. 5) were built into the model. These included both observed (Rothlisberg 1982) and hypothetical cases for comparison. The eight schemes of larval migration are numbered with increasing complexity. Schemes 1 to 3 represent non-migrating larvae. Scheme 4 has the larvae within the logarithmic layer (near-zero velocity) by day and moving just up into the water column (level 10) at night. In scheme 5 the larvae move from the near-bottom (level 10) by day to the surface at night and in scheme 6 the larvae move from level L by day to the surface by night. Larvae in scheme 7 spend the day at level 10 and move to the surface at night, spending 1 h at each of two transitional depths during both the ascent and descent. Finally, scheme 8 simulates the developing larvae's increasing capacity for vertical migration as seen in the study of Rothlisberg (1982). The three zoeal substages move only within the lower one third of the water column, the three mysis substages move through the lower two thirds of the water column and the postlarvae move throughout the entire water column. In scheme 8a, in March, with relatively warm water temperatures, the larvae spend one day in each larval substage. In October, when water temperatures are lower, each larval substage was assumed to last 2 days. The larval growth rates are based on experimental larval rearing studies (Rothlisberg, unpublished data).

The larvae were released in the model at five locations (Fig. 4, A-E), which were known adult spawning grounds. The two known periods of peak reproductive activity, October and March were tested. The differences in distance and direction between the non-migratory (exemplified by scheme 1) and migratory (scheme 6) larvae were considerable. In none of the areas did the modeled larvae, which remained stationary in the water column, move very far from their places of origin. In March, all vertically migrating larvae were transported in a northerly direction (Fig. 6); in the southeastern Gulf this was away from coastal nursery grounds, and in the north the advection had a coastward component towards the nursery grounds. In October, 6 months later, the direction was reversed, due to the $180°$ change in phase of the K1 tide with respect to the day-night cycle. The southerly advection in the eastern Gulf was towards the nursery grounds in the south and away from them in the north.

The model agrees with known postlarval recruitment dynamics, especially in the southeastern Gulf. Here in March, at the peak of egg and larval production, larvae are moved offshore; consequently, few postlarvae are delivered to the coastal nursery grounds. Six months later, in October, when relatively few females are spawning, the larvae are successfully moved inshore creating the yearly peak of postlarval immigration.

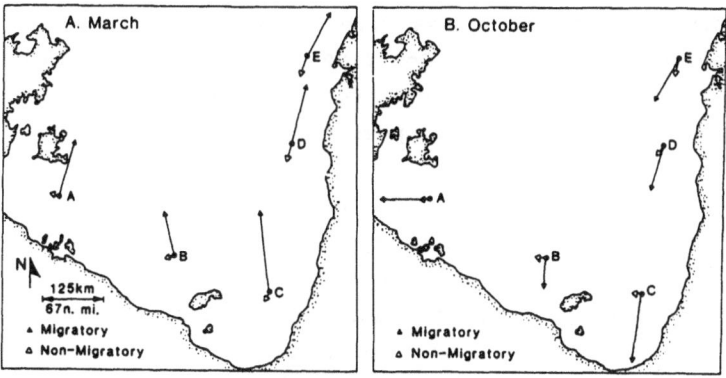

Fig. 6. Spatial variation in larval advection after 20 days, at five larval release points for both migratory and non-migratory larvae. A. March, B. October (redrawn from Rothlisberg et al. 1983).

3. DISCUSSION

The duration of the larval stage is an indication of dispersal potential, but recruitment will also be affected by the complex interplay of the seasonality and location of spawning and hatching and the hydro-meteorological regime operating, from the time of spawning through the pelagic larval phase. In addition to the strength and patterns of postlarval recruitment, the distribution and reproductive capacity of the species will ultimately be defined by habitat availability, habitat preferences and environmental tolerances of the juveniles and adults.

An example of tolerances during the larval stage was given in the P. jordani case history. The larvae apparently have a relatively narrow temperature tolerance. The optimum temperature is maintained by the normal wind and current pattern. This pattern is however, complex and includes a reversal in the direction of the longshore current direction, which affects onshore and offshore larval transport. The relative importance of the temperature tolerance of larvae and the pattern of transport cannot be assessed, but slight deviations from normal conditions, which affect both temperature and currents, markedly decrease larval survival. Some of the conclusions are based on correlations, with laboratory and limited field observations to support them. In the P. borealis fishery in the Gulf of Maine, an inverse relationship between year-class strength and sea temperature has also been established (Dow, 1981). Dow (1981) lagged the relationship back 4 years to the larval period and suggested that high temperatures influence adult migrations and dispersion at the time of egg hatching.

Pandalus jordani has a wide latitudinal distribution off the west coast of North America but the area of highest abundance and preferred habitat is relatively small. The ability of P. jordani to maintain itself in this complex and variable hydrographic milieu, in spite of its low fecundity (<3000 eggs), long larval life (about 4 months) and a short period of annual reproduction, suggests a high degree of adaptation to this upwelling ecosystem. Coastal retention of early larval stages, followed by limited offshore transport of older larvae appears to be the norm. More detailed studies of larval behavior, particularly as it pertains to the retention mechanism (e.g. Peterson et al., 1979; Wroblewski, 1982), would be warranted, especially if the perceived relationship between upwelling and larval survival does not hold up over a longer time series. Further studies of remote forcing factors might also be appropriate to give longer-term forecasts of larval survival estimates and indices of year-class strength.

Penaeus merguiensis is very widespread, occurring throughout the entire Indo-Pacific and very fecund (up to 400,000 eggs per spawning, with two or more spawnings per year). Despite having a short planktonic life (2 to 4 weeks), and a huge larval wastage at the time of maximum larval production, the species is maintaining itself in commercial abundances in the southeastern Gulf of Carpentaria. The limited dispersal and high wastage are probably offset by the high fecundity of the species and the availability of suitable juvenile habitat in the ubiquitous mangrove estuaries around the Gulf. Larval wastage of P. merguiensis may be lower in the center of its distribution, where two monsoonal periods and two peaks of postlarval recruitment are the norm (Rothlisberg et al., 1985). However, this hypothesis needs testing over the full Indo-Pacific range of this species.

In P. merguiensis successful larval transport appears to be a movement from offshore spawning grounds to nearshore estuarine nursery grounds. Seasonal changes in the direction of this transport dramatically affect postlarval recruitment patterns. Currently the model only explains the mechanisms that maintain the patterns of postlarval recruitment, not year-to-year variations in abundance. More detailed studies are needed on short-term abiotic events, such as cyclones and their associated hydro-meteorology, as well as biotic factors that may affect inter annual larval and postlarval survival. Factors that affect larval survival of penaeids have only recently been studied in detail (e.g. Emmerson, 1980; Preston, 1985) and warrant further examination in the ecological context of each species.

Perhaps the most intriguing questions about larval life span and species distribution concern the species that have phyllosoma larvae. These larvae have very long and variable life spans; however, very little is known about larval survival and growth rates because of the difficulty of culturing these forms. Growth rates can only be inferred from estimates of the age of field-caught specimens, and such estimates are difficult with widely dispersed larvae, of unknown origin, spawned by adults that reproduce over a considerable portion of the year. Against this background of a very long larval life it is relevant to compare the ranges of the species. The western rock lobster, P. cygnus, occurs only on the coast of western Australia; other species (e.g. P. ornatus and P.penicillatus), which presumably have similar planktonic larval durations, are spread throughout the entire Indo-Pacific, and in the case of P. pencillatus, extend to the eastern tropical Pacific. It is doubtful that such distances can be traversed within a single larval life span, and clearly dispersal potential, larval tolerances, and the availability of suitable postlarval, juvenile and adult habitat all affect the distribution of a species and the size of its range. An interesting corollary is the balance between dispersal and retention mechanisms required to colonize and maintain isolated island and sea mount populations of closely related scyllarid species in the genus *Jasus* throughout the southern hemisphere (see discussion in Phillips and McWilliam, 1986).

The western rock lobster, P. cygnus, though coastally distributed through its juvenile and adult stages, has a lengthy larval phase which abandons the coastal zone and is transported over 1,000 km offshore. Further studies, including the use of remote-sensing, to elucidate large-scale changes in the southeastern Indian Ocean, as well as localized coastal phenomena, are planned; they should be profitable in understanding year-to-year fluctuations in larval mortality and puerulus settlement.

4. CONCLUSIONS

Larval drift, in the three examples given, would have been a misleading term. The larvae of all three species are active vertical migrators, especially in their later stages and their distribution in the water column can significantly affect larval dispersal directions and distances. Further, to successfully complete their life cycles, the three species have shown three different larval transport trajectories: offshore-onshore for the palinurid lobster, onshore-offshore for the pandalid shrimp and onshore in the penaeid prawn. Deviations from these trajectories alter both postlarval recruitment pattern and strength. The only way to gain an appreciation for the role the larval stage plays in recruitment, dispersal and overall distribution of a species must involve detailed dynamic descriptions of the hydrographic regime, behavior and tolerances of the larvae.

REFERENCES

Bakun, A., 1973. Coastal upwelling indices, west coast of North America 1946-1971. - U.S. Dept. Comm. NOAA Tech. Rep. NMFS SSRF-671. 103 pp.

Church, J.A. & A.M.G. Forbes, 1981. Non-linear models of tides in the Gulf of Carpentaria. - Aust. J. Mar. Freshw. Res. 32: 685-698.

Church, J.A. & A.M.G. Forbes, 1983. Circulation in the Gulf of Carpentaria. Part I. Direct observations of currents in the south east corner of the Gulf of Carpentaria. - Aust. J. Mar. Freshw. Res. 34: 1-10.

Cresswell, G.R. & T.J. Golding, 1980. Observations of a south-flowing current in the southeastern Indian Ocean. - Deep-Sea Res. 27A: 449-466.

Crocos, P.J. & J.D. Kerr, 1983. Maturation and spawning of the banana prawn *Penaeus merguiensis* de Man (Crustacea: Penaeidae) in the Gulf of Carpentaria, Australia. - J. Exp. Mar. Biol. Ecol. 69: 37-59.

Emmerson, W.D., 1980. Ingestion, growth and development of *Penaeus indicus* larvae as a function of *Thallassiosira weissflogii* cell concentration. - Mar. Biol. 58: 65-73.

Dow, R.L., 1981. Environmental influences on *Pandalus borealis* in the Gulf of Maine. - In T. Frady (ed.): Proc. Inter. Pandalid Shrimp Symp. February 13-15, 1979 Kodiak, Alaska, p.351-359. Alaska Sea Grant Program. Univ. Alaska, Fairbanks. Sea Grant Rep. 81-3 June 1981. 519 pp.

Forbes, A.M.G., 1984. The contribution of local processes to seasonal hydrology of the Gulf of Carpentaria. - Oceanogr. Trop. 19: 193-201.

Forbes, A.M.G. & J.A. Curch, 1983. Circulation of the Gulf of Carpentaria. Part II. Residual circulation and mean sea level. - Aust. J. Mar. Freshw. Res. 34: 11-22.

Gunter, G., 1980. Studies in estuarine-marine dependency. - In M. Sears & D. Merriman (eds.): Oceanography: The past, pp. 474-487. Springer-Verlag, N.Y. 812 pp.

Hamon, B.V., 1965. Geostrophic currents in the southeastern Indian Ocean. - Aust. J. Mar. Freshw. Res. 16: 255-271.

Haynes, E.B., 1985. Morphological development, identification and biology of larvae of Pandalidae, Hippolytidae and Crangonidae (Crustacea, Decapoda) of the northern Pacific Ocean. - Fish. Bull. U.S. 83: 253-288.

Hjort, J., 1914. Fluctuations in the great fisheries of north Europe viewed in the light of biological research. - Rapp. P.-v. Reun. Expl. Mer 20:1-228.

Hunter, J.R., 1981. Feeding ecology and predation of marine fish larvae. - In R. Lasker (ed.): Marine fish larvae, morphology, ecology and relation to fisheries, pp. 33-79. Washington Sea Grant Program, Univ. Washington Press, Seattle. 131 pp.

Huyer, A., 1977. Seasonal variation in temperature, salinity and density over the continental shelf off Oregon. - Limnol. Oceanogr. 22: 422-453.

Jackson, G.A. & R.R. Strathman, 1981. Larval mortality from offshore mixing as a link between precompetent and competent periods of development. - Amer. Nat. 118: 16-26.

Kanciruk, P., 1980. Ecology of juvenile and adult Palinuridae (Spiny lobsters). - In J.S. Cobb & B.F. Phillips (eds.): The biology and management of lobsters Vol. II., pp. 59-96. Academic Press, N.Y. 390 pp.

Lesser, J.H.R., 1978. Phyllosoma larvae of *Jasus edwardsii* (Hutton) (Crustacea: Decapoda: Palinuridae) and their distribution off the east coast of the North Island, New Zealand. - N.Z. J. Mar. Freshw. Res. 12: 357-370.

Lukas, J., 1981. Review of the Oregon pink shrimp fishery, management strategy and research activities. - In T. Frady (ed.): Proc. Inter. Pandalid Shrimp Symp., February 13-15, 1979, Kodiak, Alaska, pp. 63- 72. Alaska Sea Grant Program, Univ. Alaska, Fairbanks. Sea Grant Rep. 81-3, June 1981. 519 pp.

May, R.C., 1974. Larval mortality in marine fishes and the critical period concept. - In J.H.S. Blaxter (ed.): The early life history of fish, pp. 3-20. Springer-Verlag, Berlin. 768 pp.

Mileikovsky, S.A., 1971. Types of larval development in marine bottom invertebrates, their distribution and ecological significance: a re-evaluation. - Mar. Biol. 10: 193-213.

Norcross, B.L. & R.F. Shaw, 1984. Oceanic and estuarine transport of fish eggs and larvae: A review. - Trans. Amer. Fish. Soc. 113: 153-165.

Parrish, R.H., C.S. Nelson & A. Bakun, 1981. Transport mechanisms and reproductive success of fishes in the California Current. - Biol. Oceanogr. 1: 175-203.

Pearcy, W.G., 1970. Vertical migration of the ocean shrimp, Pandalus jordani: a feeding and dispersal mechanism. - Calif. Fish and Game 56: 125-129.

Peterson, W.T. & C.B. Miller, 1975. Year-to-year variation in the planktology of the Oregon upwelling zone. - Fish. Bull. U.S. 73: 642-653.

Peterson, W.T., C.B. Miller & A. Hutshinson, 1979. Zonation and maintenance of copepod populations in the Oregon upwelling zone. - Deep-Sea Res. 26A: 467-494.

Phillips, B.F., 1981. The circulation of the southeastern Indian Ocean and the planktonic life of the western rock lobster. - Oceanogr. Mar. Biol. Ann. Rev. 19: 11-39.

Phillips, B.F., P.A. Brown, D.W. Rimmer & D.D. Reid, 1979. Distribution and dispersal of the phyllosoma larva of the western rock lobster Panulirus cygnus in the southeastern Indian Ocean. - Aust. J. Mar. Freshw. Res. 30: 773-783.

Phillips, B.F. & N.G. Hall, 1978. Catches of puerulus larvae on collectors as a measure of natural settlement of the western rock lobster Panulirus cygnus George. - CSIRO Aust. Div. Fish Oecanogr. Rep. 98: 1-18.

Phillips, B.F. & P.S. McWilliam, 1986. The pelagic phase of spiny lobster development. - Can. J. Fish. Aquat. Sci. 43: 2153-2163.

Phillips, B.F. & L. Olsen, 1975. Swimming behavior of the puerulus larvae of the western rock lobster. - Aust. J. Mar. Freshw. Res. 26: 415-417.

Phillips, B.F., D.W. Rimmer & D.D. Reid, 1978. Ecological investigations of the late stage phyllosoma and puerulus larvae of the western rock lobster Panulirus longipes cygnus. Mar. Biol. 45: 347-357.

Phillips, B.F. & A.N. Sastry, 1980. Larval ecology. - In J.S. Cobb & B.F. Phillips (eds.): Biology and management of lobsters Vol. II, pp. 11-57. Academic Press, N.Y. 390 pp.

Preston, N., 1985. The combined effects of temperature and salinity on hatching success and the survival, growth, and development of the larval stages o Metapenaeus bennettae (Racek & Dall). - J. Exp. Mar. Biol. Ecol. 85: 57-74.

Rabalais, N.N. & J.N. Cameron, 1983. Abbreviated development in Uca subcylindrica (Stimpson 1859) (Crustacea, Decapoda, Ocypodidae) reared in the laboratory. - J. Crust. Biol. 3: 519-541.

Rimmer, D.W., 1980. Spatial and temporal distribution of early-stage phyllosoma of the western rock lobster, Panulirus cygnus - Aust. J. Mar. Freshw. Res. 31: 485-497.

Rimmer, D.W. & B.F. Phillips, 1979. Diurnal migration and vertical distribution of phyllosoma larvae of the western rock lobster Panulirus cygnus George. - Mar. Biol. 54: 109-124.

Robinson, J., 1971. The distribution and abundance of pink shrimp (*Pandalus jordani*) off Oregon. - Invest. Rep. No. 8, Fish Comm. Ore. 48 pp.

Rochford, D.J., 1969a. Seasonal variation in the Indian Ocean along 110° E. I. Hydrological studies in the upper 500 m. - Aust. J. Mar. Freshw. Res. 20: 1-50.

Rochford, D.W., 1969b. Seasonal interchange of high and low salinity surface waters off southwest Australia. - CSIRO Aust. Div. Fish. Oceanogr. Tech. Pap. 29. 8 pp.

Rothlisberg, P.C., 1979. Combined effects of temperature and salinity on the survival and growth of the larvae of *Pandalus jordani* (Decapoda: Pandalidae). - Mar. Biol. 54: 125-134.

Rothlisberg, P.C., 1980. A complete larval description of *Pandalus jordani* Rathbun (Decapoda, Pandalidae) and its relation to other members of the genus *Pandalus*. - Crustaceana 38: 19-48.

Rothlisberg, P.C., 1982. Vertical migration and its effect on dispersal of penaeid shrimp larvae in the Gulf of Carpentaria, Australia. - Fish. Bull. U.S. 80: 541-554.

Rothlisberg, P.C., J.A. Church & A.M.G. Forbes, 1983. Modelling the advection of vertically migrating shrimp larvae. - J. Mar. Res. 41: 511-538.

Rothlisberg, P.C. & C.J. Jackson, 1982. Temporal and spatial variation of plankton abundance in the Gulf of Carpentaria, Australia 1975-1977. - J. Plankton Res. 4: 19-40.

Rothlisberg, P.C. & C.B. Miller, 1983. Factors affecting the distribution, abundance and survival of *Pandalus jordani* (Decapoda, Pandalidae) larvae off the Oregon coast. - Fish. Bull. U.S. 81: 455-472.

Rothlisberg, P.C. & W.G. Pearcy, 1977. An epibenthic sampler used to study the ontogeny of vertical migration of *Pandalus jordani* (Decapoda, Caridea). - Fish. Bull. U.S. 74: 994-997.

Rothlisberg, P.C., D.J. Staples & P.J. Crocos, 1985. A review of the life history of the banana prawn, *Penaeus merguiensis* in the Gulf of Carpentaria. - In P.C. Rothlisberg, B.J. Hill & D.J. Staples (eds.): Second Aust. Nat. Prawn Sem., pp. 125-136. NPS2, Cleveland, Australia. 368 pp.

Shanks, A.L., 1983. Surface slicks associated with tidally forced internal waves may transport pelagic larvae of benthic invertebrates and fishes shoreward. - Mar. Ecol. Prog. Ser. 13: 311-315.

Shanks, A.L., 1985. Behavioral basis of internal-wave-induced shoreward transport of megalopae of the crab *Pachygrapsus crassipes*. - Mar. Ecol. Prog. Ser. 24: 289-295.

Smith, R.L., 1974. A description of current, wind and sea level variations during coastal upwelling off the Oregon coast, July-August 1972. - J. Geophys. Res. 79: 435-443.

Staples, D.J., 1979. Seasonal migration patterns of postlarval and juvenile banana prawns, *Penaeus merguiensis* de Man, in the major rivers of the Gulf of Carpentaria, Australia. - Aust. J. Mar. Freshw. Res. 30: 143-157.

Staples, D.J., D.J. Vance & D.S. Heales, 1985. Habitat requirements of juvenile penaeid prawns and their relationship to offshore fisheries. - In P.C. Rothlisberg, B.J. Hill & D.J. Staples (eds.): Second Aust. Nat. Prawn Sem., pp. 47-54. NPS2, Cleveland, Australia. 368 pp.

Thorson, G., 1950. Reproductive and larval ecology of marine bottom invertebrates. - Biol. Rev. 25: 1-45.

Wroblewski, J.S., 1982. Interaction of currents and vertical migration in maintaining *Calanus marshallae* in the Oregon upwelling zone - a simulation. - Deep-Sea Res. 29: 665-686.

Wyrtki, K., 1962. Geopotential topographies and associated circulation in the south-eastern Indian Ocean. - Aust. J. Mar. Freshw. Res. 13: 1-17.

Wyrtki, K., 1973. Physical oceanography of the Indian Ocean. - In B. Zeitzschel & S.A. Gerlach (eds.): The biology of the Indian Ocean, pp. 18-36. Springer-Verlag, Berlin. 549 pp.
Zeldis, J.R. & J.B. Jillett, 1982. Aggregations of pelagic Munida gregaria (Fabricius) (Decapoda, Anomura) by coastal fronts and internal waves. - J. Plankton Res. 4: 839-857.

TRANSPORT OF CRAB LARVAE BETWEEN ESTUARIES AND THE CONTINENTAL SHELF

Charles E. Epifanio
College of Marine Studies, University of Delaware
Lewes, DE 19958, USA

Estuarine crabs release their larvae freely into the water column in spite of the risk of these larvae being transported out of the estuary onto the continental shelf. Many species have evolved behavioral traits that in combination with horizontal advection retain the larvae near favorable adult habitat. Leakage of these species onto the shelf results in loss of the larvae to adult population. Behavioral models resulting in retention include: 1) constant maintenance of position deep in the water column, 2) downstream advection of surface-dwelling, immature larvae, and 3) tidally rhythmic vertical migration of larvae. Larvae of some species appear to be adapted for routine dispersal onto the continental shelf. These larvae must be transported back to suitable estuarine habitat for adult existence. Models for transport back to the estuary include: 1) active horizontal swimming by mature larvae or juveniles, 2) seaward advection of surface-dwelling, immature larvae followed by landward advection of bottom-dwelling, mature larvae, 3) transport of mature larvae in convergence zones associated with landward movement of tidally induced, internal waves, and 4) wind-generated surface transport of mature larvae. A number of case studies involving well-studied estuarine species are reviewed in the paper.

1. INTRODUCTION

1.1 The problem

The life cycle of most species of estuarine and coastal crabs includes several zoeal stages and a megalopal stage (Fig. 1). Zoeal stages are generally planktonic while the megalopal stage may evidence behavioral patterns that result in either planktonic or benthic distributions.

Zoeae are capable of directional swimming at speeds about an order of magnitude slower than the subtidal seaward movement of surface water in many estuaries (Pape & Garvine, 1982; Sulkin, 1984). As most estuarine crabs are incapable of maintaining viable populations on the continental shelf, zoeae must be retained in the estuary or have some means of returning to the estuary once they have undergone development in shelf

Lecture Notes on Coastal and Estuarine Studies, Vol. 22
B.-O. Jansson (Ed.), Coastal-Offshore Ecosystem Interactions.
© Springer-Verlag Berlin Heidelberg 1988

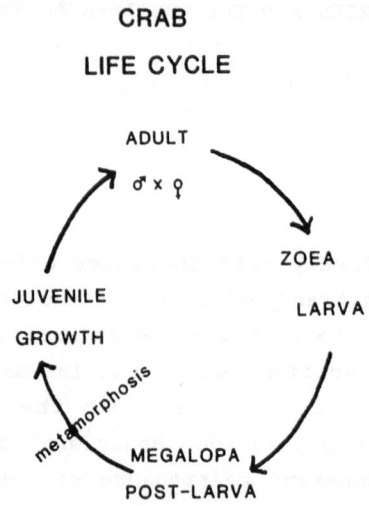

CRAB

LIFE CYCLE

ADULT

♂ x ♀

ZOEA

JUVENILE

LARVA

GROWTH

metamorphosis

MEGALOPA

POST-LARVA

Fig. 1. Life cycle of a typical brachyuran crab. Planktonic zoeal stages vary from two to eight, depending on species. The period of planktonic existence is usually less than 30 days, but may extend as long as 90 days in some species. Megalopae show planktonic and benthic behavior. This post-larval phase may last 30 days.

waters. Clearly, active swimming by zoea larvae is ineffective in maintaining horizontal position deep in the water column or undergo vertical migrations to effect retention in the estuary.

Results of field investigations suggest that various species of estuarine crab can be grouped by the degree of retention of their larvae in the estuary (Nichols & Keney, 1963; Dudley & Judy , 1973; Sandifer, 1975; Dittel and Epifanio, 1982; Johnson, 1982; Brookins and Epifanio, 1985). Retained species are prevalent in families such as the Xanthidae (mud crabs) wherein adults are incapable of swimming and live in the middle to upper reaches of the estuary. At the other extreme are swimming crabs in the family Portunidae whose larvae are normally exported from the estuary (Epifanio et al., 1984). These observed differences in the degree of retention of different species in the estuaries imply an active involvement of the larvae in their horizontal transport.

2. MODELS RESULTING IN RETENTION

There are at least three models that would result in retention of zoeae in an estuary (Fig. 2). In the first, zoeae are generally distributed in the lower regions of the water column due to passive sinking (DeWolf, 1973; Boicourt, 1982; Seliger et al., 1982). These larvae would be mixed up into the water column by turbulent processes during periods of flooding or ebbing tidal currents and would sink deeper in the water column at slack tide. As flood velocities exceed ebb velocities in bottom estuarine waters, this would provide a mechanism for transport upstream.

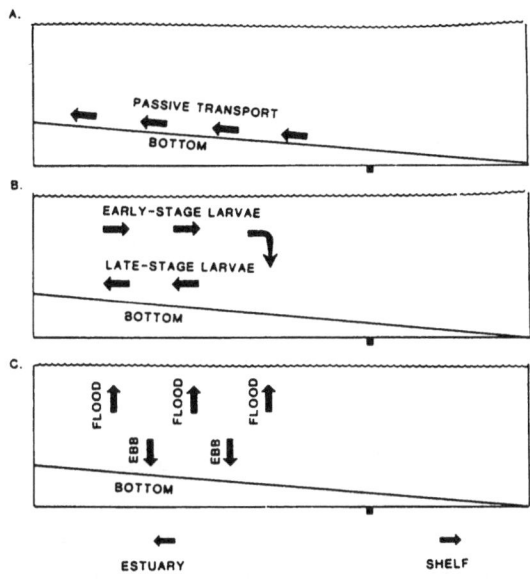

Fig. 2. Possible mechanisms for retention of larval crabs in the estuary. A. Larvae maintain a deep position throughout development and are transported upstream by nontidal flow. B. Larvae sink in water column as development proceeds. C. Larvae migrate rhythmically with tidal flow.

In the second model, zoeae take up a position high in the water column early in development, but sink to deeper levels as ontogenetic development proceeds (Bousfield, 1955). This would result in downstream dispersal during early larval life with transport back upstream as metamorphosis approached.

In the third model, larvae migrate rhythmically with tidal phases; they move up in the water column during periods of flooding tidal currents and sink deeper in the water column during periods of ebbing tidal currents (Carriker, 1951, 1961; Nelson & Perkins, 1930; Wood & Hargis, 1971: Epifanio et al., 1984). Clearly, this would result in net up stream transport of the larvae.

Model 1 is passive as it does not require directed swimming by the larvae. In contrast, models 2 and 3 represent an active process wherein vertical swimming by the zoeae controls the degree of horizontal advection. A growing body of evidence suggests that crab larvae are able to use environmental cues to regulate their vertical swimming and thus their horizontal displacement in space.

3. BEHAVIORAL BASIS OF RETENTION

In the marine and estuarine environment, the principal orienting stimuli in the vertical plane are light and gravity. Important kinetic stimuli include light, temperature, salinity, and hydrostatic pressure (Sulkin, 1984). Nevertheless, it is not likely that light is the most important cue for vertical migration of crab larvae in the natural environment. The penetration of light into turbid estuarine water is limited, and the extreme diurnal variation in intensity of light in the water column minimizes its utility as a cue for vertical movements important in retention. In contrast, gravity is a conservative stimulus for orientation throughout the water column regardless of time of day. Laboratory experiments (e.g. Sulkin, 1973; Latz & Forward, 1976; Sulkin et al., 1980, 1983; Schembri, 1982) have generally shown that early-stage zoeae display negative geotaxis (move away from the center of the earth) while late stages and megalopae display positive geotaxis (move toward center of earth). This pattern of response would yield a shallow distribution of early-stage larvae and a deep distribution of late stages and megalopae.

Further studies have shown that zoaea and megalopae respond to other stimuli in addition to gravity. Salinity appears to have minor effects on swimming and larvae are able to pass through sharp haloclines (Sulkin et al., 1980; O'Connor & Epifanio, 1985). Temperature increases swimming speed, but thermoclines do not prevent upward or downward migration (Kelly et al., 1982; MnConnaughey, 1983). It appears, then, that zoeae are able to migrate vertically regardless of the degree of stratification of the water column.

Hydrostatic pressure is more conservative than temperature or salinity, as this factor increases predictably with depth. Early-stage zoeae increase swimming speed as pressure increases (barokinesis), even in very small increments (Wheeler & Epifanio, 1978). Late-stage larvae appear to be less sensitive to pressure changes, but megalopae may show an increased sensitivity (Sulkin & Van Heukelem, 1982).

Geotaxis and barokinesis are responses to very conservative stimuli and are almost universally found in early-stage larvae. A combined response to gravity and pressure would result in upward swimming and would serve as a mechanism for depth regulation; as larvae sank deeper into the water column, barokinesis would increase and the larvae would swim to a shallower depth. Increased water depth during flooding tidal periods would also stimulate a barokinetic response in larvae and, in combination with negative geotaxis, stimulate upward swimming on flooding tidal currents. However, the tide propagates as a wave with a consequent orbital circulation imparted to the water column. A crab larva passively suspended in the water column would be entrained in this orbital circulation and would not change its depth relative to the surface. In this case, the hydrostatic-pressure environment of the crab larva would not change and, thus, would not serve as a cue to the larva that the tide was flooding. In contrast, larvae or megalopae that sink to the bottom or near bottom during ebb tidal phases would experience an increase in hydrostatic pressure during periods of flooding tidal currents and would rise in the water column to be transported upstream. The cue and mechanism for sinking or downward swimming during ebb tidal phases is unclear, however.

4. MODELS FOR RETURN OF EXPORTED LARVAE

There is considerable leakage onto the continental shelf of larvae of even those species that show retentive behavioral traits (Nichols & Keney, 1963; Dudley & Judy, 1973), and those larvae are, for the most part, lost to adult estuarine populations. However, larval exportation from the estuary appear the norm in some species, and larvae of these crabs must find their way back to an appropriate estuary to maintain continuity of the species (Fig. 3). One suggested model for transport back into an estuary requires that the larvae simply swim onshore.
Studies of swimming behavior (Phillips & Olsen, 1975) and field distributions (Phillips et al., 1978) imply that this may be an important mechanism for rock lobster larvae in Australian waters. However, measured larval swimming speeds cast doubt on this as an effective mechanism for crab larvae.

In a second model, early-stage larvae are transported out of the estuary in surface flow and enter the adjacent coastal circulation. As ontogenetic development proceeds, the larvae take up a deeper position in the water column and are transported back to the estuary in the subtidal, estuarine, bottom flow. Pape and Garvine, (1982) showed that the bottom circulation of large estuaries such as Delaware Bay extends at

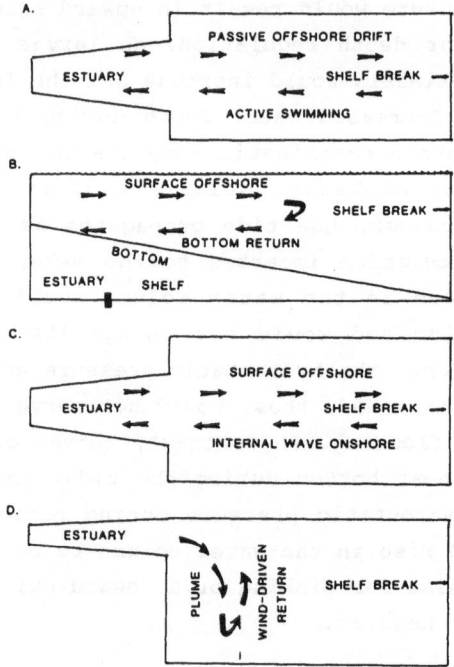

Fig. 3. Possible mechanisms for cross-shelf transport of estuarine
larvae. A. Larvae are advected offshore in surface flow and swim back
to the estuary (surface view). B. Larvae are advected offshore in
surface flow and sink to deeper depths as development proceeds
(sectional view). C. Larvae are advected offshore by surface flow and
return by internal-wave transport (surface view). D. Larvae are
transported away from estuary by surface plume and transported back to
vicinity of estuary by wind-driven transport (surface view).

least 40 km onto the shelf, and Sulkin et al. (1980) presented strong
evidence that larvae of at least one exporting species possess
behavioral traits necessary to take advantage of this circulation.
However, subsequent field sampling did not show a deepening of vertical
position of these larvae with increased age; rather, the larvae
remained in surface water throughout zoeal development (McConaugha et
al., 1983).

A third model invokes internal waves as a mechanism for rapid transport
of crab larvae back to the shore zone. As a result of the dynamics of
tidal motion at the edge of the continental shelf, packets of

shoreward-propagating internal waves are generated with each change from ebb to flood tide (Osborne & Burch, 1980; Chereskin, 1983). Associated with the onshore propagation of these waves is a set of upwelling and downwelling currents that create zones of convergence that move shoreward along with the waves. Floating material accumulates in these convergence zones and is transported rapidly shoreward along with the waves. Shanks (1983) has suggested that this phenomenon is an important transport mechanism for crab zoeae and megalopae. However, internal-wave transport requires that the larvae be located in the very surface film (neuston) regardless of developmental stages.

A final model utilizes wind-driven transport of larvae on the shelf (Johnson et al., 1984; Johnson, 1985). Here, early-stage crab larvae are carried from the parent estuary in seaward-flowing surface water. This water continues along shelf as a plume or jet adjacent to the shore. As the momentum of the plume is dissipated by friction, a wind-driven counter current is set up, providing along-shelf transport in the opposite direction for advanced-stage larvae. The model requires that larvae dwell in surface water, but not necessarily in the neuston. However, the model provides only along-shelf transport, and cross-shelf transport remains a problem.

5. CASE STUDIES

Rhithropanopeus harrisii, Uca spp., and Callinectes sapidus represent the entire spectrum of dispersal strategies of estuarine crabs. These species have been studied extensively in estuaries along the Middle Atlantic and South Atlantic Coasts of the United States where they often occur together.

5.1 Rhithropanopeus harrisii

R. harrisii is a small mud crab that lives in the oligohaline and mesohaline regions of these estuaries. Adults are incapable of swimming and probably have very restricted home ranges.

Surveys of larval abundance in mouths of major estuaries such as the Chesapeake (Sandifer, 1975; Goy, 1976) and Delaware (Dittel & Epifanio, 1982) show low incidence of R. harrisii zoeae. But adults of this species are extremely common in both estuaries and development through four zoeal and one megalopal stage requires more than two weeks - sufficient time for transport of passive particles to the lower estuary. Clearly, these larvae must be modifying their horizontal advection in a manner that minimizes flushing from the area in which they are spawned.

Lambert and Epifanio (1982) found all zoeal stages of R. harrisii retained in a small secondary estuary adjacent to the mouth of Delaware Bay. Spawning appeared to coincide with the onset of flood tide, and stage I zoeae were carried upstream by the flooding tidal currents. More detailed investigations over several consecutive tidal cycles in the Newport River, North Carolina showed a distinct, tidally rhythmic vertical migration wherein zoeae swam up in the water column curing flood-tide phases and sank to deeper depths during ebb-tide phases (Cronin, 1982). This behavior appears to be intrained in the larvae by some environmental cure, as field-caught R. harrisii larvae migrate vertically in laboratory water columns in the absence of external stimuli, while laboratory-reared larvae do not (Cronin & Forward, 1979). The cue necessary for entrainment has not been determined, although changes in hydrostatic pressure (Wheeler & Epifanio, 1978) and salinity (Latz & Forward, 1977) have been shown to elicit vertical migration of R. harrisii larvae under laboratory conditions. In any event, R. harrisii larvae clearly possess a highly developed mechanism for retention in areas of the estuary accessible to appropriate habitat.

5.2 Uca spp.

In the Middle Atlantic region, the genus Uca consists of two species of semi-terrestrial fiddler crab: U.minax and U. pugnax. U. minax lives along oligohaline marsh creeks in brackish and even freshwater marshes. U. pugnax occupies similar habitat in more saline areas of salt marsh. Juveniles and adults cannot swim and, due to their small size, are not capable of long, overland migrations. In spite of this, Uca larvae are not adapted for retention in the secondary estuaries adjacent to adult habitat.

Uca spp. larvae are hatched near the time of high tide around new and full moon assuring their downstream transport during maximum ebb flow (Wheeler, 1978; Christy, 1982). Larvae of the two species are indistinguishable by conventional techniques, and all field studies treat the species as a unit, Uca spp. Regardless (and in contrast to R. harrisii), Lambert and Epifanio (1982) found no retention of Uca zoeae in a season-long study of a secondary estuary near the mouth of Dela ware Bay.Larvae from upstream populations of U. minax were flushed during one tidal cycle to the lower regions of the secondary estuary and into Delaware Bay on the subsequent ebb phase. Larvae from the downstream populations of U. pugnas were flushed into Delaware Bay during the first ebb tide after hatching.

Early stage larvae maintain a position high in the water column resulting in their transport into the mouths of primary estuaries such as Delaware Bay (Dittel & Epifanio, 1982) and even onto the adjacent continental shelf (Christy & Stancyk, 1982). However, late stage larvae take a position deep in the water column, thus taking advantage of subtidal landward flow back into the primary estuary (Christy, 1982; Dittel & Epifanio, 1982). Movement into the marsh environments required by the adults appears to be facilitated by tidally rhythmic vertical migration (Meredith, 1982).

5.3 Callinectes sapidus

This species is a large swimming crab that occurs throughout the entire estuary as an adult. Mature males live in the oligohaline regions, extending into fresh water, while mature females migrate to the estuarine mouth for spawning. Larval development includes seven to eight zoeal stages and one megalopal stage. Hatching occurs near the time of a nocturnal high tide after which larvae migrate to the surface and are transported onto the continental shelf by the ensuing ebb tide (Provenzano et al., 1983).

Results of a three-year investigation in Delaware Bay showed the presence of stage I zoeae and megalopae; the intervening zoeal stages were not collected (Epifanio et al., 1984). Furthermore, peak abundance of stage I zoeae occurred five weeks before peak abundance of megalopae at the bay mouth. This time period approximates that required for zoeal development in the laboratory (Sulkin, 1975). Zoeal development clearly occurs in shelf waters and megalopae must possess a mechanism for returning to the estuary.

An initial hypothesis accounting for the onshore transport of megalopae was based on laboratory-based behavioral work by Sulkin et al. (1980).

In that study, the investigators demonstrated behavioral responses that would allow maintenance of position high in the water column for stage I zoeae, but deeper in the water column for stages IV and VII. The authors went on to develop a dispersal model that included transport of early-stage larvae onto the continental shelf with later return via onshore, subtidal flow of bottom water. Field evidence, however, did not support this hypothesis as zoeal stages II-VII appear to remain in surface water throughout their period of development on the inner continental shelf (Smyth, 1980; McConnaugha et al., 1983; Epifanio, unpublished data). As the preponderance of published data on flow of inshore surface water in the Middle Atlantic Bight indicated a unidirectional movement from north to south with eventual entrainment in the Gulf Stream (Bumpus, 1965, 1969, 1973; Pape & Garvine, 1982), it appeared that larvae would routinely be transported long distances - as much as 350 km (Pape & Garvine, 1982) - from their parent estuaries during the five weeks of zoeal development. Thus, the more northern estuaries would serve as sources of larvae for more southern estuaries, but there would be no mechanism for recruitment to the northernmost populations of C. sapidus, and there would be major loss to the Gulf Stream of larvae produced in Chesapeake Bay.

More recent work, however, has provided a plausible mechanism for retention of C. sapidus in the vicinity of their parent estuaries. Preliminary measurements show plumes of low-salinity surface water flowing from Delaware and Chesapeake Bays and moving to the south along the coast (Boicourt, 1981, 1982). Well offshore of these plumes, the flow of surface water is dominated by long-shore pressure gradients and moves from north to south. The region between the near-shore plume and the off-shore, pressure-dominated flow is more sensitive to forcing by wind (Fig. 4), and the prevailing southwesterly winds characteristic of the summer months in this region can impart a northward motion to the surface water (Johnson et al., 1984). Further studies in the area showed a relationship between reported distributions of blue crab larvae off Chesapeake Bay and theoretical, surface-current patterns as predicted from wind records (Johnson, 1985). A reasonable scenario for dispersal of C. sapidus larvae in the Middle Atlantic Bight involves transport of newly hatched larvae onto the continental shelf and to the south in the surface plumes of major estuaries. As these plumes dissipate due to friction, there would be mixing with the northward-flowing band of wind-driven surface water immediately offshore with consequent transport back to the general latitude of the parent estuary. Zoeae would then be transported back onshore by the predominant easterly winds that occur in early autumn (Johnson, 1985).

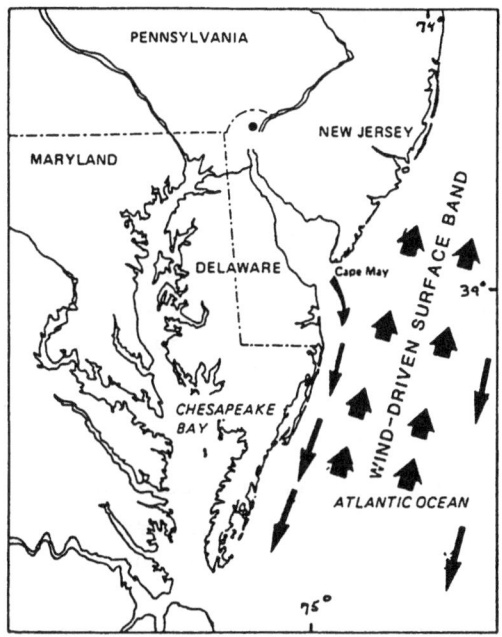

Fig. 4. Summer surface flow in lower Middle Atlantic Bight adjacent to Delaware and Chesapeake Bays. Southward flows along beach and in mid-shelf region are separated by a wind-driven northward flow important in retention of estuarine larvae near parent estuaries. Interannual varia-tion in strength of northward flow may be related to interannual varia-tion in estuarine recruitment.

Results of a comprehensive survey of megalopal distribution off the Chesapeake indicated that megalopae generally maintain a surface dis-tribution in shelf water as less than 25% of the larvae captured occurred in bottom water (Johnson, 1982). Johnson concluded from this that megalopal transport was also wind driven. However, it is doubtful that wind-driven transport at the surface would allow megalopal re invasion of large estuaries with strong seaward flow of surface water. Sulkin and Epifanio (in press) proposed an alternative hypothesis based on the portion of megalopae that Johnson found in near-bottom waters. This hypothesis assumes that megalopae move deeper in the water column as they near metamorphosis. Less mature megalopae would be transported by wind-driven surface currents, but older larvae would utilize the sub-tidal, shoreward movement of bottom waters to overcome the net seaward flow of surface waters in the estuarine mouth. This model is consistent with behavioral characteristics of megalopae in laboratory experiments (Sulkin & Van Heukelem, 1982). There is additional field evidence (Meredith, 1982; Epifanio et al., 1984) that megalopae may further accelerate their movement into estuaries by undergoing tidally rhythmic vertical migrations.

6. CONCLUSIONS

Estuarine crabs release their larvae freely into the water column in spite of the risk of the larvae being transported out of the estuary onto the continental shelf. Many species have evolved behavioral traits that allow control of horizontal advection with consequent larval retention in areas near favorable adult habitat. Leakage of these species onto the shelf generally results in loss of larvae to adult populations.

The degree of retention of estuarine zoeae varies widely among species, and some, such as the blue crab <u>Callinectes sapidus</u>, possess behavioral traits that ensure exportation onto the open shelf. This species appears to depend on wind-driven transport to retain zoeae within "striking distance" of acceptable adult habitat, and reinvasion of the estuary occurs primarily during the megalopal stage. Megalopae probably take advantage of gravity-driven estuarine circulation for the final stages of their migration back into the estuary. Blue crab zoeae that are transported to the outer continental shelf or are entrained in the Gulf Stream are probably lost to adult populations.

Species of crab that occur as adults on the continental shelf face many of the same larval dispersal problems as blue crabs, but have received far less study. Larvae released on the outer continental shelf in the Middle Atlantic Bight, for example, appear to be subjected to unidirectional flow from north to south with no known mechanism for retaining larvae in the northermost parts of the range or for forestalling entrainment in the Gulf Stream at Cape Hatteras (Rosowski, 1979). Vertical migration is an unlikely mechanism for controlling horizontal transport in these cases as sub-thermocline temperatures are generally unsuitable for larval development (Kelly et al., 1982; Van Heukelem et al., 1984). An understanding of larval transport processes on the open continental shelf will require physical measurements on a relatively fine scale. Present estimates of gross mean flow, e.g., Bumpus (1965), not sufficient for this task.

REFERENCES

Brookins, K.G., & C.E. Epifanio, 1985. Abundance of brachyuran larvae in a small coastal inlet over six consecutive tidal cycles. - Estuaries 8: 60-67.

Boicourt, W.C., 1982. Estuarine larval retention methanisms on two scales. - In V. Kennedy (es.): Estuarine comparisons, pp. 445-458. Academic Press, New York.

Boicourt, W.C., 1981. Circulation in the Chesapeake Bay entrance region: estuary-shelf interaction. - In G. Campbell, & J. Thomas (eds.): Chesapeake Bay plume study superflux, 1980, pp. 28-58. NASA Science and Technology Information Branch, Dept. Comm., NASA Conf. Publ. 2188, NOAA/NEMP III; 81 ABCDFG, 0042.

Bousfield, E.L., 1955. Ecological control of the occurrence of barnacles in the Miramichi estuary. - Nat. Mus. Can. Bull. 137: 1-69.

Bumpus, D.F., 1965. Residual drift along the bottom on the continental shelf in the Middle Atlantic Bight area. - Limnol. Oceanogr. Suppl. 10: R50-R53.

Bumpus, D.F., 1969. Reversals in the surface drift in the Middle Atlantic Bight area. - Deep Sea Res. Oceanogr. Abs. Suppl. 16: 17-23.

Bumpus, D.F., 1973. A description of the circulation on the continental shelf of the east coast of the United States. - Progress in Oceanogr. 6: 111-157.

Carriker, M.R., 1951. Ecological observations on the distribution of oyster larvae in New Jersey estuaries. - Ecol. Monogr. 21: 19-38.

Carriker, M.R., 1961. Interrelation of functional morphology, behavior and autecology in early states of the bivalve, _Mercenaria mercenaria_. - J. Elisha Mitchell Sci. Soc. 77: 168-241.

Chereskin, T.K., 1983. Generation of internal waves in Massachusetts Bay. - J. Geophys. Res. 88: 2649-2661.

Christy, J.H., 1982. Adaptive significance of semilunar cycles of larval release in fiddler crabs (Genus _Uca_): Test of an hypothesis. -Biol. Bull. 163: 251-263.

Christy, J.H., & S.E. Stancyk, 1982. Timing of larval production and flux of invertebrate larvae in a well-mixed estuary. - In V. Kennedy (ed.): Estuarine comparisons, pp. 489-503. Academic Press, New York.

Cronin, T.W., 1982. Estuarine retention of larvae of the crab _Rhithropanopeus harrisii_. - Estuar. Coastal Shelf Sci. 15: 207-220.

Cronin, T.W. & R.B. Forward, Jr., 1979. Tidal vertical migration: an endogenous rhythm in estuarine crab larvae. - Science 205: 1020-1022.

De Wolf, P., 1973. Ecological observation on the mechanisms of dispersal of barnacle larvae during planktonic life and settling. - Neth. J. Sea Res. 6: 1-129.

Dittel, A.I. & C.E. Epifanio, 1982. Seasonal abundance and vertical distribution of crab larvae in Delaware Bay. - Estuaries 5: 197-202.

Dudley, D.L. & M.H. Judy, 1973. Seasonal abundance and distribution of juvenile blue crabs in Cove Sound, N.C., 1965-68. - Estuar. Coastal Shelf Sci. 14: 51-54.

Epifanio, C.E., C.C. Valenti & A.E. Pembroke, 1984. Dispersal and recruitment of blue crab larvae in the Delaware Bay, USA. - Estuar. Coastal Shelf Sci. 18: 1-12.

Forward, Jr., R.B. Cronin, W. Thomas & D.E. Stearns, 1984. Control of diel vertical migration: Photoresponses of a larval crustacean. - Limnol Oceanogr. 29: 146-154.

Goy, J.W., 1976. Seasonal distribution and the retention of some deca-
    pod crustacean larvae within the Chesapeake Bay, Virginia. - M.S.
    Thesis, Old Dominion Univ., Norfolk, Virginia.
Johnson, D.F., 1982. A comparison of recruitment strategies among
    brachyuran crustacean megalopae of the York River, lower Chesapeake
    Bay and adjacent waters. - Ph.D. Diss., Old Dominion Univ.,
    Norfolk, Virginia.
Johnson, D.R., B.S. Hester & J.R. McConaugha, 1984. Studies of a wind
    mechanism influencing the recruitment of blue crabs in the Middle
    Atlantic Bight. - Continental Shelf Res. 3: 425-437.
Johnson, D.R., 1985. Wind-forced dispersion of blue crab larvae in the
    Middle Atlantic Bight. - Continental Shelf Res. 4: 1-14.
Kelly, P., S.D. Sulkin, & W.F. Van Heukelem, 1982. A dispersal model
    for larvae of the deep sea red crab Geryon quinquedens Smith based
    upon behavioral regulation of vertical migration in the hatching
    stage. - Mar. Biol. 72: 35-43.
Lambert, R. & C.E. Epifanio, 1982. A comparison of dispersal strategies
    in two genera of brachyuran crab in a secondary estuary. -
    Estuaries 5: 182-188.
Latz, M.I. & J.R. Forward, Jr., 1977. The effect of salinity upon
    phototaxis and geotaxis in a larval crustacean. - Biol. Bull. 153:
    163-179.
McConaugha, J.R., D.F. Johnson, A.J. Provenzano & R.C. Maris, 1983.
    Seasonal distribution of larvae of Callinectes sapidus in the
    waters adjacent to Chesapeake Bay. - J. Crust. Biol. 3: 582-591.
McConnaughey, R., 1983. The influence of thermoclines on vertical
    migration of Stage I blue crab Callinectes sapidus larvae and the
    implications to recruitment. - M.S. Thesis, Univ. Maryland, College
    Park, Maryland.
Meredith, W.A., 1982. The dynamics of zooplankton and micronekton
    community structure across a salt-marsh estuarine interface of
    lower Delaware Bay. Ph.D. Diss., Univ. Delaware, Newark, Delaware.
Nelson, T.C. & E.B. Perkins, 1930. The reactions of oyster larvae to
    currents and to salinity gradients (abstract). - Anatomical Record
    40: 288.
Nichols, P. & P.M. Keney, 1963. Crab larvae (Callinectes) in
    planktoncollections from cruises of M/V Theodor N. Gill, South
    Atlantic Coast of the United States 1953-54. - U.S. Fish. Wildl.
    Serv. Spec. Sci. Rep. Fish. 448: 1-14.
O'Connor, Nancy J. & C.E. Epifanio, 1985. The effect of salinity on the
    dispersal and recruitment of fiddler crab larvae. - J. Crust. Biol.
    5: 137-145.
Osborne, A.R. & T.L. Burch, 1980. Internal solitons in the Andaman Sea.
    - Science 208: 451-460.
Pape III, E.H. & R.W. Garvine, 1982. The subtidal circulation in
    Delaware Bay and adjacent shelf waters. - J. Geophys. Res. 87:
    7955- 7970.
Phillips, B.F. & L. Olsen, 1975. The swimming behavior of the puerulus
    stage of the Western Rock Lobster. - Aust. J. Mar. Freshw. Res. 26:
    415-417.
Phillips, B.R., D.W. Rimmer & D.D. Reid, 1978. Ecological
    investigations of the late-stage phyllosoma and puerulus larvae of
    the Western Rock Lobster Panulirum logipes cygnus. - Mar. Biol. 45:
    347-357.
Provenzano, Jr., A.J., J.R. McConaugha, K.B. Philips, D.F. Johnson & J.
    Clark, 1983. Vertical distribution of first stage larvae of the
    blue crab, Callinected sapidus, at the mouth of Chesapeake Bay. -
    Estuar. Coastal Shelf Sci. 16: 486-499.
Rosowski, M.C., 1979. The effect of temperature on growth and dispersal
    of Geryon quinquedens Smith (Brachyura; Geryonidae). M.S. Thesis.,
    Univ. Delaware, Newark, Dalaware.

Sandifer, P.A., 1979. The role of pelagic larvae in recruitment to populations of adult decapod crustaceans in the York River Estuary and adjacent lower Chesapeake Bay, Virginia. - Estuar. Coastal Mar. Sci. 3: 269-279.

Schembri, P.J., 1982. Locomotion, feeding, grooming, and the behavioral responses to gravity, light, and hydrostatic pressure in the Stage I zoea larvae of Ebalia tuberosa (Crustacea: Decapoda: Leucosiidae). - Mar. Biol. 72: 125-134.

Seliger, H.H., J.A. Boggs, R.B. Rivkin, W.H. Biggley & K.R.H. Aspden, 1982. The transport of oyster larvae in an estuary. - Mar. Biol. 71: 57-72.

Shanks, A.L., 1983. Surface slicks associated with tidally forced internal waves may transport pelagic larvae of benthic invertebrates and fishes shoreward. - Mar. Ecol. Prog. Ser. 13: 311-315.

Smyth, P.O., 1980. Callinectes (Decapoda: Portunidae) larvae in the Middle Atlantic Bight, 1975-77. Fisheries Bull. 78: 251-265.

Sulkin, S.D., 1973. Depth regulation of crab larvae in the absence of light. - J. Exp. Mar. Biol Ecol. 13: 73-82.

Sulkin, S.D., 1975. The influence of light in the depth regulation of brachyuran crabs. - Biol. Bull. 148: 333-343.

Sulkin, S.D., 1984. Behavioral basis of depth regulation in the larvae of brachyuran crabs. - Mar. Ecol. Prog. Ser. 15: 181-205.

Sulkin, S.D. & C.E. Epifanio. A conceptual model for recruitment of the blue crab, Callinectes sapidus Rathbun, to estuaries of the Middle Atlantic Bight. - In G. Jameson (ed.): Proc. Workship North Pacific Invert. Fisheries. Spec. Publ. Can. Fish. J. In press.

Sulkin, S.D., W. Van Heukelem, P. Kelly & L. Van Heukelem, 1980. The behavioral basis of larval recruitment in the crab, Callinectes sapidus Rathbun: an amendment to the concept of larval retention in estuaries. - In V. Kennedy (ed.): Estuarine comparisons. Academic Press, New York. pp. 459-475.

Sulkin, S.D., W. Van Heukelim & P. Kelly, 1983. Behavioral basis of depth regulation in the hatching and post-larval states of the mud crab Eurypanopeus depressus Hay and Shore. - Mar. Ecol. Prog. Ser. 11: 157-164.

Van Heukelem, W., M.C. Christman, C.E. Epifanio & S.D.Sulkin, 1984. Growth of juvenile eryon quinquedens Brachyura:(Geryonidae) in the laboratory. - Fish. Bull. 181: 903-905.

Wheeler, D. (1978). Semilunar hatching periodicity in the mud crab Uca pugnax (Smith). - Estuaries 1: 268-269.

Wheeler, D. & C.E. Epifanio, 1978. Behavioral response to hydrostatic pressure in larvae of two species of xanthid crabs. - Mar. Biol. 46: 167-174.

Wood, L., & W.J. Hargis, Jr., 1971. Transport of bivalve larvae in a tidal estuary. - In D. Crisp (ed.): Proceedings fourth European marine biology symposium. University Press, Cambridge. pp. 29-44.

IV. NUMERICAL MODELLING

COUPLING OF HYDRODYNAMIC AND ECOSYSTEMS MODELLING APPLIED TO TIDAL ESTUARIES

R. J. Uncles

Natural Environment Research Council,
Institute for Marine Environmental Research,
Prospect Place, The Hoe, Plymouth Pl1 3DH, U.K.

A review is given of techniques which are currently used to couple hydrodynamical and ecological models in large ecosystem simulations of tidal estuaries. Time-dependent models in one, two and three spatial dimensions are considered with reference to specific ecological models of estuarine regions. Within-tide and tidally averaged versions of fixed element and moving element one-dimensional models are the most straightforward to apply. Of these, the tidally averaged moving element model is the most satisfactory physically, but has properties which make it inconvenient for ecosystem simulations. The fixed element, tidally-averaged model is the most suitable for ecosystem simulations, although correlations between tidal flow and tidal fluctuations in the concentration of a pelagic variable are unknown and have to be neglected. This is also a difficulty with two and three-dimensional ecosystem models, which invariably use fixed spatial elements.

1. INTRODUCTION

This paper attempts to review the methods which are currently used for coupling hydrodynamical and ecological models in complex ecosystem simulations. Rather than deal with generalities, time-dependent models in one, two and three spatial dimensions are considered under separate headings. Derivations of the associated hydrodynamical equations, methods used in their numerical solutions, and properties of numerical techniques are not dealt with; these have been the subjects of other reviews (Edinger and Buchak, 1980; Liu and Leenderste, 1978; and Roach, 1982, respectively).

The role of the hydrodynamics is to advect and disperse biological and chemical components of the ecosystem through the estuary and over the course of, perhaps, several years. The problem of fundamental interest here is the very different time-scales and space-scales of hydrodynamical and ecological models. A time-step for the stable numerical solution of a

Lecture Notes on Coastal and Estuarine Studies, Vol. 22
B.-O. Jansson (Ed.), Coastal-Offshore Ecosystem Interactions
© Springer-Verlag Berlin Heidelberg 1988

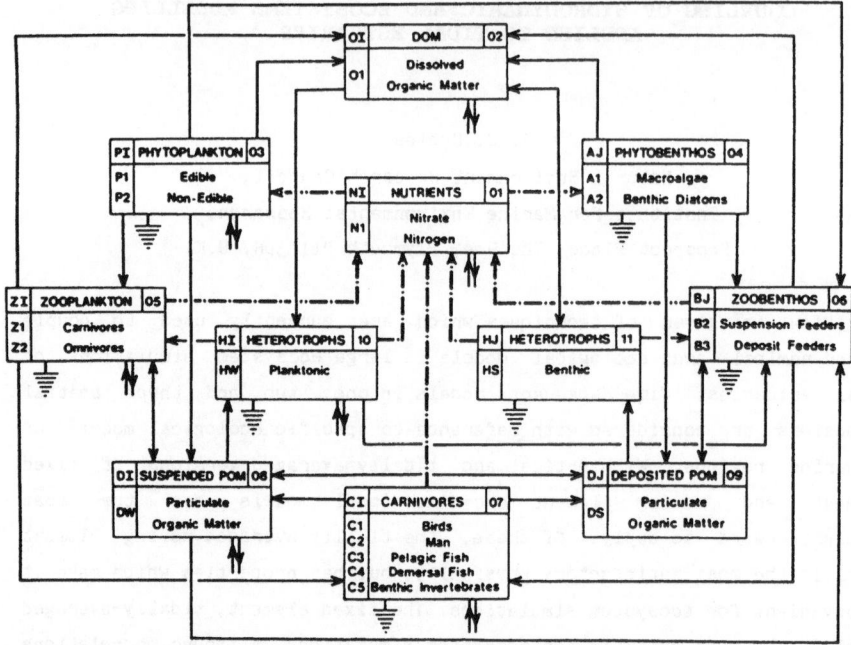

Fig. 1(a) Process flow diagram for GEMBASE with water column to left substratum to right. Solid lines represent the flow of carbon between state variables (rectangles). Broken lines represent flows of nitrate-nitrogen, the 'earth' symbols indicate losses due to respiration and the double arrows represent the transfers of material due to mixing and advection. Reproduced from Radford and Joint (1980), with permission.

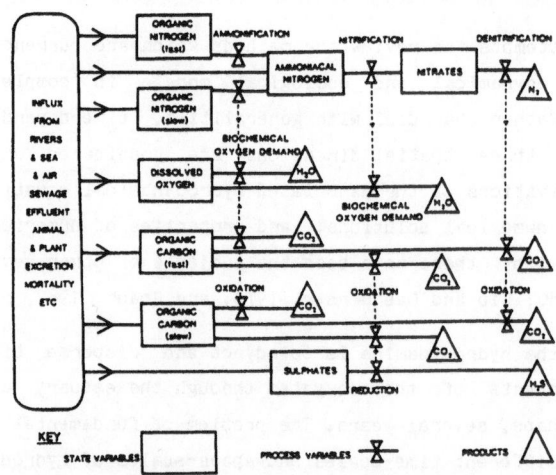

Fig. 1(b) Flow diagram of a water quality (biochemical oxygen demand) model which includes the processes by which organic materials are broken down by different groups of bacteria. Reproduced from Radford (1983), with permission.

hydrodynamical model is, typically, of the order of minutes. This is much less than the seasonal time-scales of ecological interest. The numbers of spatial elements (or grid points) necessary for accurate numerical solutions of the hydrodynamical equations in one, two and three dimensions are $(10^1 - 10^2)$, $(10^2 - 10^3)$ and $(10^3 - 10^4)$, respectively. These are between one and two orders of magnitude greater than those used in ecological models.

The reasons for this spatial coarseness in ecological models can be gleaned from Fig. 1(a), which is the flow diagram for a general ecosystem model of the Bristol Channel and Severn Estuary (GEMBASE; Radford and Joint, 1980). The number of biological state variables is an order of magnitude greater than the number of hydrodynamical state variables. Moreover, field data are required for each spatial element for all of these variables, as well as for parameters associated with their numerous interconnecting rate processes.

Despite the existence of a large literature on estuarine ecosystems, there are few well-documented examples of the coupling between hydrodynamical and ecological models. The easiest assumption for the ecologist to make is that the estuary comprises a single spatial element, and that physical transport processes are negligible (eg: Baird and Milne, 1981). In some cases, exchange processes between the ecosystem and the coastal or open-sea waters are taken into account, as in the ecosystem models of North Inlet (Summers et al., 1980), the Gulf of Maine (Campbell, in press) and the Black Sea Shelf (Belyaev, 1984). Single element models are not considered further in this paper.

Multi-element models are discussed under the heading of their spatial dimensionality, and with reference to specific ecological models. An omission here is the important and well-documented ecological model of the San Francisco Bay-Delta System (Chen and Orlob, 1975), which uses a "node-link" spatial grid system. This method is particularly useful for multi-channel systems, but does not fall easily into the categories adopted here.

Water quality models are closely related to ecological models. When an estuary becomes heavily polluted due to eutrophication, anaerobic conditions limit the variety of organisms which can exist. Under severe conditions heterotrophic processes can be modelled as a biochemical oxygen demand (BOD). The important state variable here is dissolved oxygen concentration (Fig. 1(b)). One-dimensional water quality and mixing models have been used for many years (eg: Ketchum, 1951; Stommel, 1953; Mollowney, 1973; Najarian and Harleman, 1977; O'Kane, 1980). A review of early one and two-dimensional water quality models is given for U.S. waters by Orlob (1976). A recent, brief review for U.K. waters is given by Odd (1985).

2. ONE-DIMENSIONAL MODELS

Many estuaries are much longer than they are wide, and in the presence of strong tidal currents are fairly well mixed over their cross-sections. In these cases the major spatial variability occurs along the estuarine axis, and transport of dissolved and fine suspended material can be treated as a one-dimensional process. Calculations are made in terms of cross-sectionally averaged quantities.

Very useful results can also be obtained with applications to partially mixed estuaries, provided it is understood that these models are computing estimates of cross-sectionally averaged quantities. Their important features are simplicity and rapidity of execution on the computer; they allow for transport and flushing of pelagic variables from the estuary with an accuracy which is no worse than that inherent in the modelling of most biological and chemical processes.

One-dimensional models have been used for over twenty years to solve water quality and pollution problems in estuaries. They have been used extensively for the study and prediction of salinity intrusion, oxygen balance, heat balance and, more recently, ecosystem behaviour. There are four approaches to solving the one-dimensional mass-balance equation in finite-difference form. These are the within-tide and tidally averaged versions of fixed element and moving element approximations of this equation.

2.1 Within-Tide Moving Element Model

Details of the method have been described by Mollowney (1973). The following quantities are defined; at the tidal limit ($x = 0$) the freshwater input is $Q_o(t)$. The concentration of the pelagic variable (dissolved or fine suspended particulate material) is $C(x,t)$. The sum of sources and sinks for fresh water and the pelagic variable along the axis of the estuary are $q(x,t)$ and $m(x,t)$ per unit length, respectively. The total volume of water between x and the tidal limit is $V(x,t)$.

Conservation of water is:

$$\delta_t V = Q_0 + \int_0^x q\,dx - AU \qquad\qquad 2.1(a)$$

where A is cross-sectional area and U the cross-sectionally averaged velocity. The differential of eqn. 2.1(a) is:

$$\delta_t A = q - \delta_x(AU) \qquad\qquad 2.1(b)$$

The mass-balance equation for a pelagic variable is:

$$\delta_t(AC) + \delta_x(AUC) - \delta_x(DA\delta_x C) = m \qquad\qquad 2.1(c)$$

Where D is the dispersion coefficient.

If a momentum equation is solved for U, then U, A and C can be computed simultaneously within a tidal cycle using eqns. 2.1(a)-(c). This approach is considered later. A much more efficient method has been devised for the solution of eqn. 2.1(c) using stored time-series data on U and A derived from other studies (Mollowney, 1973).

2.1.1 Solution of Mass-Balance Equation

The tidal velocity is given by:

$$\tilde{U} = U - Q/A \qquad\qquad 2.1.1(a)$$

where Q is the freshwater flow over a section:

$$Q = Q_0 + \int_0^x q\,dx \qquad\qquad 2.1.1(b)$$

Eqns. 2.1.1(a),(b) and 2.1(a) give:

$$\delta_t V + \tilde{U}A = \delta_t V + \tilde{U}\delta_x V = dV/dt = 0 \qquad\qquad 2.1.1(c)$$

where d/dt is the rate of change with time in the reference frame moving with tidal velocity $\tilde{U}$. Therefore, from eqn. 2.1.1(c) the up-estuary volume remains constant in this frame.

Substituting eqns. 2.1.1(a), (b) and (c) into eqn. 2.1(c) gives:

$$AdC/dt = \delta_x(DA\delta_x C - QC) + m \qquad\qquad 2.1.1(d)$$

In the case of salinity $C = S$ and $m = 0$ so that:

$$AdS/dt = \delta_x(DA\delta_x S - QS) \qquad\qquad 2.1.1(e)$$

If conditions are stationary then $dS/dt = 0$ and

$$DA\delta_x S = QS \qquad\qquad 2.1.1(f)$$

which enables estimates of D to be made once axial distributions of salinity have been observed.

The advantage of eqn. 2.1.1(d) over eqn. 2.1(c) is the elimination of the large and computationally difficult tidal advection term.

2.1.2 Numerical Solution

The model estuary is divided into a number of uniformly mixed elements. These are separated by interfaces at positions $x_0(t)$, $x_1(t)$, ... $x_i(t)$, ... $x_n(t)$. If the interfaces move with velocity $\tilde{U}_i$ then:

$$dx_i/dt = \tilde{U}_i \qquad\qquad 2.1.2(a)$$

and from eqn. 2.11(c):

$$d(\Delta V)_i/dt = 0 \qquad\qquad 2.1.2(b)$$

where $(\Delta V)_i$ is the volume of the ith element between interfaces at x_i and x_{i-1}.

The advective transport across x_i in the moving frame due to freshwater flows is $Q_i C_i$ (or $Q_i(C_{i+1} + C_i)/2$ with more precision). Axial dispersion is simulated by assuming equal and opposite exchanges of water between adjacent elements, E_i being the rate of volume exchange across x_i. The associated mass flux is $E_i(C_{i+1} - C_i)$ into $(\Delta V)_i$, which can be related to the dispersion coefficient by:

$$E_i = 2D_i A_i/(x_{i+1} - x_{i-1}) \qquad\qquad 2.1.2(c)$$

Eqn. 2.1.1(d) can be written:

$$(\Delta V)_i dC_i/dt = Q_{i-1}C_{i-1} + E_{i-1}(C_{i-1} - C_i)$$
$$- Q_i C_i + E_i(C_{i+1} - C_i) + M_i \qquad\qquad 2.1.2(d)$$

where M_i is the sum of sources and sinks into the ith element:

$$M_i = \int_{x_{i-1}}^{x_i} m_i \, dx$$

The solution of 2.1.2(d) can be made fully implicit and centred in time using:

$$\{\Delta V\}_i (dC_i/dt)^{n+\frac{1}{2}} = (\Delta V)_i (C_i^{n+1} - C_i^n)/\Delta t \qquad\qquad 2.1.2(e)$$

and

$$C_i^{n+\frac{1}{2}} = (C_i^{n+1} + C_i^n)/2, \qquad\qquad 2.1.2(f)$$

where Δt is the time-step and n the time-level.

With this method the equations are solved for pelagic variables at intervals of perhaps one hour throughout each tidal cycle. Information on freshwater run-off, currents and volumes can be stored as data in the program and used to simulate behaviour over spring-neap cycles. As an example of the method, calculated variations in salinity during an average tidal cycle are shown in Fig. 2.1.2(a) for several positions along the

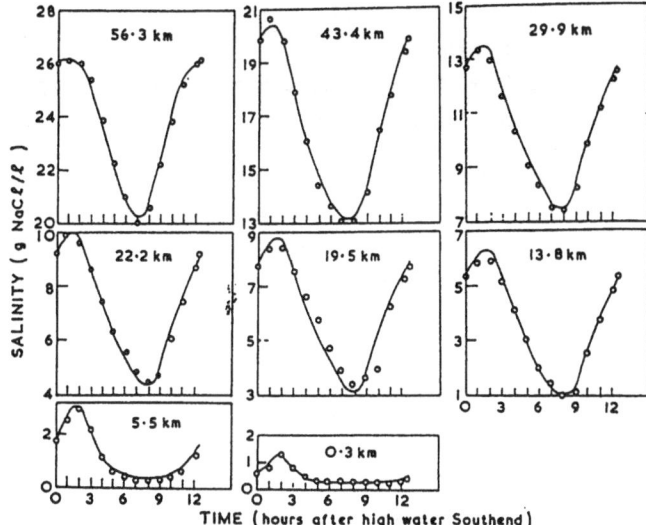

Fig. 2.1.2(a) Comparison of salinity variations during average tidal cycle in Thames Estuary, obtained from numerical model (curves) and from hydraulic model (circles). Reproduced from Mollowney (1973), with permission.

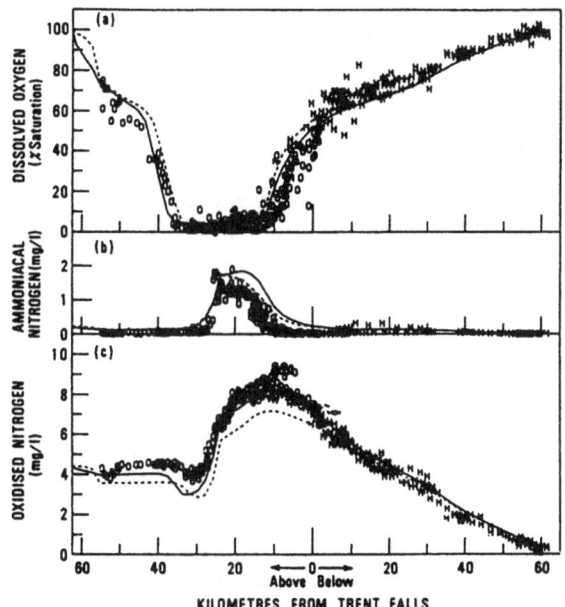

Fig. 2.1.2(b) Observed and computed distributions of some water quality variables (plotted at high water positions) in a survey of the Ouse-Humber during a summer period in 1978. Symbols denote observed data; dashed lines denote results from the steady-state model; full lines denote results from the time-dependent model. Reproduced from Gameson (1982), with permission.

Thames Estuary, UK (Mollowney, 1973; p. 76). The simulated freshwater input is $34 \ m^3s^{-1}$ at Teddington Weir ($x = 0$), and times are relative to high water at Southend ($x = 100$ km). Calculated values are compared with data from a hydraulic model (open circles).

An application of the method to the Humber Estuary, UK, is shown in Fig. 2.1.2(b) (Gameson, 1982; p. 25). Observed and computed distributions of dissolved oxygen, ammonia-nitrogen and oxidised nitrogen are shown for a spring tide survey during June 1978. Data are plotted at their high water positions against axial distance along the estuary, the estuary's mouth being 60 km below Trent Falls. The dashed lines in Fig. 2.1.2(b) show results for a steady-state simulation ($dC_i/dt = 0$ in eqn. 2.1.2(d)).

A disadvantage of the within-tide, moving element model is that an oscillating frame of reference for pelagic variables is inconvenient for interacting benthic variables, of which sediment is one of the more important. This technique may also be too detailed in that the within-tide behaviour which it simulates is much faster than the seasonal time-scale of ecological interest. A tide-averaged model is more appropriate for a study of these time-scales.

2.2 Tide-Averaged Moving Element Model

A tidal cycle average in the oscillating frame is defined to be:

$$T^{-1} \int_t^{t+T} C_i \ dt = \langle C_i \rangle \qquad \qquad 2.2(a)$$

where T is the tidal period, which may be taken as the time between successive occurrences of high, low, mid-rising or mid-falling water levels.

The equation of mass conservation in the oscillating frame (eqn. 2.1.2(d)) can be averaged over T to give:

$$(\Delta V)_i \langle dC_i/dt \rangle = \langle Q_{i-1}C_{i-1} + E_{i-1}(C_{i-1} - C_i)$$
$$- Q_iC_i + E_i(C_{i+1} - C_i) + M_i \rangle \qquad \qquad 2.2(b)$$

or $\quad (\Delta V)_i d\langle C_i \rangle/dt = \langle Q_{i-1} \rangle \langle C_{i-1} \rangle + \langle E_{i-1} \rangle (\langle C_{i-1} \rangle - \langle C_i \rangle) - \langle Q_i \rangle \langle C_i \rangle$
$$+ \langle E_i \rangle (\langle C_{i+1} \rangle - \langle C_i \rangle) + \langle M_i \rangle \qquad \qquad 2.2(c)$$

This is valid because temporal variations in the oscillating frame are small. Averaging the conservation of water volume equation (eqn. 2.1.2(b)) gives:

$$d(\Delta V)_i/dt = 0 = \langle Q_{i-1} \rangle - \langle Q_i \rangle + \langle \Delta Q_i \rangle \qquad \qquad 2.2(d)$$

where $\langle \Delta Q_i \rangle$ is the tidally averaged rate of freshwater input to the ith element. Eqn. 2.2(d) can be used to compute $\langle Q_i \rangle$ given $\langle Q_0(t) \rangle$ and evaluating $\langle \Delta Q_i \rangle$ from the known inputs and the fraction of time each element spends passing these.

The tidal average used in eqns. 2.2(a)-(d) applies in the oscillating frame. This differs from the tidal cycle average at a fixed position. The latter is a weighted time average over all elements passing through position x, whereas the former is a time average for just one element. However, it is unnecessary to oscillate the elements in order to solve eqns. 2.2(c)-(d). An initial time is chosen, such as mid-water rising at the mouth, which defines $(\Delta V)_i$ for all elements. Values of $\langle C_i \rangle$ then correspond to the tidal cycle average within an element which occupies a position between x_{i-1} and x_i at mid-tide. In steady-state the tidal cycle average is the same as the instantaneous value for a particular element, and the axial distribution at any instant of time (eg. high or low water) can be computed using a simple tidal displacement of position (eqn. 2.1.2(a)).

Values of $\langle E_i \rangle$ in eqn. 2.2(c) can be evaluated from a knowledge of salinity, S, for which $\langle M_i \rangle = 0$. Assuming a steady-state and replacing C by S in eqn. 2.2(c) gives:

$$\langle E_i \rangle = \langle Q_i \rangle \langle S_i \rangle / (\langle S_{i+1} \rangle - \langle S_i \rangle) \qquad \qquad 2.2(e)$$

Radford (1984) has applied the tide-averaged, moving element model to a water quality study of the Thames Estuary. Salinity and concentrations of dissolved oxygen and nitrogen at their mid-tide positions were simulated over the two year period 1975-1976, and compared with observations. Fig. 2.2(a) shows the observed extremes of salinity in the estuary over the period, and the curves derived from the two-year run of the model.

2.3 Within-Tide Fixed Element Model

This model uses the same elements as an associated hydrodynamical model. Typically, the equations are:
Continuity (eqn. 2.1(b)):

$$\delta_t A = B \delta_t \zeta = q - \delta_x(AU) \qquad \qquad 2.3(a)$$

where B is width and ζ the elevation of the water surface.
Momentum:

$$\delta_t U = -U \delta_x U - g \delta_x \zeta - \alpha U |U| / H - \tfrac{1}{2} g H \delta_x \ln \rho + \delta_x(NA \delta_x U)/A \qquad 2.3(b)$$

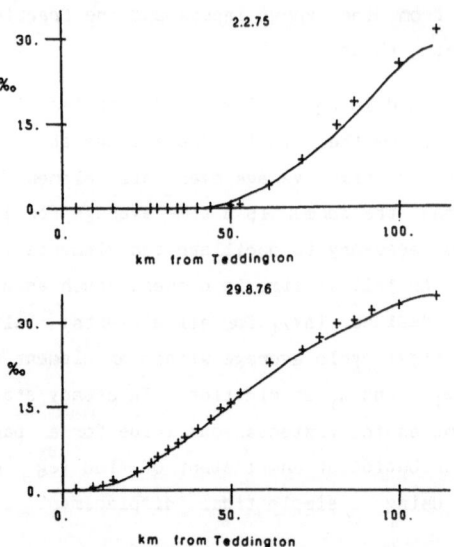

Fig. 2.2(a)   Observed (symbols) and computed (full lines) extremes of the
axial salinity distributions along the Thames Estuary during 1975 and 1976.
reproduced from Radford (1984), with permission.

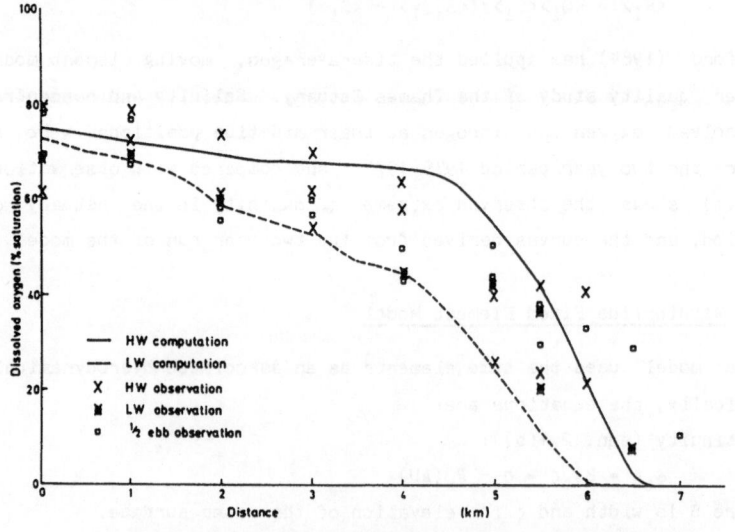

Comparison between computed and observed dissolved oxygen distributions
October 1976

Fig. 2.3(a)   Comparison of computed and observed axial distributions of
dissolved oxygen in the Shing Mun Creek (Hong Kong) during October 1976.
Distances are up-estuary from the mouth of the creek. Reproduced from
Maskell and Odd (1977), with permission.

ρ is water density, α the drag coefficient, H the sectionally-averaged depth and N the axial viscosity coefficient. Many methods are available for the solution of these equations (eg. Uncles and Jordan, 1980). The time-step, Δt, for numerical solutions is generally of the order of minutes.

The mass-balance equation for C (eqn. 2.1(c)) is solved using fixed elements. Data for U and A are either tabulated from independent calculations of eqns. 2.3(a), (b) or generated simultaneously with C. The advantages of computing independently are that eqn. 2.1(c) can often be solved with a much longer time-step than the hydrodynamical model, and that it is usual to run several ecological or water quality simulations for the same period with the same physical variables.

In a tidal flow the advective term in eqn. 2.1(c) is usually much larger than the dispersion term. This is the main motivation for using a moving element model. Relatively small errors in the numerical solution of the advection term can dominate the long-term solution of the fixed element mass-balance equation. Maskell and Odd (1977) recommend the use of the Brian and Stone technique, which is a time-centred implicit scheme. For the ith internal (non-boundary) element this has the form (see eqn. 2.1(c)):

$$\delta_t(AC)_i^{n+\frac{1}{2}} = (12\Delta t)^{-1} \sum_{k=0}^{1} \sum_{j=-1}^{+1} \varepsilon_j(-1)^{k+1}\{AC\}_{i+j}^{n+k}$$

with $\varepsilon_{-1} = 1$, $\varepsilon_0 = 4$ and $\varepsilon_1 = 1$.

$$\delta_x(AUC)_i^{n+\frac{1}{2}} = (4\Delta x)^{-1} \sum_{k=0}^{1} \sum_{j=-1}^{0} (-1)^j (AU)_{i+j}^{n+k}(C_{i+j}^{n+k} + C_{i+j+1}^{n+k})$$

$$\delta_x(AD\delta_x C)_i^{n+\frac{1}{2}} = (2\Delta x^2)^{-1} \sum_{k=0}^{1} \sum_{j=-1}^{0} (-1)^j (AD)_{i+j}^{n+k}(C_{i+j+1}^{n+k} - C_{i+j}^{n+k})$$

This method minimises damping and phase errors in the numerical solution.

The technique has been used to simulate tidal flows, salinity intrusion and oxygen balance in a tidal creek over a spring-neap period (HRS, 1977). The water quality model incorporates the effects of anaerobic mud deposits. Fig. 2.3(a) shows computed and observed dissolved oxygen in the creek at high and low water during October 1976 as functions of distance up-estuary from the mouth.

2.4 Tide-Averaged Fixed Element Model

A tidal cycle average is defined for the ith fixed element by:

$$T^{-1} \int_t^{t+T} C_i(t) \, dt = \langle C_i \rangle \qquad \qquad 2.4(a)$$

Applying this averaging to eqn. 2.1(c) gives:

$$\delta_t(\langle A \rangle \langle C \rangle) = \delta_x(D^*\langle A \rangle \delta_x \langle C \rangle - \langle Q \rangle \langle C \rangle) + \langle m \rangle$$
$$- \delta_t \langle \tilde{A} \tilde{C} \rangle \qquad \qquad 2.4(b)$$

In which temporal changes in the tidally averaged volume are ignored. Eqn. 2.4(b) is of practical use only if tidal correlations are negligible:

$$\left| \delta_t \langle \tilde{A} \tilde{C} \rangle \right| \ll \left| \delta_t(\langle A \rangle \langle C \rangle) \right| \qquad \qquad 2.4(c)$$

However, this condition is invariably assumed true rather than tested. The dispersion coefficient, D^*, is effectively defined by eqn. 2.4(b) and is not simply related to D in eqn. 2.1(c) unless tidal variations in cross-sectional area and tidal pumping of dissolved material are negligible (Uncles and Radford, 1980). Because changes in the tidally-averaged cross-sectional areas are usually very small, eqn. 2.4(b) can be written:

$$\langle A \rangle \delta_t \langle C \rangle = \delta_x(D^*\langle A \rangle \delta_x \langle C \rangle - \langle Q \rangle \langle C \rangle) + \langle m \rangle \qquad \qquad 2.4(d)$$

This is identical in form with eqn. 2.1.1(d), which is the instantaneous equation for C in a moving frame. If the tidally averaged volume of the ith fixed element between x_i and x_{i-1} is denoted by $(\Delta V)_i$:

$$(\Delta V)_i = \langle A_i \rangle (x_i - x_{i-1})$$

then eqn. 2.4(d) can be written in the same way as eqn. 2.2(c) and solved in exactly the same way:

$$(\Delta V)_i \delta_t \langle C_i \rangle = \langle Q_{i-1} \rangle \langle C_{i-1} \rangle + \langle E_{i-1} \rangle (\langle C_{i-1} \rangle - \langle C_i \rangle) - \langle Q_i \rangle \langle C_i \rangle$$
$$+ \langle E_i \rangle (\langle C_{i+1} \rangle - \langle C_i \rangle) + \langle M_i \rangle \qquad \qquad 2.4(e)$$

Coefficients $\langle E_i \rangle$ represent dispersive mixing. Although this equation is identical to that for the tidally averaged moving element model (eqn. 2.2(c)), there are important differences in the mode of application. Because the tidal average is applied at a fixed position along the axis, values of $\langle E_i \rangle$ must be determined using tidally averaged salinity data, rather than salinity data at mid-tide (or whatever tidal state is used for reference). Freshwater and other material from point sources in eqn. 2.4(e) are input to fixed elements and are not shared between those water masses which pass source points during a tidal excursion. For freshwater, input q_i (eqns. 2.1(a), (b)) is associated with the ith element:

$$\langle Q_{i-1} \rangle - \langle Q_i \rangle + \langle q_i \rangle (x_i - x_{i-1}) = 0 \qquad \qquad 2.4(f)$$

rather than eqn. 2.2(d). Local growth or decay of a pelagic variable is associated with a fixed position in space and is much easier to incorporate than in the moving element model. Incorporation of benthic variables (including sediment) is straightforward as these are always associated with elements which are fixed relative to the sea bed. The mode of numerical solution is identical to that for eqn. 2.2(b). The only major disadvantage of this method is the fact that the effects of tidal correlations are unknown. A further disadvantage which is more important for water quality than ecological modelling is the fact that a tidally-averaged model of this kind cannot predict the peaks of pollution concentrations as accurately as other models.

An intensive study of salinity and cadmium distributions in the Severn Estuary, UK, has been undertaken using the tidally averaged, fixed element model (Radford et al., 1981). Fig. 2.4(a) shows observed and computed tidally averaged cadmium for five periods during 1975-1977 as functions of distance from Maisemore Weir (x = 0). Major polluting sources and inputs from the Rivers Taff, Avon, Wye and Severn are included.

2.5 Application to Ecological Modelling

Some of the preceeding methods have been used in three large one-dimensional ecological models which are currently in development. These are ecosystem models of the Cumberland Basin, Canada (Gordon et al, in press), the Ems-Dollard, Netherlands (Ruardij and Baretta, 1982) and the Tamar Estuary, U.K. (Harris et al, 1984). A comparison of the first two ecosystems is given by Gordon and Baretta (1982). The model of the Tamar is a contaminant dispersal model with application to biology, rather than a conventional ecosystem model, but is sufficiently intricate in its treatment of chemistry and invertebrate biology to warrant inclusion here.

2.5.1 Ems-Dollard Ecosystem

This ecosystem is divided into benthic and pelagic systems. The estuary is discretized into five elements (Fig. 2.5.1(a)). The Dollard elements (1 and 2 in Fig. 2.5.1(a)) were defined on the basis of differences in species composition and activity. The remaining three were divided into roughly equal parts between element 2 and the open sea.

The model is run for tidally averaged conditions using boundary conditions on concentrations and biomasses in the sea and in the River Ems and Westerwoldse. Control variables are freshwater inputs, concentrations of detritus, silt and oxygen in the rivers and sea, and temperature throughout the region.

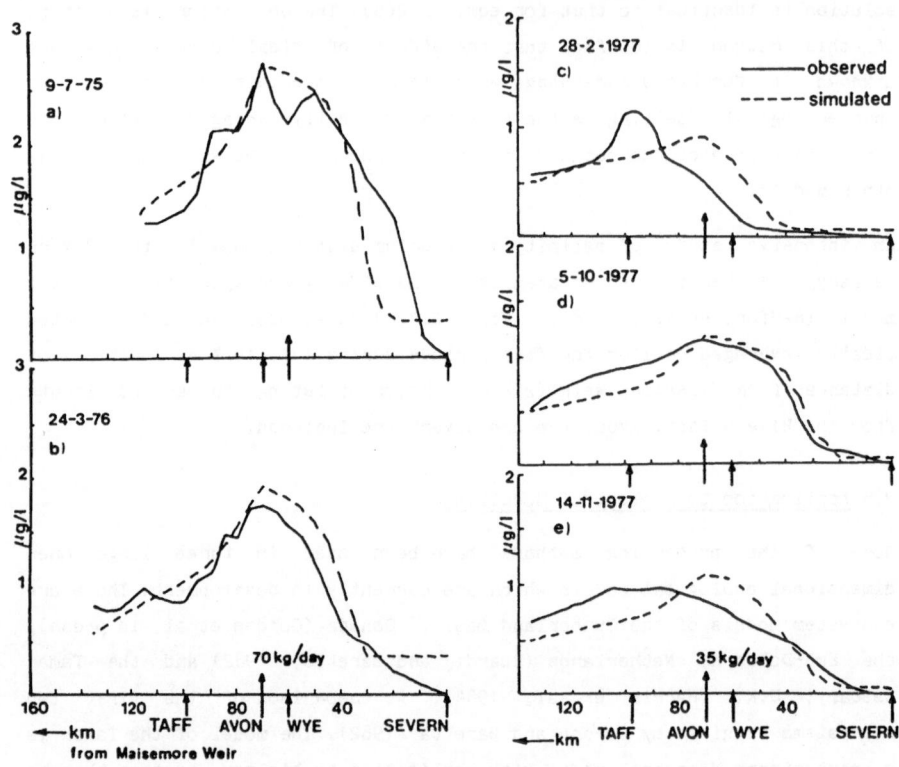

Fig. 2.4(a) Observed and simulated distributions of dissolved cadmium assuming measured input concentrations for the Taff, the Wye and the Severn but an extra input at Avonmouth. The mismatch of the peak concentrations on 28 February 1977 indicates the existence of another high and unexplained input in the region of the Taff. Reproduced from Radford et al. (1981), with permission.

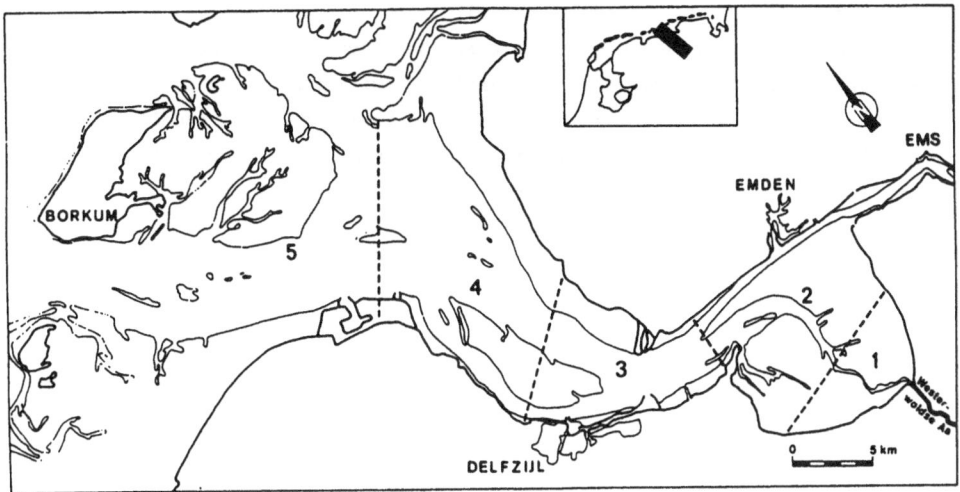

Fig. 2.5.1(a)   The division of the Ems-Dollard estuary into five spatial elements for ecological modelling. Reproduced from Ruardij and Baretta (1982), with permission.

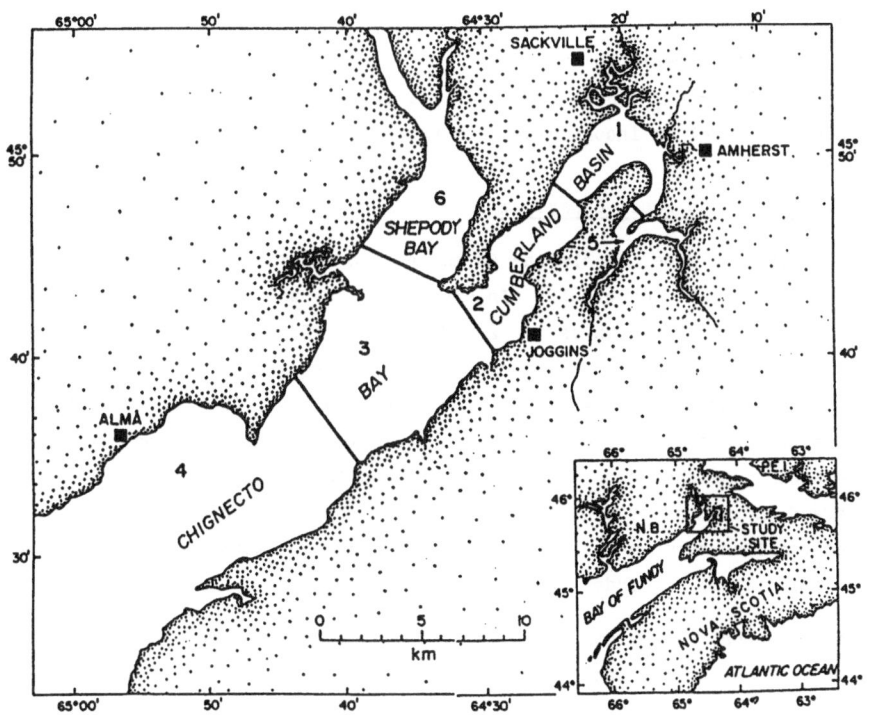

Fig. 2.5.2(a)   Chart showing elements (1-3) and boundaries (4-6) of the Cumberland Basin ecosystem model in the upper reaches of the Bay of Fundy. Reproduced from Gordon et al. (In Press), with permission.

The model assumes zero transport for all epibenthic and benthic variables. Most of the pelagic variables are transported according to the mass-balance equation (eqn. 2.1(c)). The dispersion coefficients are deduced from observed mid-tide salinity distributions. Therefore, the model can be considered a tide-averaged moving element model, although it is not strictly this because benthic variables interact only with fixed pelagic elements, and within-tide oscillations of the elements are not taken into account.

Mass-balance equations are written for the pelagic variables in each of the five elements, $i = 1$ to 5. These equations are similar to eqn. 2.2(c), except that the advection into element i from element $(i - 1)$ by the freshwater flow $\langle Q_{i-1} \rangle$ is centred in space to give:

$$\Delta V_i \delta_t \langle C_i \rangle = \langle Q_{i-1} \rangle (\langle C_{i-1} \rangle + \langle C_i \rangle)/2 + \langle E_{i-1} \rangle (\langle C_{i-1} \rangle - \langle C_i \rangle)$$
$$- \langle Q_i \rangle (\langle C_i \rangle + \langle C_{i+1} \rangle)/2 + \langle E_i \rangle (\langle C_{i+1} \rangle - \langle C_i \rangle)$$
$$+ \langle M_i \rangle \qquad\qquad 2.5.1(a)$$

Space-centring is preferable to up-wind differencing when used with such large elements. This is because the artificial numerical dispersion introduced with up-wind differencing is proportional to element size. The mixing volumes, $\langle E_i \rangle$ are computed from the steady-state version of eqn. 2.5.1(a) with C replaced by S:

$$\langle E_i \rangle = \langle Q_i \rangle (\langle S_i \rangle + \langle S_{i+1} \rangle)/2(\langle S_{i+1} \rangle - \langle S_i \rangle) \qquad\qquad 2.5.1(b)$$

A separate model is incorporated for the silt transport (Ebenhoh, 1985).

2.5.2 Cumberland Basin Ecosystem

This ecosystem model is similar in concept to the Ems-Dollard model, and is discussed by Gordon et al. (In Press). There are three interior (non-boundary) elements shown as 1, 2 and 3 in Fig. 2.5.2(a). Element length is equal to the average tidal excursion (about 16 km), and element positions are defined by their locations at low tide (Fig. 2.5.2(a)). There are three external boundaries; upper Cumberland basin (5) at the landward end (dry at low water), neighbouring Shepody Bay (6) and lower Chignecto bay (4) at the seaward end. Boundary conditions are defined by time-series data. The model predicts long-term variability (weeks to years) and within-tide, short-term variability (of the order of hours) is not included.

The transport equations for pelagic variables are based on the tide-averaged moving element equations with centred spatial differencing (formally equivalent to eqn. 2.5.1(a)). Mixing volumes are deduced from eqn. 2.5.1(a). Because the reference time is low water, all computed variables are referred to their low water positions. Although calculations are tide-averaged, within-tide oscillations of elements are taken into account when evaluating the various source terms in the equations for pelagic variables. A silt transport model is included which is the same as that for the Ems-Dollard model (Ebenhoh, 1985).

2.5.3 The Tamar Model

This model simulates the axial transport of a range of environmental toxins in a partly-mixed, macrotidal estuary in the southwest of England (Fig. 2.5.3(a)). A description of the model and some preliminary results are given by Harris et al. (1984). Aromatic hydrocarbons ranging from benzene to benzo (α) pyrene are considered, together with cadmium. The inclusion of this model here is a consequence of the fact that the chemistry of these toxins is treated with considerable realism, and the toxic effects on the mussel, Mytilus edulis, are simulated as an indication of ecological impact in the estuary.

The model is based on the one-dimensional mass-balance equation (eqn. 2.1(c)). A tide-averaged, moving element model is used, and the reference tidal state is mid-water rising. However, like the Ems Dollard model, source terms are associated with fixed elements, and there is no attempt to take into account the within-tide oscillations of elements.

The method of solution is somewhat different from the other examples. The element spacing at mid-tide is $\Delta x = 1$ km for all 30 elements. Defining:

$$\Delta_x \langle C_i \rangle = \langle C_i \rangle - \langle C_{i-1} \rangle$$
$$\langle \overline{C}_i \rangle = (\langle C_i \rangle + \langle C_{i-1} \rangle)/2$$

Then the concentration at element interface position x_{i-1} and at time $t + \Delta t/2$ is approximated by:

$$\langle C_i' \rangle = \langle \overline{C}_i \rangle - \tfrac{1}{2} \langle u_{i-1} \rangle \Delta t \, \Delta_x \langle C_i \rangle / \Delta x$$

where $\langle u_{i-1} \rangle = \langle Q_{i-1} \rangle / \langle A_{i-1} \rangle$ is the freshwater current.

Using this, the change in $\langle C_i \rangle$ over time-interval Δt is given by:

$$(\Delta V)_i (\langle C_i^{n+1} \rangle - \langle C_i^n \rangle)/\Delta t = -\langle \overline{Q}_i^{n+\frac{1}{2}} \rangle \Delta_x \langle C_i' \rangle - \langle C_i^n \rangle \Delta_x \langle Q_i^n \rangle$$
$$- \langle E_{i-1}^n \rangle \Delta_x \langle C_i^n \rangle + \langle E_i^n \rangle \Delta_x \langle C_{i+1}^n \rangle + \langle M_i \rangle \qquad 2.5.3(a)$$

The coefficients $\langle E_i \rangle$ are evaluated according to eqn. 2.5.1(b) and related to tidal range and freshwater inputs by simple regressional formulae.

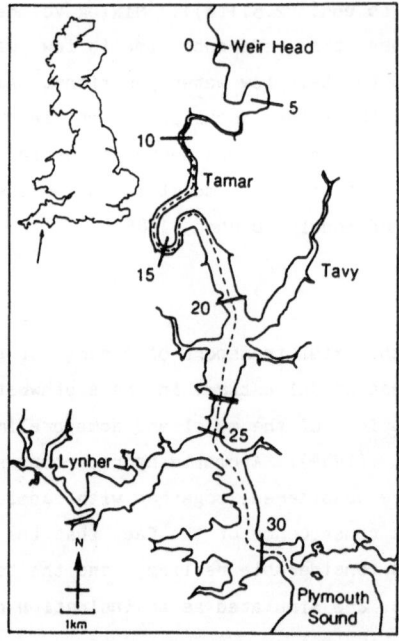

Fig. 2.5.3(a) The Tamar Estuary. Axial distances are km from the head (Weir Head). Reproduced from Harris et al. (1984), with permission.

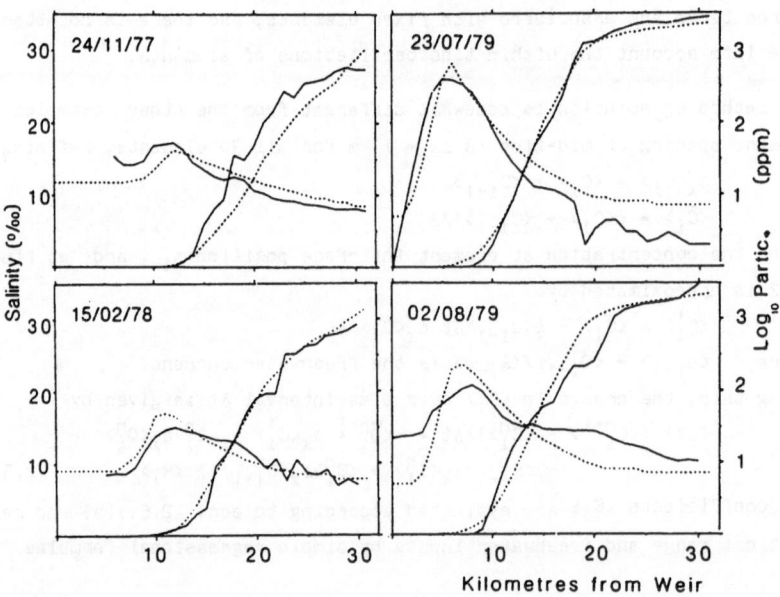

Fig. 2.5.3(b) Example comparisons of observed (full lines) and computed (dashed lines) axial distributions of salinity and suspended sediment concentrations (partic.) in the Tamar Estuary. Reproduced from Harris et al. (1984), with permission.

A mechanistic model for silt transport is not incorporated. Rather, regressional formulae are used in which concentrations of suspended sediment are related to salinity, freshwater run-off and tidal range. These formulae are given in Harris et al (1984). Transport of sediment is forced by repeating the annual cycle of movement observed during 1982.

Fig. 2.5.3(b) shows a comparison of observed and computed salinity and suspended sediment distributions along the estuary for four occasions. The model is able to reproduce both observations of salinity for high and low freshwater inputs, and the position and magnitude of the turbidity maximum. A typical run of the model for winter conditions, showing flushing times, turbidity and salinity is shown in Fig. 2.5.3(c). The model also outputs contours for dissolved and particulate forms of hydrocarbons and cadmium, as well as those for Mussel growth.

3. TWO-DIMENSIONAL MODELS

These models can be of two types. Width-averaged models compute concentrations and velocities as functions of axial distance and depth (negligible transverse variations). Depth-averaged models compute concentrations and velocities as functions of horizontal position (negligible vertical variations). Depth-averaged models have been deployed more extensively for ecological modelling studies, and these are considered first.

3.1 Depth-Averaged Models

This type of model is suited to a wide estuary which has strong axial and transverse gradients, but which is vertically well mixed. The computer time required for simulations is long, and because of this the estuary is divided into a rather small number of horizontally well mixed elements, but with significantly different ecological properties.

In order to simulate seasonal cycles for ecosystem studies it is necessary to couple the hydrodynamical and ecological sub-models. Two-dimensional, depth-averaged, hydrodynamical models have been described by many authors, but these are too detailed in space and time to be of immediate value to ecosystem models. Typically, the numerical solution of the depth-averaged hydrodynamical equations requires $10^2 - 10^3$ elements and runs with a time-step of the order of minutes (eg: Uncles, 1981, 1982; Flather and Heaps,

1975). Here, we will discuss the approaches to this coupling problem as used in the two-dimensional ecosystem studies of the Bristol Channel, UK (Uncles 1983a) and Narragansett Bay, USA (Kremer and Nixon, 1978).

3.2 Bristol Channel Ecosystem

A chart of the Bristol Channel and its discretization is shown in Fig. 3.2(a). The ecological model GEMBASE (general ecosystem model of the Bristol Channel and Severn Estuary) computes the interactions between biological and chemical variables during mixing and advection by tidal and residual currents (Radford, 1979, 1981; Radford and Joint, 1980; Joint, 1983). The biological sub-model is outlined in Fig. 1(a). The nine regions used in the GEMBASE model are named and depicted schematically in Fig. 3.2(a). Freshwater inputs to each GEMBASE region, denoted by $Q_{f,i}$ are also shown in Fig. 3.2(a). On average, the river inputs from the Severn Estuary contribute about 60% of the total flow of freshwater to the Bristol Channel, while inputs from the northern and southern coastlines of the Bristol Channel amount to about 30% and 10%, respectively. Tides are semi-diurnal with a mean tidal range of 6.6 m. Mean peak tidal currents are about 1 m s^{-1} (Uncles, 1983b).

Boundary conditions for GEMBASE are applied at the Inner Estuary (IE), and at the Celtic Sea North (CSN) and Celtic Sea South (CSS) regions (see Fig. 3.2(a)). The remaining six modelled regions were chosen as a minimum set to describe the spatially differing ecological and physical properties of the Channel. A larger number of regions would have required excessive computer time to run the model through several years of simulated time for the many numerical experiments of interest.

The problem considered here is the determination of mixing and advection of water and dissolved materials between, and through, the large GEMBASE regions shown in Fig. 3.2(a) for periods of time of the order of several years. The method of solution has the advantages of being conceptually simple, rapid to run, and capable of generating realistic rates of transport between regions, as judged from the comparison of computed and observed distributions of fresh water in the Channel. Use is made of the long-term salt balance, which, while being insufficient to define transport over the two-dimensional Bristol Channel regions shown in Fig. 3.2(a) is able to do so in the one-dimensional regions which represent the Severn

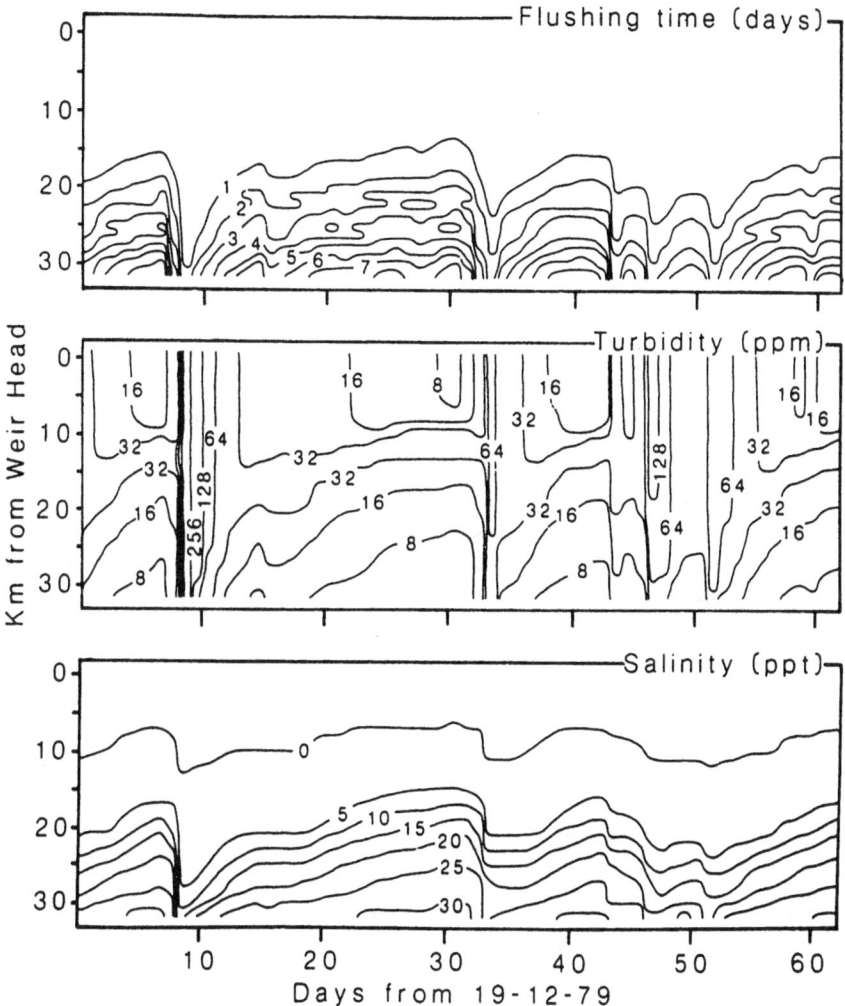

Fig. 2.5.3(c) Computed salinity, suspended sediment distributions and flushing times in the Tamar Estuary during a winter period. Reproduced from Harris et al. (1984), with permission.

Estuary.

3.2.1 Advection and Mixing

The transport of a pelagic variable across any boundary between two GEMBASE regions in Fig. 3.2(a) is considered. The instantaneous rate of transport of water, Q, across the boundary is:

$$Q = A\overline{U} \qquad\qquad 3.2.1(a)$$

where the overbar denotes an average over the boundary's area, A, and where U is the component of depth averaged velocity perpendicular to A.

The associated transport of a pelagic variable is:

$$F = A\overline{UC} = A\overline{U}\,\overline{C} + A\overline{U'C'} \qquad\qquad 3.2.1(b)$$

Here, the prime denotes a deviation from the cross-sectional average; for example, $C = \overline{C} + C'$. The residual rate of transport across A follows from eqns. 3.2.1(a),(b):

$$\langle F \rangle = \langle Q \rangle \langle \overline{C} \rangle + \{ \langle \tilde{Q}\tilde{C} \rangle + \langle A\overline{U'C'} \rangle \} \qquad\qquad 3.2.1(c)$$

where $Q = \langle Q \rangle + \tilde{Q}$ and $\overline{C} = \langle \overline{C} \rangle + \tilde{C}$ the tildes denoting cross-sectionally averaged tidal fluctuations.

The residual transport of water can be separated into two parts: one (Q_f) is due to freshwater inputs to the Channel (for example, $Q_{f,CC}$ is shown as flowing from the Central Channel South (CCS) to the Central Channel North (CCN) in Fig. 3.2(a)), and the other results from all other mechanisms, Q_R:

$$\langle Q \rangle = Q_f + Q_R$$

so that the residual transport across A is:

$$\langle F \rangle = Q_f \langle \overline{C} \rangle + Q_R \langle \overline{C} \rangle + \{ \langle \tilde{Q}\tilde{C} \rangle + \langle A\overline{U'C'} \rangle \} \qquad\qquad 3.2.1(d)$$

An empirical mixing coefficient, E, is defined by:

$$E = -\{ \langle \tilde{Q}\tilde{C} \rangle + \langle A\overline{U'C'} \rangle \}/\Delta\langle C \rangle \qquad\qquad 3.2.1(e)$$

in which $\Delta\langle C \rangle$ is the difference in $\langle C \rangle$ between the two GEMBASE regions under consideration. Equation 3.2.1(d) is:

$$\langle F \rangle = Q_f \langle \overline{C} \rangle + Q_R \langle \overline{C} \rangle - E\Delta\langle C \rangle \qquad\qquad 3.2.1(f)$$

3.2.2 Residual Flows

The freshwater flow across the region boundaries in Fig. 3.2(a) can be determined from a hydrodynamical model. As the fresh water from each source enters a region it produces a head of water which drives currents out of the Channel. These currents are extremely small compared with the tidal flows, so that flow patterns can be investigated for each input of fresh

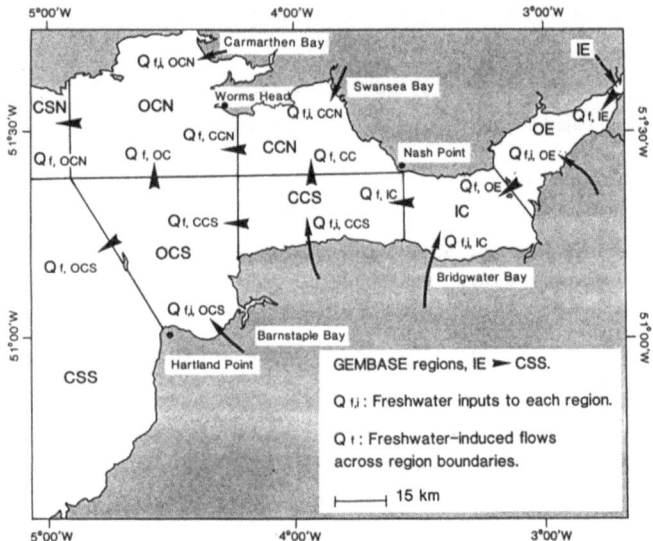

Fig. 3.2(a) Regions used in the GEMBASE model, showing freshwater inputs
to each region, $Q_{f,i}$, and freshwater-induced flows across region
boundaries, Q_f. Inner Estuary (IE); Outer Estuary (OE); Inner Channel (IC);
Central Channel South and North (CCS and CCN); Outer Channel South and
North (OCS and OCN); Celtic Sea South and North (CSS and CSN). Reproduced
from Uncles (1983a), with permission.

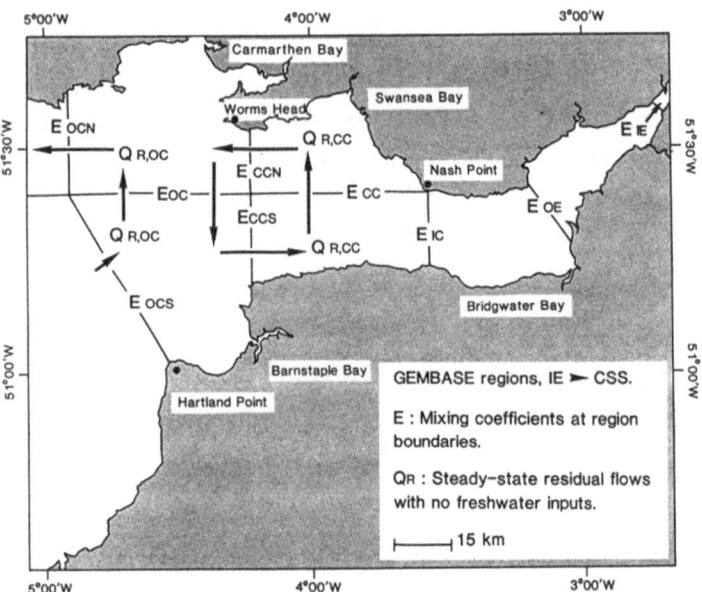

Fig. 3.2.2(a) Mixing coefficients at region boundaries, E, and steady-
state residual flows with no freshwater inputs, Q_R. Reproduced from Uncles
(1983a), with permission.

water, and then summed to take into account all inputs according to:

$$Q_f = \Gamma \cdot Q_{f,i} \qquad\qquad 3.2.2(a)$$

where Γ is a constant (9 x 9) matrix, Q_f are the flows $Q_{f,IE}$ to $Q_{f,OCS}$, and $Q_{f,i}$ are the inputs $Q_{f,i,IE}$ to $Q_{f,i,OCS}$ (see Fig. 3.2(a)).

Residual flows, Q_R, are shown in Fig. 3.2.2(a), together with mixing coefficients, E, which are defined at region boundaries. The residual flow from the Central Channel South to North is $Q_{R,CC}$, and that from the Outer Channel South to North is $(Q_{R,OC} - Q_{R,CC})$. Q_R was computed from the hydrodynamical model, and is the residual flow due to M_2 tidal nonlinearities and density currents (Uncles, 1982). A schematic picture of the large-scale residual circulation, according to the hydrodynamical model, is drawn in Fig. 3.2.2(b). Wind driven currents can also be incorporated in GEMBASE.

3.2.3 Transport Equations

The mass balance for a pelagic variable in the Central Channel South, using eqns. 3.2.1(f) and with reference to Figs. 3.2(a) and 3.2.2(a), is written:

$$
\begin{aligned}
(\Delta V)_{CCS} d\langle C_{CCS}\rangle/dt = {}& + [Q_{f,IC}(\langle C_{IC}\rangle + \langle C_{CCS}\rangle)/2 \\
& - E_{IC}(\langle C_{CCS}\rangle - \langle C_{IC}\rangle)] \\
& - [(Q_{f,CCS} - Q_{R,CC})(\langle C_{OCS}\rangle + \langle C_{CCS}\rangle)/2 \\
& - E_{CCS}(\langle C_{OCS}\rangle - \langle C_{CCS}\rangle)] \\
& - [(Q_{f,CC} + Q_{R,CC})(\langle C_{CCN}\rangle + \langle C_{CCS}\rangle)/2 \\
& - E_{CC}(\langle C_{CCN}\rangle - \langle C_{CCS}\rangle)] + \langle M_{CCS}\rangle \qquad 3.2.3(a)
\end{aligned}
$$

A similar equation holds for each GEMBASE region, and for each pelagic variable. Q_f and Q_R are computed from the hydrodynamical model. The method used to estimate E is based on the satisfaction of eqn. 3.2.3(a) for salinity, S, with $d\langle S\rangle/dt = 0 = \langle M\rangle$ for each GEMBASE region. Extensive salinity data for each region were averaged to yield long-term values. For the Outer Estuary and Inner Channel the balance is given by eqn. 3.2.1(f) with $Q_R = 0$ and $\langle F\rangle = 0$, so that:

$$E = Q_f \langle \bar{S}\rangle/\Delta\langle S\rangle \qquad\qquad 3.2.3(b)$$

yielding E_{IE}, E_{OE}, and E_{IC}. In the two-dimensional region there are six mixing coefficients, E_{CC} to E_{OCS} (Fig. 3.2.2(a)), and only four equations of the form of eqn. 3.2.3(a) - one for each region. Therefore, two values of E must be estimated directly from the hydrodynamical model, using eqn. 3.2.1(e) in conjunction with long-term averaged salinity data. The flux in eqn. 3.2.1(e) consists of a part due to tidal pumping, $\langle \widetilde{QS}\rangle$, and a part due

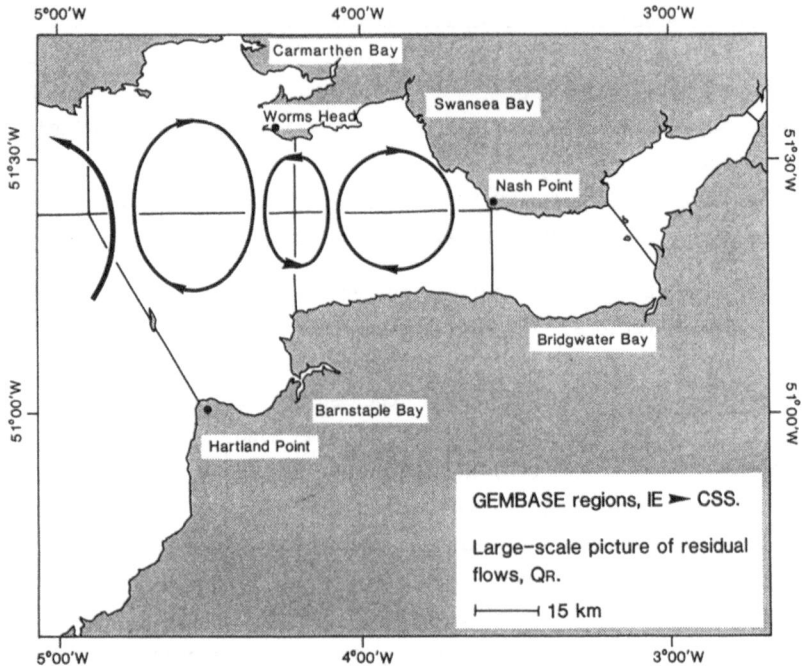

Fig. 3.2.2(b) Large-scale residual circulation, Q_R, in the Bristol Channel according to a hydrodynamical model. Reproduced from Uncles (1983a), with permission.

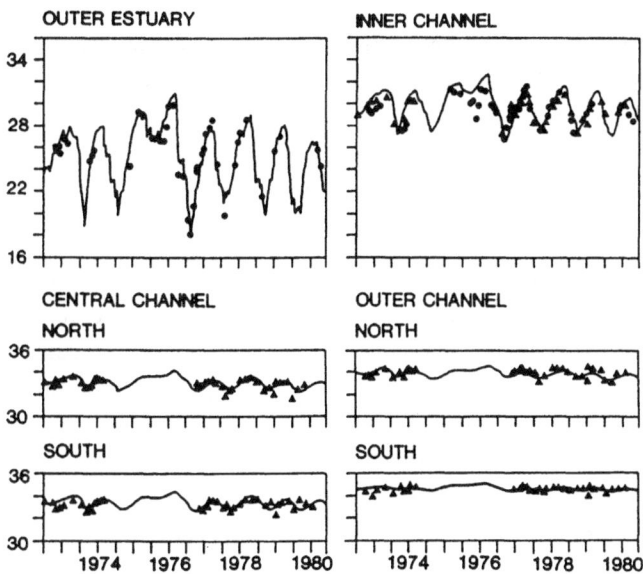

Fig. 3.2.3(a) Computed and observed salinities for the GEMBASE regions as functions of time. IMER data (Δ), other sources (o). Reproduced from Uncles (1983a), with permission.

to transverse shear in the tidal and residual currents. The effect of tidal shear can be neglected. The flux due to shear in the residual currents is:

$$\langle A\overline{U'S'}\rangle \approx \langle A\rangle\langle\overline{U'}\rangle\langle\overline{S'}\rangle \qquad\qquad 3.2.3(c)$$

Therefore eqns. 3.2.1(e) and 3.2.3(c) give (neglecting tidal pumping):

$$E \approx -\langle A\rangle\langle\overline{U'}\rangle\langle\overline{S'}\rangle/\Delta\langle S\rangle \qquad\qquad 3.2.3(d)$$

which can be computed using $\langle U'\rangle$ from the hydrodynamical model, and using $\langle S'\rangle$ and $\Delta\langle S\rangle$ from the long-term averaged salinity data.

In principle, E could be derived from eqn. 3.2.3(d) for each boundary between GEMBASE regions. However, the current shear along those boundaries which terminate on headlands is dominated by the small intense circulation patterns near the headlands, and the shear flux of salt in these circulations cannot be computed with accuracy. Also, the boundary between the Outer Channel North (OCN) and the Celtic Sea North (CSN) is very close to the seaward boundary of the hydrodynamical model, and may be subject to possible spurious boundary circulation patterns and their associated shear. Finally, in view of the very small salinity difference ($\Delta\langle S\rangle = 0.22^{o}/_{oo}$) between the Central Channel North (CCN) and the Central Channel South (CCS), it was decided to compute E_{CC} from the observed salt balance. Therefore, E_{CCN} and E_{OC} were computed directly from eqn. 3.2.3(d); the method yields a shear flux of salt into the Central Channel North (CCN) from the Outer Channel North (OCN) in Fig. 3.2(a) of $517^{o}/_{oo}$ m^3 s^{-1}, which with $\Delta\langle S\rangle = 0.98^{o}/_{oo}$ gives:

$$E_{CCN} = 5 \times 10^2 \text{ m}^3 \text{ s}^{-1}$$

The computed shear flux from the Outer Channel South (OCS) to the Outer Channel North (OCN) in Fig. 3.2(a) is $2310^{o}/_{oo}$ m^3 s^{-1}, which with $\Delta\langle S\rangle = 0.72^{o}/_{oo}$ gives:

$$E_{OC} = 3 \times 10^3 \text{ m}^3 \text{ s}^{-1}$$

Using these values, the remaining coefficients can be derived from the long-term salt balance for each region.

The results of a simulation of the salt balance using the preceeding model are shown in Fig. 3.2.3(a). The agreement between observed and computed data is generally good, not only for the mean values of salinity for each region, but also for the seasonal variations. To achieve this measure of agreement it was necessary to increase E_{CCS} by 50% over the value computed using the long-term salt balance, and to increase E_{IC} by 25%. Such trial and error adjustment is justified in view of the neglect of tidal pumping in the application of eqn. 3.2.1(e), and, more importantly, the neglect of

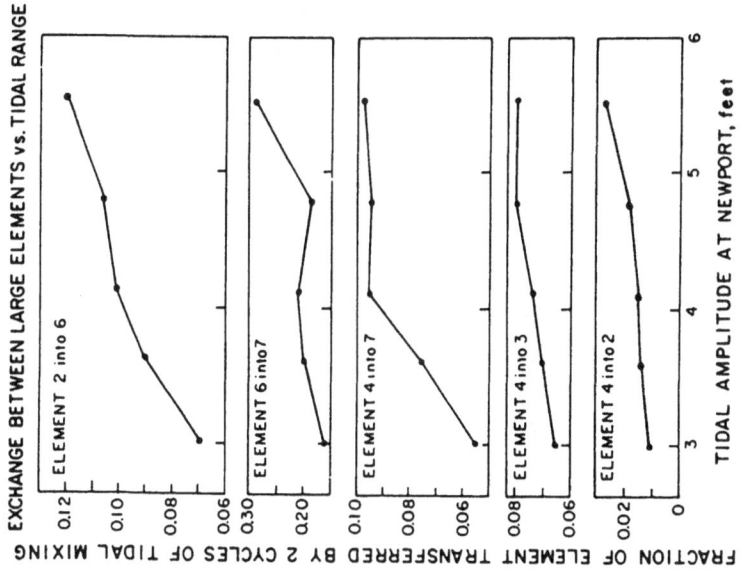

EXCHANGE BETWEEN LARGE ELEMENTS vs. TIDAL RANGE

Fig. 3.3.1(a) Representative exchanges of water between large spatial elements of Narragansett Bay, computed as a function of tidal amplitude at Newport. Reproduced from Kremer and Nixon (1978), with permission.

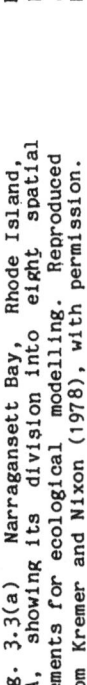

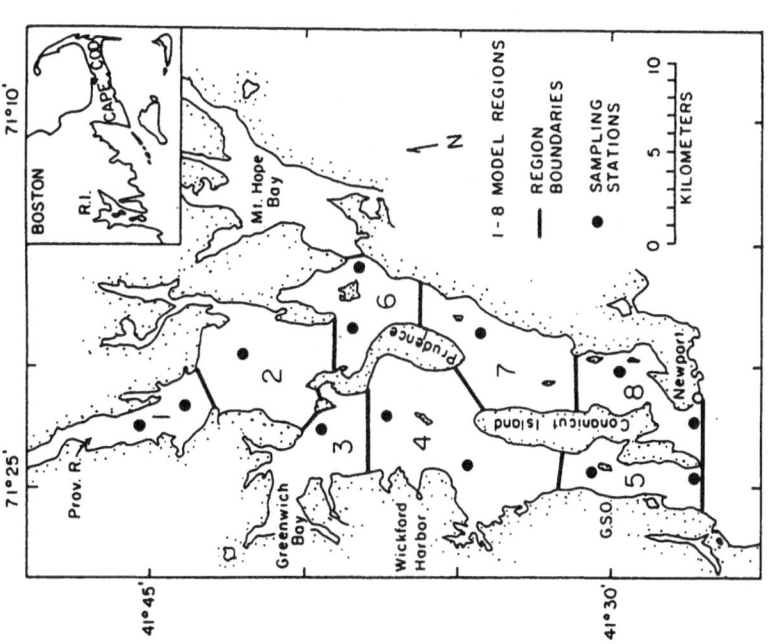

Fig. 3.3(a) Narragansett Bay, Rhode Island, USA, showing its division into eight spatial elements for ecological modelling. Reproduced from Kremer and Nixon (1978), with permission.

correlations between Q_f and $<S>$ in taking the long-term average of eqn. 3.2.3(a) for the salt balance. The agreement between observed and computed mean values of salinity for each region might be expected in view of the fact that these data were used in the process of defining mixing coefficients for the model. Nevertheless, the fact that seasonal variations are also reproduced with reasonable accuracy implies that the transport model has a substantial, although much simplified, physical basis.

3.3 Narragansett Bay Ecosystem

A chart of this estuarine region and its discretization into eight vertically mixed, homogeneous elements is shown in Fig. 3.3(a). The ecological model is less complex than that for the Bristol Channel; only the phytoplankton, zooplankton and major nutrients are treated with mechanistic detail (Kremer and Nixon, 1978)

The spatial elements in Fig. 3.3(a) are coupled by a mixing model which is rather different in concept to that for the Bristol Channel ecosystem model.

3.3.1 Mixing Model

The basis of this approach is again a fine-element, within-tide, hydrodynamical numerical model of the region. Depth-averaged velocities and elevations are computed for five infinitely repeating tidal ranges between neap and spring tides. Computed velocities for each spring to neap tidal state are then used to drive a fine-element, mass-balance model for a conservative, pelagic tracer. All of the fine-elements occupying one of the large elements in Fig. 3.3(a) are assigned an initial tracer concentration of unity. All other fine-elements are assigned an initial tracer value of zero. After two tidal cycles the net daily transport to each large element is computed as a fraction of the original mass of tracer.

The calculations are repeated for all five tidal ranges, and the data for fractional transfers tabulated against tidal range for use in the ecological model. This technique enables the daily-averaged transport of water between any two elements to be calculated. Representative exchanges of water between large spatial elements of Narragansett Bay are shown in Fig. 3.3.1(a) as functions of tidal range at Newport.

If two adjoining elements, i and j, have concentrations $\langle C_i \rangle$ and $\langle C_j \rangle$ of a pelagic variable, and if they exchange net water volume with each other at a daily rate $\langle Q_i \rangle$ and $\langle Q_j \rangle$, then the net mass transport into element i from element j is:

$$\langle F_i \rangle = \langle Q_j \rangle \langle C_j \rangle - \langle Q_i \rangle \langle C_i \rangle$$
$$= \langle Q_j \rangle (\langle C_j \rangle - \langle C_i \rangle) - (\Delta \langle Q \rangle) \langle C_i \rangle \qquad 3.3.1(a)$$

where $\Delta \langle Q \rangle = \langle Q_i \rangle - \langle Q_j \rangle$.

If $\langle Q_i \rangle > \langle Q_j \rangle$, then the mixing volume (equivalent to E in eqns. 3.2.1(e) and 3.2.3(a)) is $\langle Q_j \rangle$ and the residual flow from element i to j is $\Delta \langle Q \rangle$ (equivalent to Q_R in eqn 3.2.1(f)). If $\langle Q_j \rangle > \langle Q_i \rangle$, then rewriting eqn. 3.3.1(a) as:

$$\langle F_i \rangle = \langle Q_i \rangle (\langle C_j \rangle - \langle C_i \rangle) - (\Delta \langle Q \rangle) \langle C_j \rangle \qquad 3.3.1(b)$$

shows that the mixing volume is now $\langle Q_i \rangle$ and the residual flow from element j to i is $- \Delta \langle Q \rangle$. Therefore, this process of transferring water masses is essentially equivalent to an up-wind differencing of the ecological transport equations. The inclusion of all elements in the daily transfer process (rather than just immediate neighbours) is equivalent to an implicit solution of the transport equations with a time-step of one day.

3.4 Width-Averaged Models

In deep, narrow estuaries, transverse variations can be neglected compared with variations through the water column and along the axis. A width-averaged transport model for the tidally averaged pelagic variable is then appropriate:

$$\delta_t (\langle B \rangle \langle C \rangle) + \nabla . (\langle B \rangle \underline{\langle V \rangle} \langle C \rangle) - \nabla . (\langle B \rangle \underline{\langle D \rangle} \nabla \langle C \rangle)$$
$$= \delta_z (\langle B \rangle \langle \Omega \rangle \langle C \rangle) + \langle m \rangle \langle B \rangle \qquad 3.4(a)$$

with $\nabla = (\delta_x, \delta_z)$ and $\underline{\langle V \rangle} = (U, W)$.

The settling velocity of fine particulate material is $\langle \Omega \rangle$, and $\langle m \rangle$ is the sum of all other tidally averaged sources and sinks per unit volume.

Numerical models of the within-tide, width-averaged hydrodynamical equations have been developed (eg: Hamilton, 1975; Wang and Kravitz, 1980), as have models of the tidally-averaged hydrodynamics which are more suitable for water quality and ecosystem studies (eg: Nihoul et al., 1979; Festa and Hansen, 1976; Hansen and Rattray, 1965). A feature of these tidally-averaged models is the assumption of steady-state conditions. Therefore, time-averaging is taken to be sufficiently long that the circulation is in equlibrium with the forcing functions of freshwater inputs, wind-stress and density (salinity) gradients. This is not ideal because often the dominant vertical residual circulation is due to axial density gradients (gravitational circulation), and these have an equilibrium time-scale which is approximately equal to that for the salinity field (the estuarine flushing time), which can be of order weeks to months. The tidally-averaged, time-dependent model considered in the section on three-dimensional models would therefore seem more appropriate for a study of width-averaged ecosystems than these steady-state models.

3.5 Applications

There does not appear to be a conventional two-dimensional ecosystem model which considers width-averaged conditions. Winter et al. (1975) investigate the relationships between the growth of phytoplankton and climatic and hydrodynamic conditions in Puget Sound, USA, which is a temperate fjord with marked tides. The steady-state, tidally averaged circulation, $\langle \underline{V} \rangle$, is determined using the analytical methods of Hansen and Rattray (1965). Eqn. 3.4(a) is solved for plant chlorophyll using the computed values of $\langle \underline{V} \rangle$ and assuming negligible axial gradients ($\delta_x = 0$). Therefore, the equation is reduced to its one-dimensional form in the vertical dimension, and is solved using methods similar to those outlined for the one-dimensional, tidally averaged fixed element model.

A two-layer, two-element ecosystem model of southern Danish Waters has been presented by Gargas and Jacobsen (1984). The equations are equivalent to eqn. 3.2.3(a), with one mass balance equation for each element and with steady-state conditions assumed. Given a further equation for water volume conservation, this provides three equations in three unknowns (the horizontal flows through the upper and lower layers, and the vertical volume exchange between them. Entrainment is not incorporated). Knowing salinity and freshwater inputs over the area and throughout the year therefore enables flows and mixing to be defined for the ecosystem model.

Two-dimensional, two-layer, steady-state models have been discussed by Officer (1980) and applied to the problems of turbidity maxima formation and silica uptake in estuaries.

The numerical circulation model described by Festa and Hansen (1976) has been used as a basis for understanding dissolved silica distributions in San Francisco Bay (Peterson et al., 1978), and turbidity maxima in

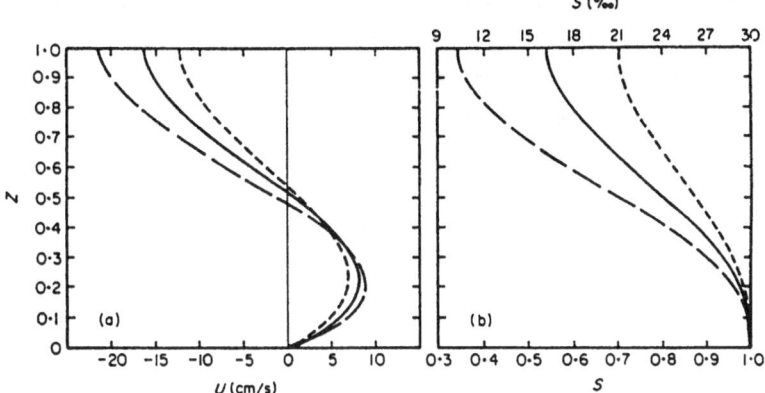

Fig. 3.5(a)   Variations of computed (a) current velocity and (b) salinity profiles at the seaward boundary for three values of the freshwater inputs. Reproduced from Festa and Hansen (1976), with permission.

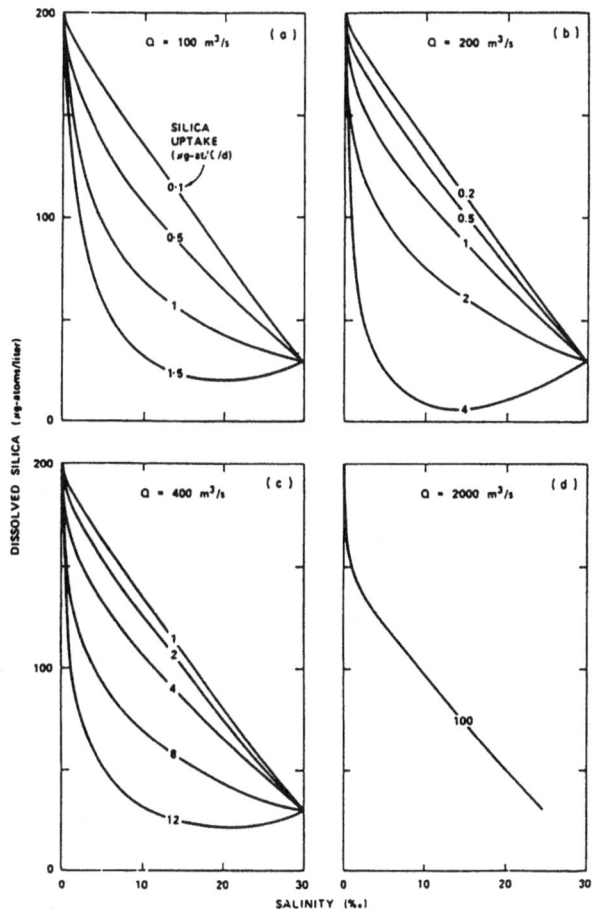

Fig. 3.5(b)   Computed silica-salinity covariance for different freshwater inputs (Q) and silica uptake rates. Reproduced from Peterson et al. (1978), with permission.

partially mixed estuaries (Festa and Hansen, 1978). In these models, $\langle B \rangle$ = constant, and eqn. 3.4(a) is written (Festa and Hansen, 1978):

$$\delta_t\langle C \rangle = - J(\Psi,C) + \langle D_x \rangle \delta_{xx}\langle C \rangle + \langle D_z \rangle \delta_{zz}\langle C \rangle$$
$$+ \langle \Omega \rangle \delta_z \langle C \rangle + \langle m \rangle \qquad\qquad 3.5(a)$$

in which J is the Jacobian and Ψ is a stream function:

$$\langle U \rangle = -\delta_z \Psi \text{ and } \langle W \rangle = \delta_x \Psi$$

Eqn. 3.5(a) is solved at element centres ($x_i = i\Delta x$, $z_j = j\Delta z$):

$$(\langle C_{ij}^{n+1} \rangle - \langle C_{i,j}^{n-1} \rangle)/2\Delta t = - J_{i,j}^n(\Psi,\langle C \rangle)/4\Delta x \Delta z$$
$$+ \langle D_x \rangle (\langle C_{i+1,j}^n \rangle + \langle C_{i-1,j}^n \rangle - \langle C_{i,j}^{n+1} \rangle - \langle C_{i,j}^{n-1} \rangle)/\Delta x^2$$
$$+ \langle D_z \rangle (\langle C_{i,j+1}^n \rangle + \langle C_{i,j-1}^n \rangle - \langle C_{i,j}^{n+1} \rangle - \langle C_{i,j}^{n-1} \rangle)/\Delta z^2$$
$$+ \langle \Omega \rangle (\langle C_{i,j+1}^n \rangle - \langle C_{i,j-1}^n \rangle)/2\Delta z + \langle m_{i,j}^n \rangle \qquad 3.5(b)$$

where $J_{i,j}^n(\Psi,\langle C \rangle)$ is solved in conservative form (Arakawa, 1966). Using $\langle \Omega \rangle$ = 0 and C = S in eqn. 3.5(b) gives an equation for the salt balance. Typical solutions are given in Festa and Hansen (1976,1978). Variations of velocity and salinity at the seaward boundary for three values of freshwater run-off are shown in Fig. 3.5(a).

The method has been used by Peterson et al. (1978) to investigate the uptake of silica ($\langle \Omega \rangle$ = 0 and $\langle m \rangle$ < 0) in north San Francisco Bay. The uptake is most clearly represented by plotting computed silica against computed salinity. This is shown in Fig. 3.5(b) for four values of the freshwater inputs.

4. THREE-DIMENSIONAL MODELS

This type of model is suited to an estuarine region which has strong gradients in both the vertical and horizontal dimensions. The computer time required for simulations is very long, and because of this the estuary is divided into a rather small number of horizontally well mixed segments of comparable size, but with significantly different ecological properties.

The modelled area is further sub-divided into layers which discretize each horizontal segment. This vertical layering can be such that the depth of each layer changes throughout a tidal cycle. In this technique, each layer maintains a fixed fraction of the water column as the surface elevation changes with time (eg: Owen, 1980). Alternatively, the layers can remain fixed whilst the water surface moves through them (eg: Hamilton, 1975).

In order to simulate seasonal cycles for ecosystem studies it is necessary to solve the tidally averaged equations. Three-dimensional, within-tide models have been described by several authors, but these are too detailed and time-consuming for ecosystem studies. Here, we will focus on the solution of the tide-averaged equations, and in particular on the rather general techniques used for an ecosystem model of Tolo Harbour, Hong Kong. This work was undertaken by Hydraulics Research, Wallingford, UK, in collaboration with the Institute for Marine Environmental Research (HR, 1983 a, b).

4.1 Tolo Channel Ecosystem

A chart of the Tolo Channel and its discretization is shown in Fig. 4.1(a). The ocean boundary is Mirs Bay. The region is divided into 13 horizontal segments. These are further layered into between one and four elements, depending on the local depth of water. The element numbering is shown for each segment in Fig. 4.1(a), as well as the residence time of water in each element (computed using a numerical model for conditions during December 1979 (HR, 1983 a)). The segmentation was chosen from considerations of the natural geometry and water quality; generally, the segments are large compared with a tidal excursion and contain the strongest horizontal residual circulations.

The tides in the region are mixed semi-diurnal and diurnal, with a mean tidal range of about 1.5 m, and a maximum tidal range of about 2.6 m. Tidal currents reach a maximum of 0.2 m s^{-1} in the central channel, and are as slow as 0.05 m s^{-1} in bays. There is negligible inflow of fresh water during the cool dry season between November and March, and the region is unstratified. During the hot, wet season a series of flashy spates occurs. The water column then stratifies because of the buoyancy of fresh water and the inadequate mixing produced by the weak tidal currents. The region has large inputs of natural effluents which exert a strong BOD. For this reason a water quality model is solved in conjunction with an ecosystem model (see Figs. 1(a),(b)).

The approach taken in modelling the Tolo Channel ecosystem is to first compute residual velocities and elevations over the period of interest (typically one year) using a tidally averaged hydrodynamical model. These velocities and elevations are then used to drive the advection-dispersion equations for the combined water quality and ecosystem models. The hydrodynamical model is described first.

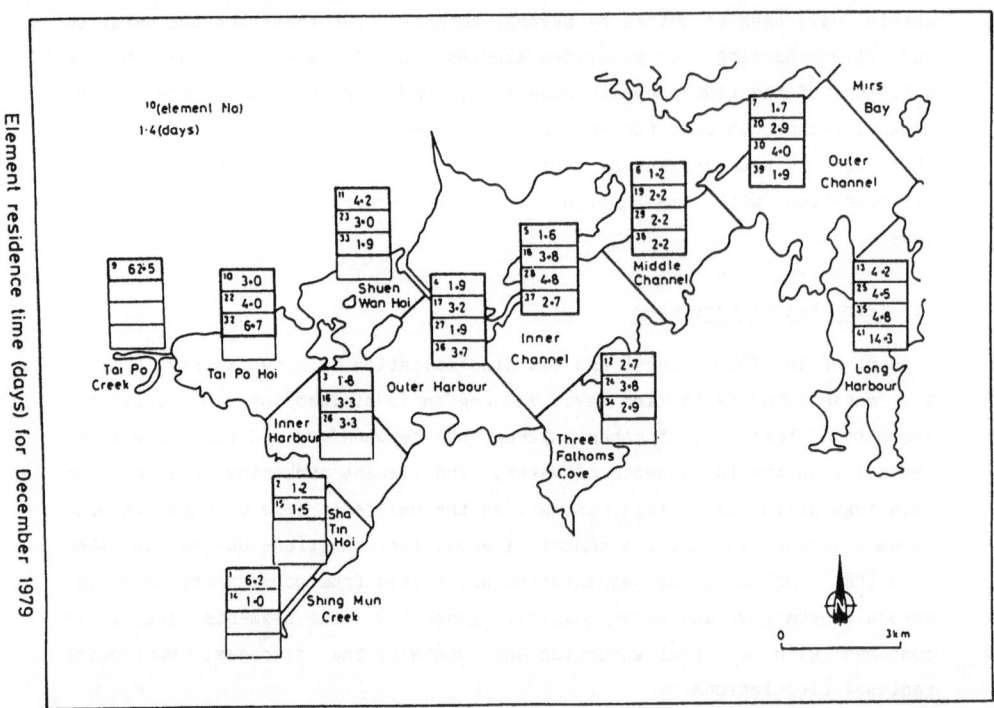

Element residence time (days) for December 1979

Fig. 4.1(a) Discretization of Tolo Channel showing computed residence times (in days) during a winter period of the spatial elements. Reproduced from HR (1983a), with permission.

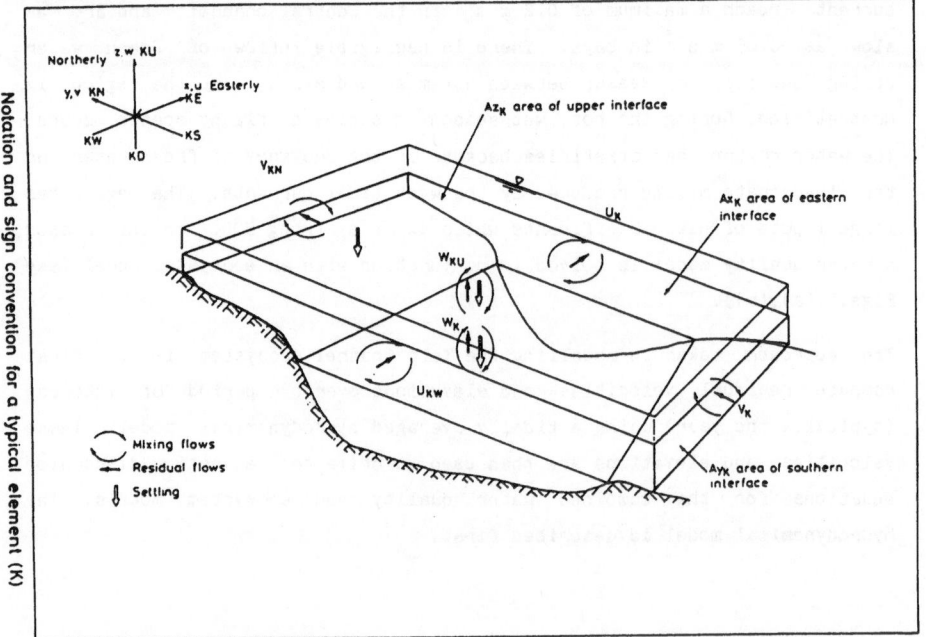

Notation and sign convention for a typical element (K)

Fig. 4.3(a) Notation and sign convention for a typical element, k. Reproduced from HR(1983b), with permission.

4.2 <u>Hydrodynamical Model</u>

The tidally-averaged model requires the freshwater inputs, rainfall and rate of evaporation for each segment of Fig. 4.1(a). Salinity and mean water level are specified at the ocean boundary (Mirs Bay). Temperature is specified throughout the region and is used with salinity to compute water density. Freshwater flows and wind-stress vary on a daily basis, whereas temperature, rate of evaporation, oceanic salinity and water level vary on a seasonal basis (if at all).

The tidally averaged equations of motion neglect Coriolis force. They are derived from the within-tide equations and are subject to the usual simplifying assumptions (HR, 1983 a). The equations are:
Conservation of (incompressible) water volume:
$$\nabla . \langle \underline{V} \rangle = 0 \qquad\qquad 4.2(a)$$
Conservation of momentum (x and y directions):
$$\langle \rho \rangle (\delta_t \langle U \rangle + \delta_z \langle U \rangle \langle W \rangle) = -\delta_x \langle P \rangle + \delta_z \langle \tau_x \rangle \qquad\qquad 4.2(b)$$
$$\langle \rho \rangle (\delta_t \langle V \rangle + \delta_z \langle V \rangle \langle W \rangle) = -\delta_y \langle P \rangle + \delta_z \langle \tau_y \rangle \qquad\qquad 4.2(c)$$
Hydrostatic Balance:
$$\delta_z \langle P \rangle + \langle \rho \rangle g = 0 \qquad\qquad 4.2(d)$$
where $\quad \underline{V} = (U,V,W), \quad \nabla = (\delta_x, \delta_y, \delta_z)$
and $\langle \tau_x \rangle$ and $\langle \tau_y \rangle$ are the tidally averaged shear stresses.

Coriolis force is neglected because it can play only a minor role when considered with the system of segmentation and length-scales in the Channel (Fig. 4.1(a)). Horizontal derivatives in the advection and shear-stress terms are neglected compared with vertical derivatives.

A salt-balance equation is solved. Salinity is required both for determination of density and for use by the ecosystem model:
$$\delta_t \langle S \rangle + \nabla . \langle S \rangle \langle \underline{V} \rangle - \nabla . \langle \underline{F} \rangle = 0 \qquad\qquad 4.2(e)$$
with $\langle \underline{F} \rangle$ the diffusive flux.
The density is defined from (temperature T in °C):
$$\langle \rho \rangle = 1000 + (0.797 - 0.001875 \langle T \rangle) \langle S \rangle$$
$$- 1000 \times (0.562(\langle T \rangle - 4)/277)^{1.85} \text{ kg m}^{-3}$$
for $12°C < \langle T \rangle < 34°C$.

4.3 <u>Numerical Model</u>

Layers of thickness ΔZ are used to form elements within the segments of

Fig. 4.1(a). The widths of a segment perpendicular to the x and y directions are B_y and B_x. The areas of an element's interfaces perpendicular to directions x, y, z are A_x, A_y and A_z (Fig. 4.3(a)). Western and northern interfaces are denoted by subscripts W and N. Upper and lower interfaces by subscripts U and D. Fig. 4.3(a) shows the Kth element. In all of the following equations the K subscript is omitted - it being understood that quantities refer to the Kth element.

Element velocities are defined as averages over interfaces; for example:
$$\langle \overline{U} \rangle = \int_{A_x} \langle U \rangle dA_x, \quad \langle \overline{V} \rangle = \int_{A_y} \langle V \rangle dA_y.$$
Integrating over A_x ($\int dydz$) in eqn. 4.2(b) for an interior element (for which A_x and B_y are constant) gives:
$$\langle \overline{\rho} \rangle \left\{ \delta_t A_x \langle \overline{U} \rangle + (\langle U \rangle \langle W \rangle B_y)_U - (\langle U \rangle \langle W \rangle B_y)_D \right\}$$
$$= -A_x \delta_x \langle \overline{P} \rangle + (\langle \tau_x \rangle B_y)_U - (\langle \tau_x \rangle B_y)_D \qquad 4.3(a)$$
Integrating over A_y in eqn. 4.2(c) for an interior element (for which A_y and B_x are constant) gives:
$$\langle \overline{\rho} \rangle \left\{ \delta_t A_y \langle \overline{V} \rangle + (\langle V \rangle \langle W \rangle B_x)_U - (\langle V \rangle \langle W \rangle B_x)_D \right\}$$
$$= -A_y \delta_y \langle \overline{P} \rangle + (\langle \tau_y \rangle B_x)_U - (\langle \tau_y \rangle B_x)_D \qquad 4.3(b)$$
The density over the interface is taken to be uniform, $\langle \rho \rangle = \langle \overline{\rho} \rangle$.

In a surface element the water levels and therefore the areas vary with time. However, the volume conservation equation can be used to derive (HR, 1983 a):
$$\langle \overline{\rho} \rangle \delta_t \langle \overline{U} \rangle = -\delta_x \langle \overline{P} \rangle + \left\{ (\langle \tau_x \rangle B_y)_U - (\langle \tau_x \rangle B_y)_D \right\} / A_x \qquad 4.3(c)$$
and $\qquad \langle \overline{\rho} \rangle \delta_t \langle \overline{V} \rangle = -\delta_y \langle \overline{P} \rangle + \left\{ (\langle \tau_y \rangle B_x)_U - (\langle \tau_y \rangle B_x)_D \right\} / A_y \qquad 4.3(d)$
where $\langle \tau_x \rangle$ and $\langle \tau_y \rangle$ at the surface (U - level) are components of wind-stress.

For an element at the horizontal, model bed $\langle U \rangle_D = \langle V \rangle_D = \langle \overline{W} \rangle = 0$, and eqns. 4.3(a),(b) are:
$$\langle \overline{\rho} \rangle \left\{ \delta_t A_x \langle \overline{U} \rangle + (\langle U \rangle \langle W \rangle B_y)_U \right\}$$
$$= -A_x \delta_x \langle \overline{P} \rangle + (\langle \tau_x \rangle B_y)_U - (\langle \tau_x \rangle B_y)_D \qquad 4.3(e)$$
and $\qquad \langle \overline{\rho} \rangle \left\{ \delta_t A_y \langle \overline{V} \rangle + (\langle V \rangle \langle W \rangle B_x)_U \right\}$
$$= -A_y \delta_y \langle \overline{P} \rangle + (\langle \tau_y \rangle B_x)_U - (\langle \tau_y \rangle B_x)_D \qquad 4.3(f)$$
with $\qquad \langle \tau_x \rangle_D = \alpha \langle \overline{\rho} \rangle \langle \overline{U} \rangle |U_T| \qquad\qquad 4.3(g)$
and $\qquad \langle \tau_y \rangle_D = \alpha \langle \overline{\rho} \rangle \langle \overline{V} \rangle |V_T| \qquad\qquad 4.3(h)$
where U_T and V_T are peak tidal velocities, and α is a frictional drag coefficient.

Values of <U> and <V> at upper and lower interfaces in eqns. 4.3(a), (b) can be estimated by interpolating $\langle \bar{U} \rangle$ and $\langle \bar{V} \rangle$. The hydrostatic balance (eqn. 4.2(d)) can be integrated through the height of a surface element interface to give (for the surface layer):

$$\delta_x \langle \bar{P} \rangle = g \langle \bar{\rho} \rangle \delta_x \zeta + \tfrac{1}{2} g \Delta Z \delta_x \langle \bar{\rho} \rangle \qquad 4.3(i)$$

Where ζ is surface elevation. This can then be used in a step-wise fashion to derive the pressure force within a lower element:

$$\delta_x \langle \bar{P} \rangle = \delta_x \langle \bar{P} \rangle_U + \tfrac{1}{2} g \left[(\Delta Z \delta_x \langle \bar{\rho} \rangle) + (\Delta Z \delta_x \langle \bar{\rho} \rangle)_U \right] \qquad 4.3(j)$$

where subscript U now refers to the upper (overlying) element.

The vertical velocities can be found from conservation of water volume, which for an interior element is (Fig. 4.3(a)):

$$(\langle \bar{U} \rangle A_x) - (\langle \bar{U} \rangle A_x)_W + (\langle \bar{V} \rangle A_y)_N - (\langle \bar{V} \rangle A_y)$$
$$+ (\langle \bar{W} \rangle A_z)_U - (\langle \bar{W} \rangle A_z) = 0 \qquad 4.3(k)$$

At the horizontal, model bed $\langle \bar{W} \rangle = 0$, so that eqn. 4.3(k) determines $\langle \bar{W} \rangle$ for all interior elements in a segment.

4.3.1 Salinity Model

Applying the salt balance (eqn. 4.2(e)) in conservative form to an elemental volume, ΔV, gives (see Fig. 4.3(a)):

$$L(\langle \bar{S} \rangle) = \delta_t(\langle \bar{S} \rangle \Delta V) + (\langle \bar{S} \rangle \langle \bar{U} \rangle A_x) - (\langle \bar{S} \rangle \langle \bar{U} \rangle A_x)_W$$
$$+ (\langle \bar{S} \rangle \langle \bar{V} \rangle A_y)_N - (\langle \bar{S} \rangle \langle \bar{V} \rangle A_y)$$
$$+ (\langle \bar{S} \rangle \langle \bar{W} \rangle A_z)_U - (\langle \bar{S} \rangle \langle \bar{W} \rangle A_z)$$
$$- \Big[(\langle F_x \rangle A_x) - (\langle F_x \rangle A_x)_W$$
$$+ (\langle F_y \rangle A_y)_N - (\langle F_y \rangle A_y)$$
$$+ (\langle F_z \rangle A_z)_U - (\langle F_z \rangle A_z) \Big] = 0 \qquad 4.3.1(a)$$

This applies to all elements, where in the surface or bed interfaces:

$$\langle F_z \rangle = \langle \bar{S} \rangle \langle \bar{W} \rangle$$

The salinity is defined at the centre of the cell in Fig. 4.3(a), and $\langle \bar{S} \rangle$ is a volume average. The area averages $\langle \bar{S} \rangle$ correspond to averages over A_x, A_y or A_z according to whether they are mulitplied by A_x, A_y or A_z in eqn. 4.3.1(a); they are determined by interpolation of $\langle \bar{S} \rangle$ values at element centres.

4.3.2 Mixing Coefficients

The usual assumption is made to quantify horizontal mixing; in the x direction:

$$\langle F_x \rangle = A_x D_x \delta_x \langle \overline{S} \rangle \qquad\qquad 4.3.2(a)$$

where D_x is defined in an empirical way from:

$$D_x = A_x D_{x1}/A + \beta |\langle \overline{U} \rangle| \Delta x \qquad\qquad 4.3.2(b)$$

where A is the total cross-sectional area and D_{x1} is the axial dispersion coefficient deduced from a one-dimensional salt balance. The second term is an attempt to quantify vertical shear dispersion in the currents. Δx is the x-distance between element centres, and β a scaling factor. A similar equation applies for D_y.

Vertical density gradients reduce the transport of mass and momentum through the water column. The Tolo Channel model uses a mixing length to describe the interfacial shear stress and mass flux (concentration $\langle C \rangle$):

$$\langle \tau_x \rangle = \langle \overline{\rho} \rangle \, l_m^2 |\delta_z \langle \overline{U} \rangle| \delta_z \langle \overline{U} \rangle$$

and
$$\langle F_z \rangle = \langle \overline{\rho} \rangle \, l_m l_c |\delta_z \langle \overline{U} \rangle| \delta_z \langle \overline{C} \rangle$$

In homogeneous conditions:

$$l_m = l_c = l_{mo} = 0.4 \, z(1 - z/H)^{\frac{1}{2}}$$

where H is the total depth of a segment. In stably stratified conditions l_m and l_c are functions of the Richardson number, Ri:

$$Ri = -g\delta_z \langle \overline{\rho} \rangle \, / \, \langle \overline{\rho} \rangle |\delta_z \langle \overline{U} \rangle|^2$$

and the mixing length varies with Ri as:

$$l_m/l_c = (1 + \gamma Ri)^{-\frac{1}{2}} \qquad\qquad 4.3.2(c)$$

where the value of γ depends on the extent of the time-averaging (Odd and Rodger, 1978).

4.3.3 Choice of Parameters and Simulation

The friction factor (eqns. 4.3(g),(h)) and other parameters defined in eqns. 4.3.2(a)-(c), were determined by trial and error fitting of results from the tide-averaged three-dimensional model with those from a detailed, within-tide, two-dimensional (x-z-t) model over a five day, fluvial spate period.

The tidally averaged model was run for one year (1982). Computed and observed salinities for three levels in the water column for two segments are shown in Fig. 4.3.3(a). Data for velocities and salinity are stored for subsequent use by the transport equations in the combined water quality and ecosystem models.

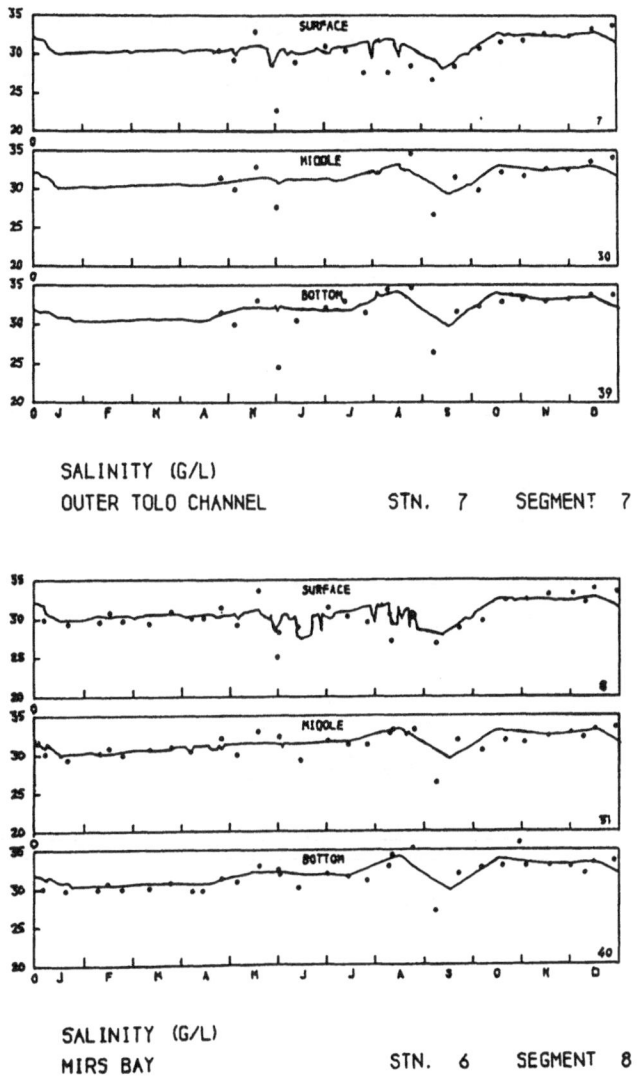

SALINITY (G/L)
OUTER TOLO CHANNEL   STN. 7   SEGMENT 7

SALINITY (G/L)
MIRS BAY   STN. 6   SEGMENT 8

Fig. 4.3.3(a) Computed and observed salinity profiles during 1982 in
segments 7 and 8 of the Tolo Channel ecological model. Reproduced from HR
(1983a), with permission.

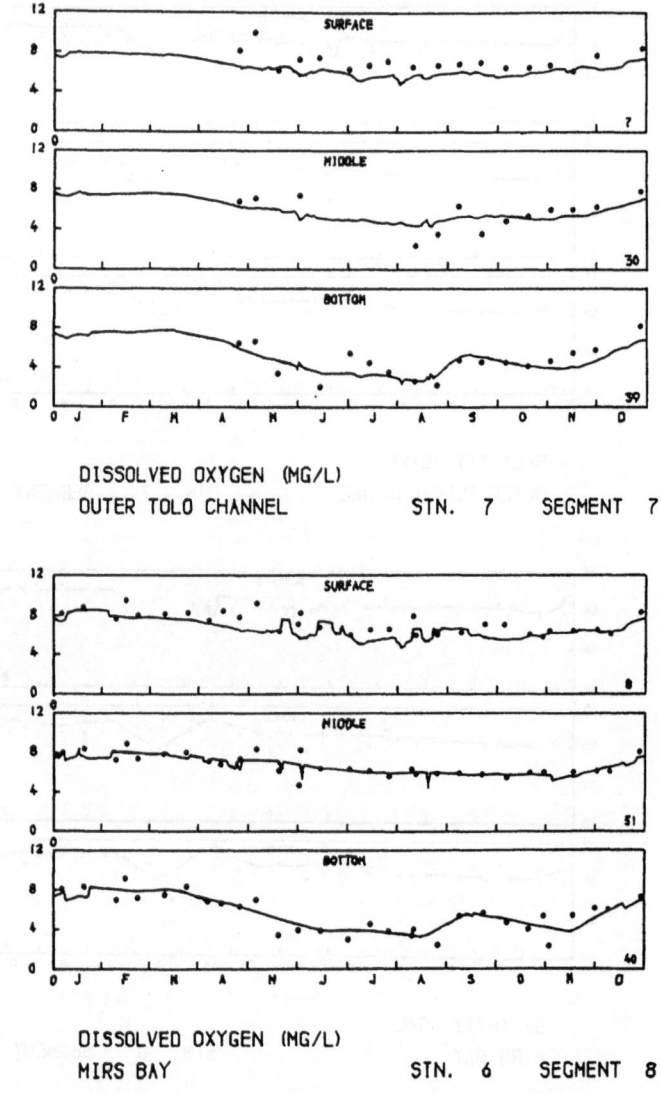

DISSOLVED OXYGEN (MG/L)

OUTER TOLO CHANNEL   STN. 7   SEGMENT 7

DISSOLVED OXYGEN (MG/L)

MIRS BAY   STN. 6   SEGMENT 8

Fig. 4.4(a)   Computed and observed dissolved oxygen profiles during 1982
in segments 7 and 8 of the Tolo Channel ecological model.
Reproduced from HR(1983b), with permission.

4.4 Transport Equations

All pelagic variables are assumed to have a transport equation of the form:

$$\delta_t\langle C\rangle + \nabla.\langle \underline{V}\rangle\langle C\rangle - \nabla.\langle\underline{F}\rangle = \delta_z\langle\Omega\rangle\langle C\rangle + \langle m\rangle \qquad 4.4(a)$$

where $\langle\Omega\rangle$ is the settling velocity of fine particulate material, and $\langle m\rangle$ is the sum of all tidally averaged sources and sinks described by the water quality and ecological models for substance C.

The left-hand-side of eqn. 4.4(a) is identical in structure to that for the salt balance (eqn. 4.2(e)). Applying eqn. 4.4(a) to an elemental volume gives a difference equation in which the left-hand-side is identical in structure with that for salinity (eqn. 4.31(a)):

$$L(\langle\bar{C}\rangle) = \langle\bar{m}\rangle\Delta V + (\langle\Omega\rangle\langle\bar{C}\rangle A_z)_U - (\langle\Omega\rangle\langle\bar{C}\rangle A_z) \qquad 4.4(b)$$

where $\langle\bar{C}\rangle$ is an average of $\langle C\rangle$ over A_z and is determined by interpolation of $\langle\bar{C}\rangle$.

Eqn. 4.4(b) is solved using a fully implicit, time-centred scheme (with time-step, Δt, equal to 1/8 day) in which:

$$\delta_t(\langle\bar{C}\rangle\Delta V)^{n+\frac{1}{2}} = \left[(\Delta V\langle\bar{C}\rangle)^{n+1} - (\Delta V\langle\bar{C}\rangle)^n\right]/\Delta t \qquad 4.4(c)$$

and $\qquad \langle\bar{C}\rangle^{n+\frac{1}{2}} = (\langle\bar{C}\rangle^{n+1} + \langle\bar{C}\rangle^n)/2 \qquad\qquad 4.4(d)$

Eqns. 4.4(a)-4.4(d) are the basis of the calculation of transport for the oxygen balance, suspended particulates and other ecological variables over the region and through time.

The model was run for 1982 using observed forcing functions and boundary conditions. Loading of nutrients from streams and outfalls were specified. Fig. 4.4(a) shows simulated and observed dissolved oxygen concentrations in Mirs Bay and Outer Tolo Channel (Fig. 4.1(a)). The model simulated all the main features of the oxygen balance except levels of supersaturation in the surface layers which were caused by photosynthesis.

5. Final Remarks

The main practical features of one-dimensional and two-dimensional coupled hydrodynamical and ecological models are the simplicity with which the physics is treated, and the associated rapidity of execution on the computer. Simplicity is a non-trivial attribute. Ecological modelling is undertaken by multi-disciplinary teams, the members of which need to run these models for their own purposes without becoming involved with either the physical oceanography or the details of hydrodynamical modelling. Such models allow for the physical transport and flushing with an accuracy which is sufficient for the requirements of the ecological model.

Coupled models share a common design feature. A fine-element hydrodynamical model is first used to define tidal and long-term flows across element boundaries within the ecological model. Computed or observed salinity data are then used to determine mixing across these boundaries. In the case of tidally-averaged, fixed element, one-dimensional models the conservation of water volume equation is sufficient to define these long-term flows.

In some cases it is possible to parameterize results for the flows and mixing across boundaries in terms of tidal range, freshwater inputs and wind-stress. This is ideal because it provides maximum freedom for the ecological modeller - all of the physical processes can be incorporated into one subroutine, and any period of time or set of conditions can be investigated once the forcing functions are specified.

The alternative procedure, and the one used for the three-dimensional ecological model reviewed here, is to run a tidally-averaged hydrodynamical model over a period of particular interest, and to store the computed data for subsequent access by the ecological model. Therefore, numerical experiments with modified freshwater inputs or tidal ranges require separate runs of the hydrodynamical model. However, there are circumstances when this should be done anyway, such as when proposed major engineering works are likely to have a large effect on the estuary's circulation.

ACKNOWLEDGEMENTS

I am grateful to J.A. Stephens and T.Y. Woodrow for assistance with the literature review and manuscript preparation. This work forms part of the Physical Processes programme of the Institute for Marine Environmental Research, a component body of the Natural Environment Research Council (NERC). It was partly supported by the Department of the Environment on Contract No. DGR 480/48.

REFERENCES

Arakawa, A. 1966. Computational design for long-term numerical integration of the equations of fluid motion. Two-dimensional incompressible flow. Part 1. J. Comp. Phys., 1, 119-143.

Baird, D. and H. Milne. 1981. Energy flow in the Ythan Estuary, Aberdeenshire, Scotland. Estuar. Coastal Shelf Sci., 13, 455-472.

Belyaev, V.I. 1984. Simulation of functioning of a complex ecosystem. Ecological Modelling, 26, 9-15.

Campbell, D.E. In Press. Process variability in the Gulf of Maine - a macroestuarine environment. In: Estuarine Variability (Ed: D.A. Wolfe). Academic Press Inc.

Chen, C.W. and G.T. Orlob. 1975. Ecological Simulatation for Aquatic Environments. In: Systems Analysis and Simulation in Ecology, Vol 3 (Ed: B. Patten). Academic Press, 475-588.

Ebenhoh, W. 1985. A dynamic model of the redistribution of suspended silt in an estuary. BOEDE Publ. en Versl., 2, 21 pp.

Edinger, J.E. and E.M. Buchak. 1980. Numerical hydrodynamics of estuaries. In: Estuarine and Wetland Processes (Ed: P. Hamilton and K.B. Macdonald). Plenum Press. 115-146.

Festa, J.F. and D.V. Hansen. 1976. A two-dimensional numerical model of estuarine circulation: The effects of altering depth and river discharge. Estuar. Coastal Mar. Sci., 4, 309-323.

Festa, J.F. and D.V. Hansen. 1978. Turbidity maxima in partially mixed estuaries: A two-dimensional numerical model. Estuar. Coastal Mar. Sci., 7, 347-359.

Flather, R.A. and N.S. Heaps. 1975. Tidal computations for Morecambe Bay. Geophys. J. R. Astron. Soc., 42, 489-517.

Gameson, A.L.H. 1982. Dissolved-Oxygen Regime. In: The quality of the Humber Estuary (Ed: A.L.H. Gameson). Yorkshire Water Authority, Leeds, U.K. 15-26.

Gargas, E. and T.S.Jacobsen. 1984. A box model of the biological/chemical turnover in the southern part of the Danish Waters. Limnologica (Berlin), 15, 263-275.

Gordon, D.C. Jr. and J.W. Baretta. 1982. A preliminary comparison of two turbid coastal ecosystems: The Dollard (Netherlands - FRG) and the Cumberland Basin (Canada). Hydrobiol. Bull., 16, 225-267.

Gordon, D.C., P.D. Keizer, G.R. Daborn, P. Schwinghamer and W.L. Silvert. In Press. Adventures in holistic ecosystem modelling: the Cumberland Basin ecosystem model. Netherlands J. Sea Res.

Hamilton, P. 1975. A numerical model of the vertical circulation of tidal estuaries and its application to the Rotterdam Waterway. Geophys. J. R. Astron. Soc., 40, 1-21.

Hansen, D.V. and M. Rattray, Jr. 1965. Gravitational circulation in straits and estuaries. J. Mar. Res., 23, 105-122.

Harris, J.R.W., A.J. Bale, B.L. Bayne, R.F.C. Mantoura, A.W. Morris, L.A. Nelson, P.J. Radford, R.J. Uncles, S.A. Weston and J. Widdows. 1984. A preliminary model of the dispersal and biological effect of toxins in the Tamar Estuary, England. Ecological Modelling, 22, 253-284.

HR. 1983a. Tolo Harbour water quality model (TASGM: a three-dimensional gravitational circulation model). Hydraulics Research Ltd., U.K. Report No. Ex 1108. 30 pp.

HR. 1983b. Tolo Harbour water quality model (QUEST2: a three-dimensional water quality and ecosystem model). Hydraulics Research Ltd., U.K. Report No. Ex 1140.

HRS. 1977. Sha Tin New Town, Hong Kong. (Mathematical model studies of pollution in Shing Mun Tidal Creek). Hydraulic Research Ltd., U.K. Report No. Ex802. 9 pp.

Joint, I.R. 1983. Development of an ecosystem model of a turbid estuary. Can. J. Fish. Aquat. Sci., 40 (Sup. 1), 341-348.

Ketchum, B.H. 1951. The exchanges of fresh and salt waters in tidal estuaries. J. Mar. Res., 10, 18-37.

Kremer, J. and S.W. Nixon. 1978. A coastal marine ecosystem. Ecol. Stud., 24, Springer-Verlag. 250 pp.

Liu, S.-K. and J.J. Leenderste. 1978. Multidimensional numerical modelling of estuaries and coastal seas. In: Advances in Hydroscience (Ed: V. T. Chow), Vol 11. Academic Press. 95-164.

Maskell, J.M. and N.V.M. Odd. 1977. A mathematical model of the oxygen balance in a well-mixed estuary. Hydraulics Research Ltd., U.K. Report No. It 171. 52 pp.

Mollowney, B.M. 1973. One-dimensional models of estuarine pollution. In: Mathematical and hydraulic modelling of estuarine pollution, Water Pollution Res. Tech. Paper No. 13 (HMSO), 106-113.

Najarian, T.O. and D.R.F. Harleman. 1977. A real time model of nitrogen-cycle dynamics in an estuarine system. Prog. Wat. Tech., 8, 323-345.

Nihoul, J.C.J., F.C. Ronday, J. Smitz and G. Billen. 1979. Hydrodynamic and water quality model of the Scheldt Estuary. In: Marsh-estuarine systems simulation (Ed: R.F. Dame). USC Press, Columbia, SC. 71-82.

Odd, N.V.M. 1985. Modelling the dispersal of effluents in tidal waters. Hydraulics Research Ltd., U.K. Report No. SR 47. 29 pp.

Odd, N.V.M. and Rodgers, J.G. 1978. Vertical mixing in stratified tidal flows. J. Hyd. Div. ASCE., 104, Hy3, Proc Paper 13599.

Officer, C.B. 1980. Box models revisited. In: Estuarine and wetland processes (Ed: P. Hamilton and K.B. MacDonald). Plenum, N.Y. 65-114.

O'Kane, J.P. 1980. Estuarine Water-Quality Management. Pitman Pub. Ltd., Lond., 155 pp.

Orlob, G.T. 1976. Estuarial Models. In: Systems Approach to Water Management (Ed: A.K. Biswas). McGraw-Hill, Lond., 253-293.

Owen, A. 1980. A three-dimensional model of the Bristol Channel. J. Phys. Oceanogr., 10, 1290-1302.

Peterson, D.H., J.F. Festa and T.J. Conomos. 1978. Numerical simulation of dissolved silica in the San Francisco Bay. Estuar. Coastal Mar. Sci., 7, 99-116.

Radford, P.J. 1979. Some aspects of an estuarine ecosystem model-GEMBASE. In: state of the art in ecological modelling (Ed: S.E. Jorgensen). ISBN 87 87257 17 -3. 301-322.

Radford, P.J. 1981. Modelling the impact of a tidal power scheme on the Severn Estuary ecosystem. Paper presented at the International Symposium - Energy and ecological modelling (Louisville, Kentucky, April 20-23, 1981). 13 pages.

Radford, P.J. 1983. Systems modelling of estuaries. In: Practical procedures for estuarine studies (ed: A.W. Morris). Natural Environment Research Council, Swindon, U.K. 239-262.

Radford, P.J. 1984. Effects on the water quality of the Thames Estuary. In: Predicted effects of proposed changes in patterns of water abstraction on the ecosystems of the lower River Thames and its tidal estuary. Natural Environment Research Council, Swindon, U.K. 97-179.

Radford, P.J. and I.R. Joint. 1980. The application of an ecosystem model to the Bristol Channel and Severn Estuary. Water. Pollut. Control, 79 244-250.

Radford, P.J., R.J. Uncles and A.W. Morris. 1981. Simulating the impact of technological change on dissolved Cadmium distribution in the Severn Estuary. Water Res. 15, 1045-1052.

Roache, P.J. 1982. Computational fluid mechanics. Hermosa Publishers, USA. 446 pp.

Ruardij, P. and J.W. Baretta. 1982. The Ems-Dollard ecosystem modelling workshop. BOEDE Publ. en Versl., 2, 46 pp.

Stommel, H. 1953. Computation of pollution in a vertically mixed estuary. Sewage and Ind. Wastes, 25, 1065-1071.

Summers, J.K., W.M. Kitchens, H.N. McKellar, Jr. and R.F. Dame. 1980. A simulation model of estuarine subsystem coupling and carbon exchange with the sea. II. North Inlet model structure, output and validation. Ecological Modelling, 11, 101-140.

Uncles, R.J. 1981. A numerical simulation of the vertical and horizontal M_2 tide in the Bristol Channel and comparisons with observed data. Limnol. Oceanogr., 26, 571-577.

Uncles, R.J. 1982. Computed and observed residual currents in the Bristol Channel. Oceanol. Acta, 5, 11-20.

Uncles, R.J. 1983a. Modelling tidal stress, circulation, and mixing in the Bristol Channel as a prerequisite for ecosystem studies. Can. J. Fish. Aquat. Sci., 40 (Sup. 1), 8-19.

Uncles, R.J. 1983b. Hydrodynamics of the Bristol Channel. Mar. Pollut. Bull., 15, 47-53.

Uncles, R.J. and M.B. Jordan. 1980. A one-dimensional representation of residual currents in the Severn Estuary and associated observations. Estuar. Coastal Mar. Sci., 10, 39-60.

Uncles, R.J. and P.J. Radford. 1980. Seasonal and spring-neap tidal dependence of axial dispersion coefficients in the Severn – a wide, vertically mixed estuary. J. Fluid Mech., 98, 703-726.

Wang, D.-P. and D.W. Kravitz. 1980. A semi-implicit two-dimensional model of estuarine circulation. J. Phys. Oceanogr., 10, 441-454.

Winter, D.F., K. Banse and G.C. Anderson. 1975. The dynamics of phytoplankton blooms in Puget Sound, a fjord in the northwestern United States. Mar. Biol., 29, 139-176.

V. COASTAL-OFFSHORE INTERACTIONS

V COASTAL-OFFSHORE INTERACTIONS - AN EVALUATION OF PRESENTED EVIDENCE

B-O. Jansson, A.D. McIntyre, S.W. Nixon, M.M. Pamatmat, B. Zeitzschel, J.J. Zijlstra

During the workshop the individual papers were discussed in the context of the overall theme. In addition, group discussions within the different sections, led by a working group member aimed at summarizing and evaluating the presented evidence of coastal-offshore interaction. Related future research was also discussed although many of the suggestions were found to have a strong general bearing. Since not all of the participants have submitted an article their constructive participation is mainly traced in this summary of the proceedings.

1. WATER EXCHANGE

Depending on the type of shallow coastal system, the principal mixing zone is situated either in inshore basins, around inlets, or offshore on the coastal shelf with tide, wind and buoyancy as main agents and affected by topography and the earth's rotation. In areas with strong tides, most mixing takes place in inshore basins and inlets, in the case of large rivers most mixing takes place on the continental shelf. The exchange of nutrients and pollutants between inshore and offshore regions does not depend solely on the exchange of water. Due to adsorption they are coupled to processes of flocculation, consolidation and erosion. The direction of the transport may be affected by the asymetry of the tidal cycle. In cases of persistent coastal currents, mixing with the offshore primarily takes place through cyclonic eddies.

<u>Recommendations for future research</u> include studies on frontal structures using infrared satellite images and development of a microwave radar for registration of seasurface current distributions. An adequate description of turbulent exchange properties in buoyancy flow is needed, and requires further research. Models should be able to deal with the transport of buoyant and sedimentary particles or vertically migrating organisms. Emphasis should be placed on flocculation, consolidation and erosion properties of fine sediment and the adsorption and desorption of nutrients and pollutants.

Lecture Notes on Coastal and Estuarine Studies, Vol. 22
B.-O. Jansson (Ed.), Coastal-Offshore Ecosystem Interactions.
© Springer-Verlag Berlin Heidelberg 1988

2. MASS BALANCE STUDIES

A significant part of the Workshop's activity was directed to considering a number of coastal habitats and ecosystems, in particular the mass balances between them and offshore regions. The focus was on ecosystems dominated by salt marshes, mangroves, and tidal mud flats as well as regions below low water mark including coral reefs, submerged macrophyte beds, fjord systems, drowned valleys and polluted bays.

There has been considerable controversy in the literature regarding the relationship between these coastal systems and adjacent sea areas. Some investigators have postulated substantial export of nutrients and/or organic material to the offshore, giving the coastal ecosystems an essential role in the larger ecology of near- and offshore systems. This was subjected to critical review by the Workshop. The conclusions drawn in the separate reviews are summarized in Table 1.

Table 1. Degree of exchange of organic matter and nutrients between coastal systems and the ocean.

EVIDENCE SOURCE	PRESENT VIEW
Stable isotope studies	Most terrigenous material stays in the nearshore
Tidal flats	Mostly import of organic matter Heavy internal cycling
Salt marshes	Exports of organic matter to nearshore. Large internal cycling of nutrients
Mangroves	Export of organic matter to nearshore
Fjords	Well mixed fjords are nutrient limited, stagnant are sinks
Coral reefs	Self-maintained; very limited metabolic interaction with surrounding ocean
Global view of riverine-estuarine transports	Oceanic flows dominate

Perhaps the most confident statement can be made in relation to coral reefs which were recognised in the extreme case as shoal water ecosystems having no coupling with land. Although normally surrounded by oligotrophic ocean, reefs are characterised by high productivity.

Their conspicuous metabolic performance is accomplished by retention and recycling of organic and inorganic material. The potential of coral reefs for net organic export is insignificant in quantity compared with either their internal cycling or with the net metabolic activity of the photic zone of the surrounding ocean. While coral reefs, like other ecosystems influenced by man are threatened by local pollution, direct destruction or over-exploitation, there is no clear evidence that the coral reef ecosystem is threatened on a global scale.

For submerged macrophyte beds the evidence is also probably sufficiently unequivocal to justify the statement that they are insignificant exporters of material on the scale of continental shelves. They may, however, be important on a local scale in oligotrophic situations as conservative reservoirs of material which is slowly released from detritus seasonally. In certain circumstances over-exploitation of predators in kelp beds may destabilise the community structure but there may be natural longterm "flip-flop" fluctuations (e.g. Nova Scotian lobsters and urchins, Californian sea-otters). Large scale exploitation of macrophyte beds (e.g. for methane gas energy, alginic acid, fertilizers) may destroy the habitat for certain animals (e.g. lobsters, rock-lobsters, certain fish, intertidal birds).

It is more difficult to reach conclusions about mangrove ecosystems. Productivity studies and mass-balance calculations have established that in general mangroves export large quantities of detritus to estuarine waters which, given the immense energy from tides, precipitation and run-off in some tropical areas, are seasonally extended to near-shore areas. However, there are few quantitative studies for sediments and nutrients, especially considering the great diversity of mangrove systems from one region to another, and we do not understand if these wetlands are a source or sinks, particularly in the case of organic material. Also, the mass balance approach is not sufficient in itself to address the issue of the importance of mangroves to coastal and inshore interests such as fisheries. Mere net exchange or outwelling does not necessarily indicate utilization by higher trophic levels. It is relevant that mangroves are extensively exploited by man for fuel, charcoal, chipwood, pulpwood, tanin, building material and fence posts. Mangroves are also destroyed by other human activities such as clearing for fish or shrimp pond culture, housing development, highway construction, agriculture and in relation to oil spills. There has been an alarming loss of these ecosystems to tropical estuaries, but it is far from clear just how this impact can be evaluated.

Salt marshes also present problems of evaluation in the context of this Workshop. They are highly productive in both gross and net terms, and like coral reefs are characterised by a high rate of internal recycling. While there is a general tendency for marsh/estuarine systems to export organic carbon, this is likely to be of only local importance with respect to the receiving ocean. The qualitative differences in the carbon exchanges between the macrophyte dominated estuarine regions and the adjacent waters may be important, since many systems appear to import high quality phytoplankton carbon and export lower quality detritus, which might also apply for mangrove systems.

In general, whether in salt marshes, mangroves, or tidal flats, the picture is of substantial internal recycling together with some degree of flux with adjacent systems. What determines whether a system has excess mineralization or production is determined not so much by the size of the cycle as by the hydrodynamic processes, including net sediment transport. Thus, while it will be useful to make broad generalizations about typical examples, in practice, each area should be examined for its own specific characteristics before its flux can be determined.

Recommendations for future research. Among the various problems within the different coastal systems suggested for further studies some major ones with bearing on the theme of the workshop can be summarized:

The stable isotope technique in conjunction with physical and biological information is recommended for tracing the flows of terrigenous organic matter. For tidal flats there is a great need for comparative studies of organic matter budgets in temperate and tropical regions.

The role of salt marshes in supporting coastal fisheries by exporting organic matter has still to be assessed. Mass balance studies of sediments and nutrients in mangrove systems are needed.

More inter-disciplinary research has to be done to unify the details of the important source/sink processes on larger spatiotemporal scales.

3. ACTIVE TRANSPORTS

Some of the Workshop's attention was focused on the transport - passive or active - of living animals across the coastal-offshore boundary and on the effects of such transports on the ecosystem in the two areas. In contrast to the situation concerning transport of nutrients and organic matter, about which opinions tend to diverge, there appears to be a general consensus about the importance of migrations between the coastal and offshore regions for offshore fish and crustacean species occupying a coastal nursery. In spite of their great commercial interest, few species have been comprehensively studied to establish the importance of their coastal-offshore distribution and movements. Usually the distribution of the juvenile stage has only partly been studied, mostly only in the estuary, thus failing to establish the presence of an exclusively coastal nursery.

Recommendations for further studies. The survival value of a coastal nursery and the consequences of its alteration, pollution, elimination etc. has been insufficiently studied. For instance fjords as nursery areas for fish overwintering in lower latitudes is an important concept which has not yet been explored. Such studies are of importance since in many parts of the world the structure of coastal areas is altered by technological perturbations, while the quality is reduced by pollution, so that the continued existence of the species involved in coastal-offshore migrations, could be at stake. At present our knowledge of the specific requirements of juvenile crustacea and fish in the coastal nurseries is in most cases insufficient to advise about the effects of man-made changes in the coastal area.

It was recognized, however, that in view of the renewed interest in the fate of early, pre-recruit stages of commercially important species of fish and crustacea as formulated by e.g.IOC, studies of post-larval stages, also of species occupying a coastal nursery, will probably increase. In this context the International Recruitment Experiment Programme (IREP) should also be mentioned.

On the topic of coastal-offshore migrations little knowledge is available on the problem of how larvae of species which use coastal nurseries manage to reach the coastal areas from the offshore spawning sites. There are indications that in several cases, larvae or post-larvae, which have an insufficient locomotory capacity to reach the coastal area under their own power, display complicated and varying behavioural patterns on their way to the coast. Attempts to model larval transport, using information on behavioural patterns and three-dimensional water transport models should be highly rewarding, although some early attempts seem to indicate that existing physical models are often of insufficient detail.

Finally, it would appear from some preliminary estimates that offshore migrating of animals are insignificant transporters of organic matter as compared to transports due to physical processes.

4. NUMERICAL MODELLING

In our efforts to understand the flux of material between the coast and the ocean the physical oceanographic processes must be a main organizational tool. Mathematical models linking observations and theories, physical processes and biological processes constitute a powerful device for dynamically synthesizing the relevant information. For many areas, models describing physical dynamics already exist, but progress in understanding physical dynamics exceeds that of biological dynamics.

Recommendations. Future physical-biological models will benefit from the use of moored instruments and from satellites. The investigation of inshore/offshore ecosystem coupling requires a close collaboration between biologists, chemists and physical oceanographers. Numerical hydrodynamic models yielding information on flow structure and its time variability are essential tools for these studies. In particular, relevant flow structures are fronts, topographically and density-induced circulations, long internal wave motions, directional changes of residual bottom currents and bottom stress distribution induced by waves and currents. The time variability should include seasonal, meteorological and tidal time scales, and the instability of coastal currents. From comparing time-dependent, coupled hydrodynamical-ecological models in one, two, and three dimensions it appears that the fixed element, tidally-averaged model is the most suitable for ecosystem simulations.

5. CONCLUSIONS

Of the coastal-offshore interactions on a global scale, outwelling is not considered quantitatively significant, at least on the time scale of years to decades. There are individual sites where export could be important but the impact is essentially local.

The riverine and oceanic import of material to the coastal areas is intensively recycled there and in many cases the productivity of a coastal system is to a greater extent determined by recycling than by inputs from outside (Fig. 1).

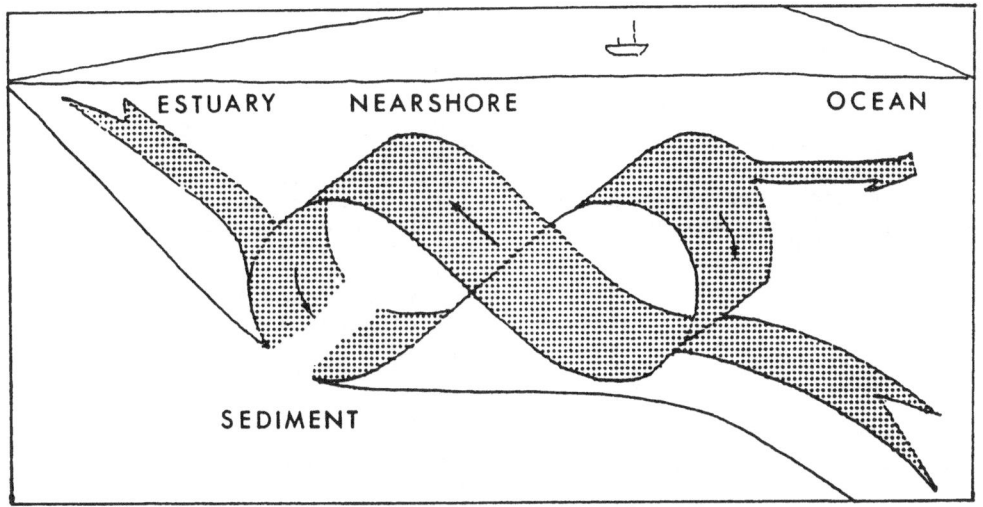

Fig. 1. On a global scale the oceanic processes eliminate the coastal areas and outwelling is of minor importance. The coastal zone is mostly characterized by an intensive internal recycling, here symbolized by the "eternity loop" with riverine, oceanic and sediment connections.

This does not suggest that there is no general interaction between coastal and offshore regions. Indeed the importance of "information" exchange must be highlighted. For example, the inshore areas are essential in constituting a crucial part in the life-support systems of offshore populations by providing nursery grounds and, occasionally feeding areas.

The value of inshore areas in themselves as recreational sites, bird sanctuaries etc. is recognized.

SUBJECT INDEX